Dairy Products: Processing Technology and Sensory Properties

Dairy Products: Processing Technology and Sensory Properties

Guest Editors

Michele Faccia
Giuseppe Natrella

MDPI

Basel • Beijing • Wuhan • Barcelona • Belgrade • Novi Sad • Cluj • Manchester

Guest Editors
Michele Faccia
Department of Soil, Plant and Food Sciences
University of Bari
Bari
Italy

Giuseppe Natrella
Department of Soil, Plant and Food Sciences
University of Bari
Bari
Italy

Editorial Office
MDPI AG
Grosspeteranlage 5
4052 Basel, Switzerland

This is a reprint of the Special Issue, published open access by the journal *Foods* (ISSN 2304-8158), freely accessible at: https://www.mdpi.com/journal/foods/special_issues/68C37DM1G2.

For citation purposes, cite each article independently as indicated on the article page online and as indicated below:

Lastname, A.A.; Lastname, B.B. Article Title. *Journal Name* **Year**, *Volume Number*, Page Range.

ISBN 978-3-7258-3549-2 (Hbk)
ISBN 978-3-7258-3550-8 (PDF)
https://doi.org/10.3390/books978-3-7258-3550-8

Contents

Preface . . . . **vii**

Michele Faccia and Giuseppe Natrella
Dairy Products: Processing Technology and Sensory Properties
Reprinted from: *Foods* **2024**, *13*, 2551, https://doi.org/10.3390/foods13162551 . . . . **1**

Onyeka M. Ikele, Chigoziri T. Ogu, Xiuping Jiang and George A. Cavender
Evaluation of Different Lactic Acid Bacteria as Starter Cultures for Nono—A West African Fermented Dairy Product
Reprinted from: *Foods* **2024**, *13*, 3030, https://doi.org/10.3390/foods13193030 . . . . **5**

Antonella Santillo, Maria Giovanna Ciliberti, Mariangela Caroprese, Agostino Sevi and Marzia Albenzio
Fatty Acids Profile and Consumers' Preferences of Pecorino Cheese Manufactured from Milk of Sheep Supplemented with Flaxseed and *Ascophyllum nodosum*
Reprinted from: *Foods* **2024**, *13*, 2165, https://doi.org/10.3390/foods13142165 . . . . **17**

Biljana Trajkovska, Gjore Nakov, Sari Thachappully Prabhat and Prarabdh C. Badgujar
Effect of Blueberry Pomace Addition on Quality Attributes of Buttermilk-Based Fermented Drinks during Cold Storage
Reprinted from: *Foods* **2024**, *13*, 1770, https://doi.org/10.3390/foods13111770 . . . . **32**

Artur Mykhalevych, Magdalena Buniowska-Olejnik, Galyna Polishchuk, Czesław Puchalski, Anna Kamińska-Dwórznicka and Anna Berthold-Pluta
The Influence of Whey Protein Isolate on the Quality Indicators of Acidophilic Ice Cream Based on Liquid Concentrates of Demineralized Whey
Reprinted from: *Foods* **2024**, *13*, 170, https://doi.org/10.3390/foods13010170 . . . . **45**

Idiana de Macêdo Barbosa, Katya Anaya, Cláudia Souza Macêdo, Robson Rogério Pessoa Coelho, Claudio Cipolat-Gotet, Emerson Gabriel dos Santos Oliveira Silva, et al.
Characterization of Physicochemical and Sensory Properties of Cheeses Added with Bovine Colostrum
Reprinted from: *Foods* **2023**, *12*, 4474, https://doi.org/10.3390/foods12244474 . . . . **67**

Loubna Abou el qassim, Beatriz Martínez, Ana Rodríguez, Alberto Dávalos, María-Carmen López de las Hazas, Mario Menéndez Miranda and Luis J. Royo
Effects of Cow's Milk Processing on MicroRNA Levels
Reprinted from: *Foods* **2023**, *12*, 2950, https://doi.org/10.3390/foods12152950 . . . . **81**

Theofilos Massouras, Evangelia Zoidou, Zinovia Baradaki and Marianna Karela
Physicochemical, Microbiological and Sensory Characteristics of White Brined Cheese Ripened and Preserved in Large-Capacity Stainless Steel Tanks
Reprinted from: *Foods* **2023**, *12*, 2332, https://doi.org/10.3390/foods12122332 . . . . **91**

Giuseppe Natrella, Giuseppe Gambacorta and Michele Faccia
Application of Commercial Biopreservation Starter in Combination with MAP for Shelf-Life Extension of Burrata Cheese
Reprinted from: *Foods* **2023**, *12*, 1867, https://doi.org/10.3390/foods12091867 . . . . **110**

Giuseppe Natrella, Giuseppe Gambacorta, Giacomo Squeo and Michele Faccia
Impact of Milk Thermization on the Quality Characteristics of P.D.O. "Canestrato Pugliese" Ovine Hard Cheese
Reprinted from: *Foods* **2023**, *12*, 1080, https://doi.org/10.3390/foods12051080 . . . . **124**

Marta Albisu, Sonia Nieto, Olaia Martínez, María Ángeles Bustamante, Luis Javier R. Barron and Ana Isabel Nájera
Optimization of Modified Atmosphere Packaging for Sheep's Milk Semi-Hard Cheese Wedges during Refrigerated Storage: Physicochemical and Sensory Properties
Reprinted from: *Foods* **2023**, *12*, 849, https://doi.org/10.3390/foods12040849 . . . . **140**

Noelia Gil, Gisela Quinteros, Monica Blanco, Shafirah Samsuri, Nurul Aini Amran, Patrico Orellana-Palma, et al.
Vacuum-Assisted Block Freeze Concentration Studies in Cheese Whey and Its Potential in Lactose Recovery
Reprinted from: *Foods* **2023**, *12*, 836, https://doi.org/10.3390/foods12040836 . . . . **156**

Preface

The dairy sector plays a crucial role in the global food industry, with cheese being one of the most widely consumed foods worldwide. Its production is continuously evolving to meet the needs of increasingly conscious consumers, who demand safer products with enhanced nutritional properties, made without chemicals through sustainable processing methods that minimize waste for the sake of the environment.

To achieve these goals, advanced technologies, quality control strategies, sensory evaluation methods, and the valorization and recovery of nutrients from by-products have significantly influenced both cheese production and consumer perception.

This Special Issue, titled Dairy Products: Processing Technology and Sensory Properties, is intended for the scientific community and industry professionals engaged in dairy research, production, and quality assessment. It aims to provide new insights to face the modern challenges of the sector.

The papers in this issue cover a wide range of topics, involving many issues such as the development of functional cheeses, fortified with blueberry pomace and bovine colostrum; the effects of animal diet on the chemical composition of cheese; the impact of processing on the quality and traceability of dairy products; an investigation into non-coding RNAs as potential biomarkers and bioactive components in milk and dairy products; the effects of thermization on the chemical and sensory characteristics of a P.D.O. Apulian cheese; and the production of dairy ingredients with enhanced functional and organoleptic qualities through the optimization of the block freeze concentration technique. Additionally, some studies focused on shelf life extension strategies.

The outcomes of each paper integrate recent scientific findings with practical applications, providing interesting insights for researchers and producers.

I would like to extend my sincere gratitude to all the experts and researchers who contributed to this Special Issue. In particular, I am profoundly grateful to Professor Michele Faccia, who is an invaluable source of inspiration. His vast and profound scientific knowledge combined with his kind and supportive nature make him not only an outstanding mentor but also a guiding figure and motivator for students, researchers, and colleagues.

Giuseppe Natrella
Guest Editor

foods

MDPI

Editorial

Dairy Products: Processing Technology and Sensory Properties

Michele Faccia * and Giuseppe Natrella

Department of Soil, Plant and Food Sciences, University of Bari, Via Amendola 165/A, 70126 Bari, Italy; disspa@pec.uniba.it
* Correspondence: michele.faccia@uniba.it; Tel.: +39-080-5442939

In developed countries, the dairy sector is going through a highly challenging phase as a consequence of changes in consumers' expectations and the spread of new cultural approaches to food. Among the most challenging requirements are producing products with a more balanced chemical composition, valorizing or improving their nutraceutical properties, extending shelf life to reduce food waste, and finding new tools to enhance traceability and safety. Of course, all these goals must be reached without impairing foods' sensory characteristics, which remain a constraint in food choices. This Special Issue comprises several interesting contributions to the field authored by researchers from 11 different countries. The contents of these papers can be grouped into three categories: the development and consumer acceptance of innovative products with improved nutritional characteristics, the effect of processing on quality and traceability, cheese shelf life, and preservation conditions.

In Article 1, an interesting functional dairy product was developed consisting of a buttermilk-based fermented drink fortified with blueberry pomace, a fruit byproduct. This is an innovation creation given that the literature related to this research largely focuses on the fortification of buttermilk with industrial or self-produced ingredients but not with byproducts [1–3]. The use of bovine colostrum in cheesemaking was studied in Article 2, with the aim of developing fresh and matured cheeses rich in bioactive compounds. Colostrum differs significantly from milk and contains higher concentrations of bioactive compounds, such as immunoglobulins, enzymes, vitamins, and growth factors, but its presence in milk may cause problems in industrial processes [4]. After chemical, microbiological, and sensory evaluation, the authors found the presence of colostrum is compatible with the production of fresh cheese at a 75:25 ratio. The authors of Article 3 produced Pecorino Cheese from the milk of sheep whose diet had been supplemented with flaxseed and algae (*Ascophyllum nodosum*) and registered a higher content of unsaturated fatty acids and lower levels of atherogenic and thrombogenic indexes than in the control cheese. This improvement in nutritional quality did not have a detrimental impact on the cheese's sensory attributes.

Four research groups contributed to this Special Issue by publishing papers dealing with the effect of processing on the quality and traceability of dairy products. In Article 4, the possibility of using concentrated demineralized liquid whey in the formulation of ice cream was investigated. This study also considered two processing variables: the addition of lactase to the whey concentrate to hydrolyze lactose and fortification with whey protein isolate. The obtained products were subjected to a series of laboratory analyses, with a particular focus on the physical characteristics. The results were thoroughly discussed, and the effect of all the variables was evaluated by highlighting the pros and cons of the different formulations. An interesting study was conducted on the effect of processing at the level of microRNAs in milk and milk-related products (Article 5). MiRNAs are noncoding RNAs that are present in milk and might have bioactive effects in humans. Research on these molecules as possible biomarkers of the dairy system, diet, and animal health status is rapidly increasing worldwide [5–7], but more information is needed about the effect of milk processing on their presence. In this study, the levels of seven microRNAs were

Citation: Faccia, M.; Natrella, G. Dairy Products: Processing Technology and Sensory Properties. *Foods* **2024**, *13*, 2551. https://doi.org/10.3390/foods13162551

Received: 8 August 2024
Revised: 12 August 2024
Accepted: 14 August 2024
Published: 16 August 2024

assessed in raw milk and three derived products: microwaved milk, yogurt, or cheese. The results demonstrated that milk treatments tended to decrease the level of all the miRNAs to some extent, even though the concentration effect that took place during cheesemaking counterbalanced this decrease. In conclusion, raw milk and cheese may contain similar concentrations of miRNAs, which are higher than those of yogurt and microwaved milk.

The effect of applying milk thermization in the production of Canestrato Pugliese, a Protected Designation of Origin (P.D.O.) ovine hard cheese, was assessed in Article 6. According to the official production protocol, this cheese can only be manufactured from raw milk, with a high risk of defects that might undermine the profitability of the cheesemaking process [8,9]. The results obtained in this study demonstrated that heat treatment did not lead to remarkable differences in the gross composition with respect to cheese made from raw milk, but caused different microbiological profiles that led to differences in proteolysis during ripening. The sensory analysis revealed that the thermized cheese lost some of its typical sensory characteristics, which was likely due to a reduction in indigenous microbiota populations. It was concluded that milk thermization could only be applied with the development and use of an autochthonous starter.

Article 7 focused on the application of vacuum-assisted block freeze concentration (BFC) to concentrate different types of whey. In BFC, the liquid is completely frozen, and the temperature at the core of the product is set to below its freezing point. Subsequently, the block is thawed, and the concentrated fraction is then separated from the ice fraction [10,11]. During the process, a vacuum pressure can be applied. The study first investigated the influence of the initial concentration, time, and vacuum pressure on the concentration index and solute yield; then, the optimal time and vacuum pressure conditions were applied to three different types of whey. Additionally, the effect of vacuum-assisted BFC on lactose content was also studied, and the results suggested that lactose tends to remain in the concentrated phase rather than in the ice. In this way, it is possible to recover, in a single step, at least 70% of the lactose initially contained in the whey.

Finally, three papers focused on the shelf life of cheese and its preservation conditions. The effect of two different types of preservation containers (stainless steel tanks—SST, and tin containers—TC) on the cheeses' physicochemical, microbiological, and textural characteristics was investigated in Article 8, along with the volatile profile of white brined cheese. The results of this study showed that the material and capacity of the ripening–preservation containers did not statistically significantly affect the physicochemical, textural, microbiological, and sensory characteristics of the white brine cheeses. The authors concluded that stainless steel tanks can be used by cheese factories with a significant focus on repackaging, as an SST container keeps dairy products fresher at lower temperatures, as well as having the advantages of being reusable and highly resistant to corrosion.

Article 9 focused on the shelf-life extension of Burrata cheese, a fresh pasta filata cheese that is becoming very popular worldwide. This cheese is very similar to mozzarella and is increasingly manufactured at an industrial level. The same strategies proposed for extending the shelf life of mozzarella [12–14] have also been tested on burrata, but with poor results. In this study, the combination of a commercial bioprotective starter and modified-atmosphere packaging (MAP) was evaluated as a strategy to delay the spoilage of a product's quality. The main outcome of the research was that a synergy between the modified atmosphere and bioprotective starter, in conjunction with good-quality raw materials and good manufacturing practices, can significantly improve the microbiological stability of burrata without the use of chemical additives. Finally, Albisu et al. (Article 10) evaluated the influence of different types of packaging conditions—air, vacuum, and four different modified atmospheres—on the quality of semihard Idiazabal sheep's milk cheese ripened for 56 days under refrigerated conditions. MAP was found to be the most effective preservation technique when compared to air- and vacuum-packaging treatments. Air-packaged cheeses presented a moldy flavor by day 35, whereas vacuum packaging led to a paste-like appearance and holes after 14 days. MAP mixtures with a CO_2 concentration between 50/50 and 80/20% CO_2/N_2 (v/v) were found to ensure sensory quality and stability in the distribution of these raw sheep-milk cheese wedges.

In summary, the Special Issue "Dairy Products: Processing Technology and Sensory Properties" has supplied some new information to help combat modern challenges faced by the dairy sector and has demonstrated that innovative research in this sector is very active.

Author Contributions: Conceptualization, M.F. and G.N.; writing—original draft preparation, M.F.; writing—review and editing, M.F. and G.N. All authors have read and agreed to the published version of the manuscript.

Conflicts of Interest: The authors declare no conflicts of interest.

List of Contributions

1. Trajkovska, B.; Nakov, G.; Prabhat, S.T.; Badgujar, P.C. Effect of Blueberry Pomace Addition on Quality Attributes of Buttermilk-Based Fermented Drinks during Cold Storage. *Foods* **2024**, *13*, 1770. https://doi.org/10.3390/foods13111770.
2. Barbosa, I.d.M.; Anaya, K.; Macêdo, C.S.; Coelho, R.R.P.; Cipolat-Gotet, C.; Silva, E.G.d.S.O.; Araújo, N.G.; Chagas, B.M.E.d.; Oliveira, J.P.F.d.; Boari, C.A.; et al. Characterization of Physicochemical and Sensory Properties of Cheeses Added with Bovine Colostrum. *Foods* **2023**, *12*, 4474. https://doi.org/10.3390/foods12244474.
3. Santillo, A.; Ciliberti, M.G.; Caroprese, M.; Sevi, A.; Albenzio, M. Fatty Acids Profile and Consumers' Preferences of Pecorino Cheese Manufactured from Milk of Sheep Supplemented with Flaxseed and Ascophyllum nodosum. *Foods* **2024**, *13*, 2165. https://doi.org/10.3390/foods13142165.
4. Mykhalevych, A.; Buniowska-Olejnik, M.; Polishchuk, G.; Puchalski, C.; Kamińska-Dwórznicka, A.; Berthold-Pluta, A. The Influence of Whey Protein Isolate on the Quality Indicators of Acidophilic Ice Cream Based on Liquid Concentrates of Demineralized Whey. *Foods* **2024**, *13*, 170. https://doi.org/10.3390/foods13010170.
5. Abou el qassim, L.; Martínez, B.; Rodríguez, A.; Dávalos, A.; López de las Hazas, M.-C.; Menéndez Miranda, M.; Royo, L.J. Effects of Cow's Milk Processing on MicroRNA Levels. *Foods* **2023**, *12*, 2950. https://doi.org/10.3390/foods12152950.
6. Natrella, G.; Gambacorta, G.; Squeo, G.; Faccia, M. Impact of Milk Thermization on the Quality Characteristics of P.D.O. "Canestrato Pugliese" Ovine Hard Cheese. *Foods* **2023**, *12*, 1080. https://doi.org/10.3390/foods12051080.
7. Gil, N.; Quinteros, G.; Blanco, M.; Samsuri, S.; Amran, N.A.; Orellana-Palma, P.; Schwinden, E.; Hernández, E. Vacuum-Assisted Block Freeze Concentration Studies in Cheese Whey and Its Potential in Lactose Recovery. *Foods* **2023**, *12*, 836. https://doi.org/10.3390/foods12040836.
8. Massouras, T.; Zoidou, E.; Baradaki, Z.; Karela, M. Physicochemical, Microbiological and Sensory Characteristics of White Brined Cheese Ripened and Preserved in Large-Capacity Stainless Steel Tanks. *Foods* **2023**, *12*, 2332. https://doi.org/10.3390/foods12122332.
9. Natrella, G.; Gambacorta, G.; Faccia, M. Application of Commercial Biopreservation Starter in Combination with MAP for Shelf-Life Extension of Burrata Cheese. *Foods* **2023**, *12*, 1867. https://doi.org/10.3390/foods12091867.
10. Albisu, M.; Nieto, S.; Martínez, O.; Bustamante, M.Á.; Barron, L.J.R.; Nájera, A.I. Optimization of Modified Atmosphere Packaging for Sheep's Milk Semi-Hard Cheese Wedges during Refrigerated Storage: Physicochemical and Sensory Properties. *Foods* **2023**, *12*, 849. https://doi.org/10.3390/foods12040849.

References

1. Mudgil, D.; Barak, S. Development of functional buttermilk by soluble fibre fortification. *Agro Food Ind. Hi Tech* **2016**, *27*, 44–47.
2. Rose, H.; Bakshi, S.; Kanetkar, P.; Lukose, S.J.; Felix, J.; Yadav, S.P.; Gupta, P.K.; Paswan, V.K. Development and Characterization of Cultured Buttermilk Fortified with *Spirulina plantensis* and Its Physico-Chemical and Functional Characteristics. *Dairy* **2023**, *4*, 271–284. [CrossRef]
3. Mudgil, D.; Barak, S.; Darji, P. Development and characterization of functional cultured buttermilk utilizing *Aloe vera* juice. *Food Biosci.* **2016**, *15*, 105–109. [CrossRef]
4. Sánchez-Macías, D.; Herrera-Chávez, B.; Quevedo-Barreto, L.; Maldonado-Bonifaz, A.; González-Castillo, Á.; Mesa, A.J.T. Colostrum in cheese milk: Effects on physicochemical and microbiological characteristics of milk, whey and fresh cheese. *Int. Dairy J.* **2024**, *155*, 105957. [CrossRef]

5. Schanzenbach, C.I.; Kirchner, B.; Ulbrich, S.E.; Pfaffl, M.W. Can milk cell or skim milk miRNAs be used as biomarkers for early pregnancy detection in cattle? *PLoS ONE* **2017**, *12*, e0172220. [CrossRef] [PubMed]
6. Miretti, S.; Lecchi, C.; Ceciliani, F.; Baratta, M. MicroRNAs as biomarkers for animal health and welfare in livestock. *Front. Vet. Sci.* **2020**, *7*, 578193. [CrossRef] [PubMed]
7. Özdemir, S. Identification and comparison of exosomal microRNAs in the milk and colostrum of two different cow breeds. *Gene* **2020**, *743*, 144609. [CrossRef] [PubMed]
8. Albenzio, M.; Corbo, M.R.; Rehman, S.U.; Fox, P.F.; De Angelis, M.; Corsetti, A.; Sevi, A.; Gobbetti, M. Microbiological and biochemical characteristics of Canestrato Pugliese cheese made from raw milk, pasteurized milk or by heating the curd in hot whey. *Int. J. Food Microbiol.* **2001**, *67*, 35–48. [CrossRef] [PubMed]
9. Piombino, P.; Pessina, R.; Genovese, A.; Lisanti, M.T.; Moio, L. Sensory profiling, volatiles and odor-active compounds of Canestrato pugliese PDO cheese made from raw and pasteurized ewes' milk. *Ital. J. Food Sci.* **2008**, *20*, 225–237.
10. Sánchez, J.; Hernández, E.; Auleda, J.M.; Raventós, M. Freeze concentration technology applied to dairy products. *Food Sci. Technol. Int.* **2011**, *17*, 5–13. [CrossRef] [PubMed]
11. Prestes, A.A.; Helm, C.V.; Esmerino, E.A.; Silva, R.; da Cruz, A.G.; Prudencio, E.S. Freeze concentration techniques as alternative methods to thermal processing in dairy manufacturing: A review. *J. Food Sci.* **2022**, *87*, 488–502. [CrossRef] [PubMed]
12. Michele, F.; Luisa, A.; Marianna, M.; Amalia, C.; Matteo Alessandro, D.N. The effect of incorporating calcium lactate in the saline solution on improving the shelf life of fiordilatte cheese. *Int. J. Dairy Technol.* **2013**, *66*, 373–381. [CrossRef]
13. Costa, C.; Lucera, A.; Conte, A.; Zambrini, A.V.; Del Nobile, M.A. Technological strategies to preserve burrata cheese quality. *Coatings* **2017**, *7*, 97. [CrossRef]
14. Sinigaglia, M.; Bevilacqua, A.; Corbo, M.R.; Pati, S.; Del Nobile, M.A. Use of active compounds for prolonging the shelf life of mozzarella cheese. *Int. Dairy J.* **2008**, *18*, 624–630. [CrossRef]

foods

MDPI

Article

Evaluation of Different Lactic Acid Bacteria as Starter Cultures for Nono—A West African Fermented Dairy Product

Onyeka M. Ikele [1,2], Chigoziri T. Ogu [1], Xiuping Jiang [2] and George A. Cavender [2,*]

[1] Department of Applied Microbiology and Brewing, Nnamdi Azikiwe University, Awka 420110, Nigeria; oikele@clemson.edu (O.M.I.); ct.ogu@unizik.edu.ng (C.T.O.)
[2] Department of Food, Nutrition and Packaging Sciences, Clemson University, Clemson, SC 29634, USA; xiuping@clemson.edu
* Correspondence: gcavend@clemson.edu

Abstract: Nono is a traditional cultured dairy product consumed across West Africa. In this study, five cultures isolated from Nigerian-produced nono and three purified lactic acid bacteria from the USDA-NRRL were examined for use in preparing nono starter cultures. Isolated cultures were characterized using microbiological and biochemical tests, including 16s rDNA sequencing to identify the genotype. Each isolated strain was cultured and inoculated into UHT milk (1% v/v) and allowed to ferment for 24 h at 25 °C. Fermented products were evaluated for pH, moisture content, water activity, and viscosity, and their descriptive sensory properties were noted. The isolate that resulted in sensory properties most similar to traditional nono was then used as the primary strain for subsequent starter culture blends made with the NRRL cultures. These blends were used for the fermentation of nono and compared with commercial nono samples. Isolates obtained from nono were as follows: *Lactobacillus fermentum*, *Lactobacillus paracasei*, and, surprisingly, *Lactobacillus rhamnosus*, which has not been previously reported as a part of the nono microflora. There was no significant difference in the physical parameters of nono made from the individual indigenous isolates and a similar pattern was observed for the organisms from NRRL, except that their total titratable acidity and viscosities were significantly ($p < 0.05$) higher than those of the indigenous organisms. Compounded starter made with *L. rhamnosus* and NRRL cultures was then used to make nono that showed significantly ($p < 0.05$) different pH and viscosity values than commercially purchased nono, while sensory evaluation showed that nono made from the new starter culture had a high overall consumer acceptance score.

Keywords: cultured dairy; nono; sensory evaluation; starter culture; lactic acid bacteria; fermented milk

Citation: Ikele, O.M.; Ogu, C.T.; Jiang, X.; Cavender, G.A. Evaluation of Different Lactic Acid Bacteria as Starter Cultures for Nono—A West African Fermented Dairy Product. *Foods* **2024**, *13*, 3030. https://doi.org/10.3390/foods13193030

Academic Editor: Barbaros Özer

Received: 20 August 2024
Revised: 10 September 2024
Accepted: 22 September 2024
Published: 24 September 2024

1. Introduction

Nono is a fermented dairy product made in different parts of Nigeria and West Africa, especially in settlements inhabited by the Fulani tribe. Nono production is known to be a craft and also a source of livelihood for the Fulani women who hand-milk the cows and ferment the milk into nono; they are popularly called 'milkmaids'. Nono holds a cultural heritage for the Fulani tribe, who are known to be traditionally pastoral nomadic farmers who reside in different parts of West Africa [1–3]. The culture of nomadic pastoralism made it possible for the consumption of nono to spread from the Fulanis to other tribes in the region where they reside at a given time. On the other hand, nomadic pastoralism poses a disadvantage to the safety and quality of nono made by these women, since there are no standard fermentation facilities available and no standardized product-processing methods. Other disadvantages are product inconsistencies between the different nono batches made, as well as the presence of microbial contaminants and pathogens in the finished product.

Nono is essentially made through hand milking, overnight boiling of the milk, cooling and fermentation, storage, and vending. As much as the boiling step is a critical control point in its production process, contamination occurs at the fermentation step, stemming from the process of back-slopping with improperly preserved cream (pre-ferment) from the previous fermentation batch [2]. This cream (*Manshanu*) is purported to be the starter culture for the fresh fermentation process; however, it is usually preserved in a wooden calabash kept in a hut. The cream has been found to be heavily laden with pathogens and fecal contaminants [3–8], evidently due to the association of nono with houseflies. This poses a food safety risk t consumers; therefore, methods to provide a safe and wholesome product are paramount. Likewise, variations in product consistency abound as a result of differences in the microbial composition of each cream used for back-slopping.

Nono has been reported to be a beverage rich in carbohydrates, proteins, and minerals [2,9,10]. It contains free fatty acids, lactose, calcium, sodium, potassium, magnesium, iron, and zinc [2,5]. It is also known to be rich in probiotic microorganisms, especially those of the lactic acid bacteria group. Nono is produced through lactic acid fermentation, and different studies carried out on nono have shown the presence of lactic acid bacteria, viz., *Lactobacillus brevis*, *Lactobacillus casei*, *Lactobacillus fermentum*, *Lactobacillus plantarum*, *Lactobacillus bulgaricus*, *Lactobacillus senioris*, *Lactobacillus helveticus*, and *Streptococcus thermophilus*, *Lactocossus lactis*, *Leuconostoc pseudomesenteroides*, and *Lactobacillus cremoris* [8,11–17].

However, these listed microorganisms do not all appear in one product; instead, they occur in variations in the different nono products examined. These organisms are known to drive the fermentation process, impacting the physical and sensory properties of nono—alongside the possible metabolic contributions of contaminants and pathogens like *Staphylococcus aureus*, *Alcaligenes faecalis*, *Clostridium sporogenes*, *Salmonella*, and *Escherichia coli* [18–20]. Thus, this study sought to evaluate different lactic acid bacteria reported to be indigenous to nono, for the sole purpose of creating a beneficial consortium that serves as the best starter culture with positive consumer acceptance and, by extension, solves the food safety problem.

2. Materials and Methods

2.1. Materials

Shelf-stable UHT milk (Horizon Organic, USA) was sourced from a local supermarket. Microbial media (MRS agar and MRS broth) were sourced from VWR Avantor, Radnor, PA, USA. pH buffers, indicators, and chemicals (ethanol and sodium hydroxide) were also sourced from VWR Avantor, USA.

2.2. Isolation, Characterization, and Identification of Indigenous Lactic Acid Bacteria in Nono

Ten nono samples were pooled and subjected to 1 in 10-fold serial dilution in phosphate-buffered saline (pH 6.8). Subsequently, 0.1 mL of the 10^{-3} tube from each sample was cultured on MRS agar plates and incubated at 30 °C and 5% CO_2 in an anaerobic incubator for 24–48 h. This was carried out according to the modified methods of Fagbemigun [20]. Isolates obtained from the cultured plates were separated into pure cultures based on their colony morphologies and thereafter subjected to a Gram stain, catalase test, and oxidase test for preliminary identification.

Subsequently, 16s rDNA sequencing was used to identify the isolates to the species level at Zymobiomics, Orange, CA, USA. The DNA samples were prepared for targeted sequencing with the Quick-16S ŒPlus NGS Library Prep Kit (Zymo Research, Irvine, CA, USA). These primers were custom-designed by Zymo Research to provide the best coverage of the 16S gene while maintaining high sensitivity, and the actual sequencing was performed by the aforementioned commercial lab (Zymobiomics, Irvine, CA, USA) as part of their commercial offerings.

Three lactic acid bacterial samples—*Lactobacillus plantarum*, *Lactobacillus casei*, and *Lactococcus lactis*—from fermented milk were also obtained from the USDA-NRRL culture collection and used as extraneous isolates for the product formulation.

2.3. Nono Production

2.3.1. Standardization of Starter Culture Isolates

The method reported by Ikele et al. [21] was used to standardize the starter cultures used for the fermentation process. Briefly, a 0.25 mL aliquot of pure culture isolates (10^5 cfu/mL) of each lactic acid bacterium was incubated in 25 mL of MRS broth without agitation at 30 °C for 24 h and then used as an inoculum to begin the fermentation process.

2.3.2. Fermentation Protocol

A modified method described by Adesokan [12] was used for nono production. For the lab fermentation procedure, 1% (v/v) of each indigenous isolate in De Man, Rogosa, and Sharpe (MRS) broth was inoculated into UHT-pasteurized whole milk (Grade A organic, Horizon Organic, Broomfield, CO, USA) in situ, sealed, and incubated at 30 °C for 24 h.

All production trials for analytical/instrumental analyses were performed in triplicate.

2.4. Physiochemical Analysis

Determinations of pH, moisture content, water activity, viscosity, total titratable acidity, and color were performed on the samples of nono. The pH was determined using a digital pH meter (pHenomenal, VWR, Randor, PA, USA). The pH electrode was immersed in 10 mL of the sample until a stable reading was obtained, and the values were recorded.

To determine moisture content, the samples were placed dropwise onto an aluminum pan provided by the manufacturer of a halogen moisture analyzer (model number 677723 Schuler Scientific, Englewood, CA, USA) and allowed to run to completion using the built-in sensing feature. Water activity determination was likewise carried out using a specialized instrument, in this case an Aqualab water activity meter (model number 1100843 Aqualab, Pullman, WA, USA), with the samples being loaded into disposable sample cups before initiating the measurement cycle.

Viscosity was measured using a rotary viscometer (model number 126408 Produstrial, Fredon, NJ, USA). For each measurement, the viscometer probe was immersed in a 15 mL aliquot of the sample before rotation was initiated, and the sample was allowed to reach a stable reading before the values were recorded.

Total titratable acidity determination was carried out according to the titration method described by Fabro [22]. Briefly, a 20 mL aliquot of fermented milk samples was added to 40 mL of distilled water that had been boiled and cooled, along with 2 mL of phenolphthalein solution as an indicator (prepared by dissolving 1% phenolphthalein in 95% ethanol). The mixture was then titrated with 0.1 M NaOH until a pink color change was observed.

To determine instrumental color, a 10 mL aliquot of each fermented milk sample was dispensed into the lid of individual 100 mm Petri dishes and covered with the inverted dish body, which was then placed onto the white calibration tile provided by the colorimeter manufacturer. A calibrated handheld colorimeter (model number CR400, Konica Minolta, Ramsey, NJ, USA) was then used to determine the color through the Petri dish.

Preliminary Sensory Property Screening

Test nono samples made from each of the indigenous isolates were examined for characteristic appearance, taste, aroma, and texture by the research team. Isolates which exhibited characteristic sensory properties were selected as choice isolates for the starter culture formulation study.

2.5. Evaluation of Effects of Best Starter Culture Consortium on Physical and Sensory Properties of Nono

The choice isolate from the indigenous cultures was used in bi- and multiple-culture cocktails with the extraneous fermented milk cultures from the USDA-NRRL at 1% (v/v), at 30 °C for 24 h, for the fermentation process. The physical and sensory parameters of the fermented products were measured as previously described. Unfermented fresh milk and a commercially purchased nono were used as the control for this experiment.

Sensory evaluation of the finished product was carried out by ten untrained panelists (who are frequent nono consumers) from Nnamdi Azikiwe University, Awka, Nigeria, using a 9-point hedonic test according to the modified methods of Dafur [19]. They characterized the nono on attributes of appearance, taste, aroma, and texture.

Institutional oversight was provided by Nnamdi Azikiwe University, under their existing approval for food tastings. Participant consent was obtained via oral consent in accordance with the published oral consent guidelines provided by the Department of Applied Microbiology and Brewing, Nnamdi Azikiwe University, Awka, Nigeria. Panelists who did not provide consent or withdrew consent were excluded from this study and any data regarding their responses were destroyed.

2.6. Statistical Analyses

The data obtained were analyzed as means with analyses of variance (ANOVA) using statistical software (GraphPad Prism, version 10.3.1). Results were deemed significant if $p \leq 0.05$.

3. Results

3.1. Isolation, Characterization, and Identification of Indigenous Lactic Acid Bacteria in Nono

Five *Lactobacillus* isolates (A–E) were isolated from the examined nono samples on the basis of their colony morphologies, and their biochemical characteristics are shown in Table 1. Molecular characterization of the isolates through sequence blast identified them as *Lactobacillus fermentum* (three isolates), *Lactobacillus paracasei*, and *Lactobacillus rhamnosus* (Table 1); a presumptive microbial identity heat map is also shown in Figure 1.

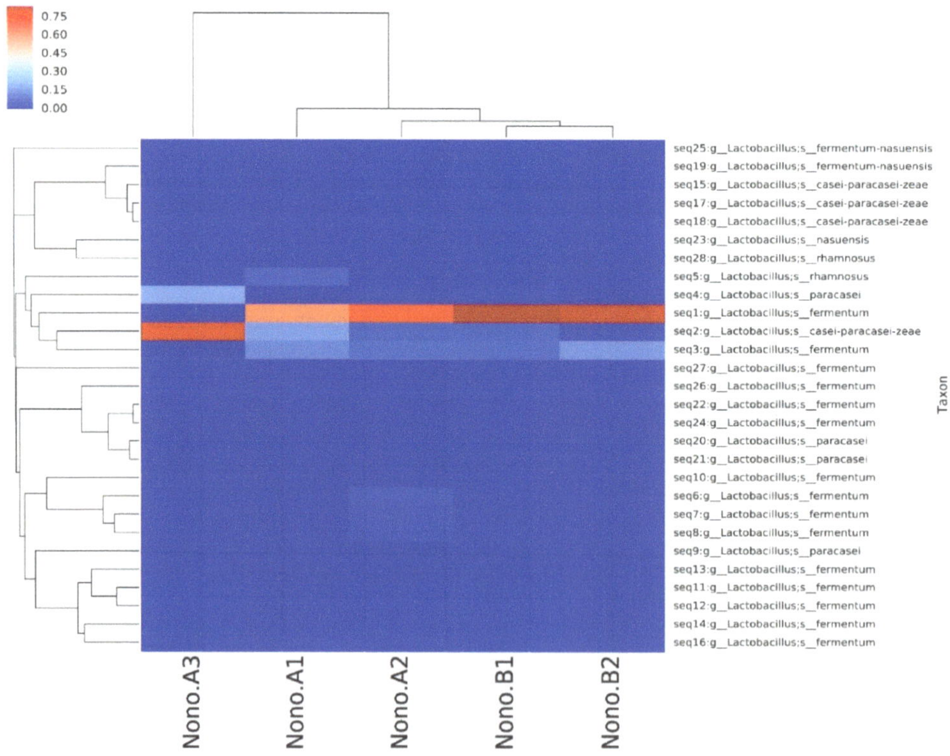

Figure 1. 16s rDNA amplicon mapping of isolates.

Table 1. Identification of indigenous isolates from nono.

Isolate Groups	Colony Morphology		Biochemical Tests		Presumptive Organisms	16s rDNA Identity
		Gram stain	Catalase	Oxidase		
A	Glistening, punctiform whitish colonies with entire margins and smooth appearance	Positive rods	Negative	Negative	*Lactobacillus* sp.	*L. fermentum*
B	Punctiform milkish colonies with entire margins and smooth appearance.	Positive rods	Negative	Negative	*Lactobacillus* sp.	*L. paracasei*
C	Circular milkish colonies with glistening appearance	Positive cocco-bacilli	Negative	Negative	*Lactobacillus* sp.	*L. fermentum*
D	Circular whitish colonies with slimy appearance	Positive rods	Negative	Negative	*Lactobacillus* sp.	*L. fermentum*
E	Punctiform, milkish colonies with slimy appearance	Positive cocco-bacilli	Negative	Negative	*Lactobacillus* sp.	*L. rhamnosus*

3.2. Assessment of Effects of Starter Culture Isolates on Physical and Sensory Properties of Nono

Indigenous starter culture isolates exhibited a low acid pH with no significant ($p > 0.05$) difference between cultures. For most of the isolates, no difference was found in moisture content (except for *L. rhamnosus*), water activity (except for *L. paracasei*), color, and viscosity values compared with those of the nono fermented with each indigenous isolate. For the extraneous starter cultures, there were no significant differences in the pH (except for *L. casei*), moisture content, and color (except for *L. plantarum*) when compared to the indigenous cultures. However, nono produced from these extraneous cultures had significantly ($p < 0.05$) higher total titratable acidity (3.0–3.8 g/L) and viscosity values (5.5 mpa.s) when compared to the indigenous cultures, as shown in Figures 2–7. Screening of the sensory capacities of indigenous and extraneous isolates is shown in Table 2.

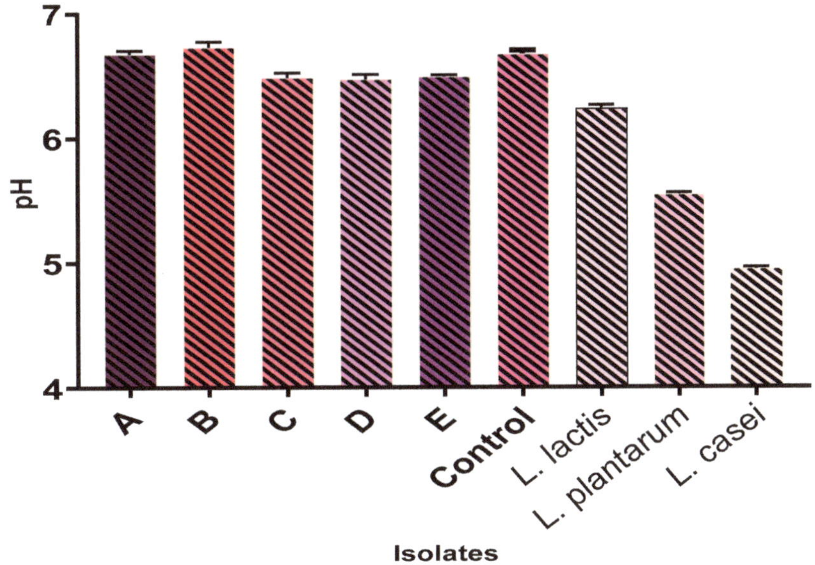

Figure 2. pH values of each starter culture candidate after fermentation. A: *Lactobacillus fermentum* 1; B: *Lactobacillus paracasei*; C: *Lactobacillus fermentum* 2; D: *Lactobacillus fermentum* 3; E: *Lactobacillus rhamnosus*.

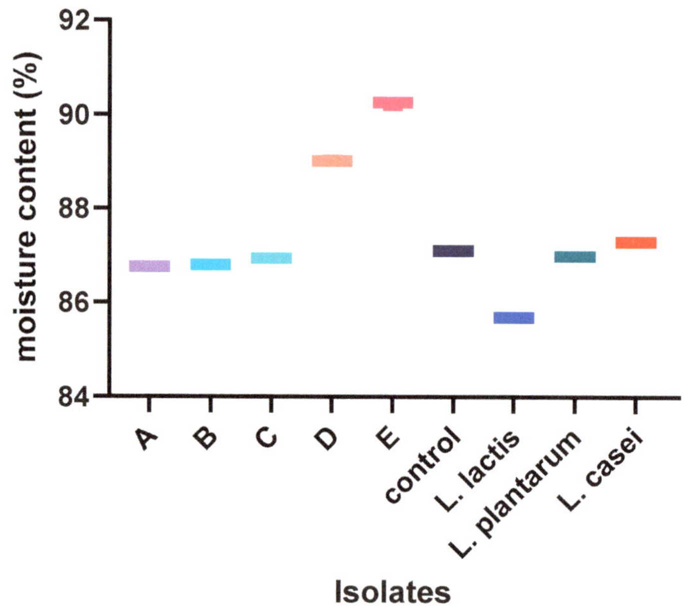

Figure 3. Moisture content values of each starter culture candidate after fermentation. A: *Lactobacillus fermentum* 1; B: *Lactobacillus paracasei*; C: *Lactobacillus fermentum* 2; D: *Lactobacillus fermentum* 3; E: *Lactobacillus rhamnosus*.

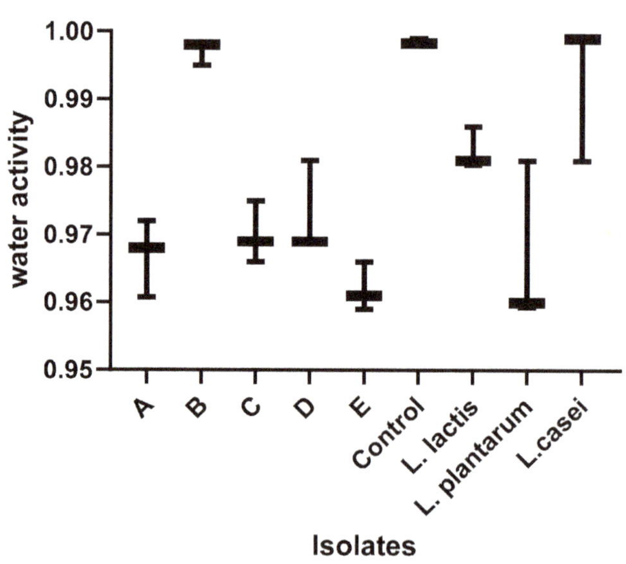

Figure 4. Water activity values of each starter culture candidate after fermentation. A: *Lactobacillus fermentum* 1; B: *Lactobacillus paracasei*; C: *Lactobacillus fermentum* 2; D: *Lactobacillus fermentum* 3; E: *Lactobacillus rhamnosus*.

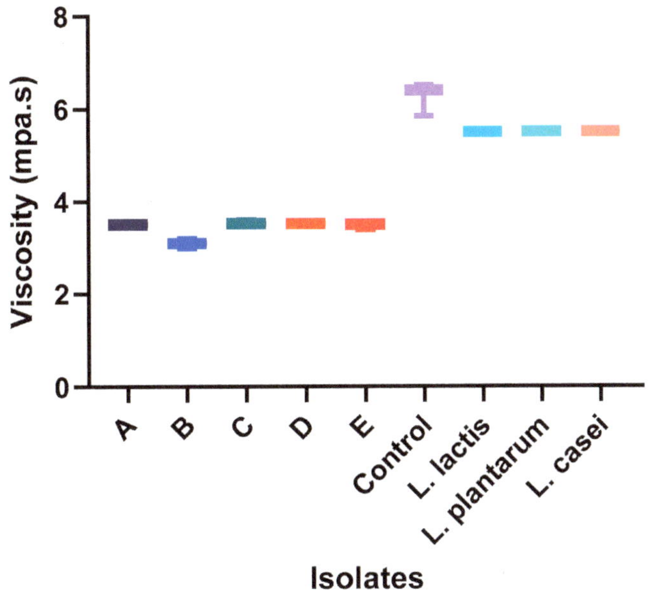

Figure 5. Viscosity values of each starter culture candidate after fermentation. A: *Lactobacillus fermentum* 1; B: *Lactobacillus paracasei*; C: *Lactobacillus fermentum* 2; D: *Lactobacillus fermentum* 3; E: *Lactobacillus rhamnosus*.

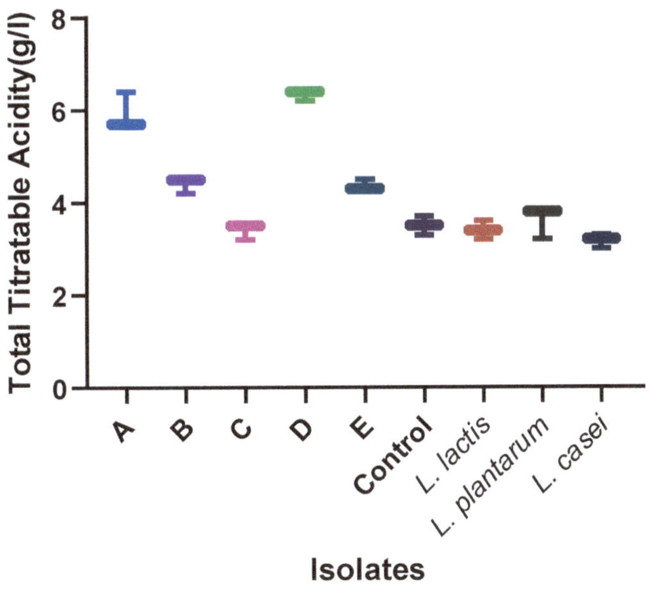

Figure 6. Total titratable acidity of each starter culture candidate after fermentation. A: *Lactobacillus fermentum* 1; B: *Lactobacillus paracasei*; C: *Lactobacillus fermentum* 2; D: *Lactobacillus fermentum* 3; E: *Lactobacillus rhamnosus*.

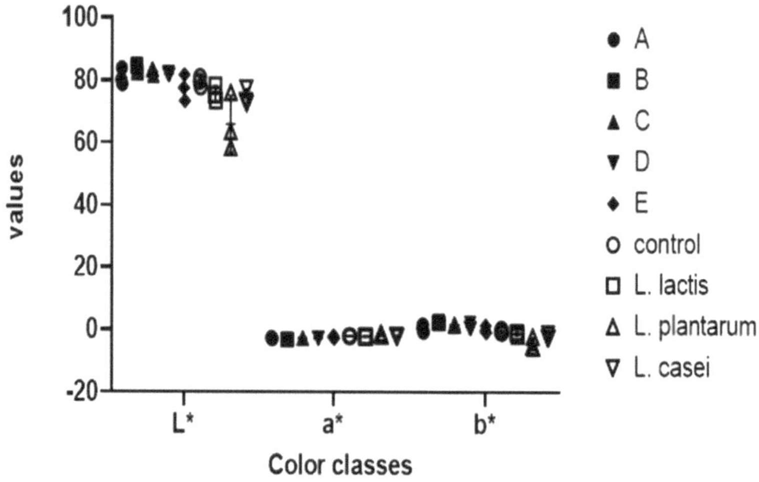

Figure 7. Color values of each starter culture candidate after fermentation. A: *Lactobacillus fermentum* 1; B: *Lactobacillus paracasei*; C: *Lactobacillus fermentum* 2; D: *Lactobacillus fermentum* 3; E: *Lactobacillus rhamnosus*.

Table 2. Preliminary screening of the sensory properties of nono made from individual isolates.

Isolates	Sensory Property of Fermented Product
Lactobacillus fermentum 1	Nono aroma, fresh milk taste, and white color
Lactobacillus paracasei	Nono taste only
Lactobacillus fermentum 2	Fresh milk taste only
Lactobacillus fermentum 3	Nono aroma, fresh milk taste, and white color
Lactobacillus rhamnosus	Nono appearance, taste, and aroma
Lactobacillus lactis	Yogurt aroma and taste
Lactobacillus plantarum	Fresh milk taste and yogurt aroma
Lactobacillus casei	Fresh milk taste and yogurt aroma

3.3. Evaluation of Effects of Best Starter Culture Consortium on Physical and Sensory Properties of Nono

Lactobacillus rhamnosus exhibited the best sensory quality in the fermented product that typified that of the regular nono and was made the choice isolate. This isolate was then used in co-culture with the extraneous isolates as bi-cultures and mixed cultures. The mixed culture displayed the physical properties (Table 3) and sensory properties (Table 4) most consistent with conventional nono.

Table 3. Evaluation of the effects of the best starter culture consortium on the physical properties of nono.

Samples	pH	Moisture Content (%)	Water Activity	Viscosity (mpa.s)	Total Titratable Acidity	Color		
						L*	a*	b*
Fresh milk	6.66 ± 0.02 ^{b}	87.12 ± 0.01 ^{a}	0.99 ± 0.00 ^{a}	5.33 ± 0.29 ^{a}	3.50 ± 0.20 ^{a}	79.52 ± 1.58 ^{a}	−1.99 ± 0.05 a	−0.38 ± 0.63 ^{a}
Commercial nono	6.56 ± 0.05 ^{b}	87.51 ± 0.25 ^{a}	0.99 ± 0.00 ^{a}	165.3 ± 2.52 ^{c}	2.93 ± 0.12 ^{a}	85.59 ± 1.45 ^{a}	−3.19 ± 0.09 ^{b}	4.09 ± 0.28 ^{c}
Starter culture mix	4.49 ± 0.02 ^{a}	87.30 ± 0.02 ^{a}	1.00 ± 0.01 ^{a}	113.7 ± 7.64 ^{b}	5.50 ± 0.30 ^{b}	83.45 ± 2.52 ^{a}	−3.02 ± 0.16 ^{b}	2.73 ± 0.44 ^{b}
p-value	0.002	0.144	0.066	0.003	0.003	0.104	0.007	0.0017

Mean values in the same column with different letters are significantly different ($p \leq 0.05$).

Table 4. Sensory analyses of test nono samples made from the starter culture mix.

Attributes	Mean ± S.D	*p*-Value
Appearance	8.04 ± 0.70	0.584
Taste	8.37 ± 0.69	0.343
Aroma	7.49 ± 0.56	0.145
Texture	7.05 ± 0.89	0.015
Overall acceptance	7.65 ± 0.47	0.009

4. Discussion

The present study sought to formulate a starter culture suitable for nono production with the goal of its standardization and improved safety. The lactic acid cultures isolated and used in this study have been partly reported by [12,15,20] as part of the microflora of nono. However, this study identified *Lactobacillus rhamnosus*, which has not been previously reported by different authors [8,11–17] as part of the microflora of nono. This study isolated *Lactobacillus fermentum*, *Lactobacillus paracasei*, and *Lactobacillus rhamnosus*. *L. fermentum* was the most commonly occurring isolate, which corresponds with the reports of [15,20], who also reported *L. fermentum* in nono. Three milk-fermenting lactic acid cultures not isolated from this study, but reported to be a part of the nono microflora [12,20], were also obtained from the USDA-NRRL culture collection center and incorporated as a part of the starter culture consortium in order to conduct a fair assessment on the fermentation capacities of each microbe reported to be found in nono.

The suitability of these organisms as starter culture candidates was assessed in terms of the physical and sensory properties they imparted on the fermented product, while Adesokan [12] chose the suitability of their own isolates for starter culture formulation solely on the basis of diacetyl production capacity. For the indigenous isolates used in this study, there was no significant ($p > 0.05$) difference observed in the pH, moisture content, and viscosity when they were used individually as starter cultures. The pH values of nono made from each isolate ranged from 6.4 to 6.7, which partly corresponds with the reports of Adisokan et al. [12] and Dafur et al. [19] and differs entirely from the reports of Abdulrahman et al. [1], Omola et al. [2], and those of Nebedum and Obiakor [9]. Adesokan [12] also reported a moisture content range of 80.7- 87.11 for nono made from their starter culture experiment, which corresponds partly with that of this study. When compared to the fresh whole milk used for the control experiment, these indigenous isolates impacted the product and significantly ($p < 0.05$) increased viscosity and total titratable acidity in the fermented milk. According to Bachmann [23], viscosity is a key physical property that affects the textural property of fermented milk products, stemming from the microbial binding of water through exopolysaccharide formation during the fermentation process. Obioha [15,24] reported that these microbial exopolysaccharides produced during fermentation increase water retention in the fermented milk product/nono; this may explain the results of the present study, which showed no significant ($p > 0.05$) difference in the moisture content values of the test nono products and the unfermented milk sample. Of the extraneous organisms used in the starter culture, *Lactobacillus casei* and *L. plantarum* showed decreased pH values (4.95 and 5.46, respectively), but they were not statistically significant ($p < 0.05$) compared to the pH values of other starter culture candidates. Li [25] reported that lactic acid bacteria are known for medium acidification and the coagulation of milk during fermentation. These extraneous cultures (*L. lactis*, *L. plantarum*, and *L. casei*) produced a significantly ($p < 0.05$) viscous product compared to that of the indigenous cultures when used singly as starter cultures. Therefore, they were used in the starter culture consortium to achieve the desired texture for nono, in combination with the best indigenous isolate which exhibited the desired sensory characteristics of nono (*L. rhamnosus*). *Lactobacillus rhamnosus* has not been previously reported by authors as a part of the nono microflora, which makes it one of the key findings in this study.

The final starter culture consortium used for the fermentation of nono was a 1% combination of *L. rhamnosus*, *L. plantarum*, *L. casei*, and *L. lactis*; this consortium yielded a product very similar to the popularly consumed commercial product, both in terms of physical and sensory properties. Comparing the nono fermented with this starter culture to the commercially purchased nono used as a control, there was a significant ($p < 0.05$) difference in the pH, total titratable acidity, viscosity, and a* and b* dairy colors. The starter culture consortium had a significantly lower pH (4.471), which varies from the results of [12], who reported a pH range of 5.5–7.9. The low pH seen in this study is expected because *Lactococcus lactis* and *Lactobacillus casei* are known acidifiers in milk fermentation [20]. The starter culture consortium exhibited a mean total titratable acidity of 5.50 g/L, which differs from the findings of both Abdulrahman [1] and Omola [3]. This could be further explained as a possible result of differences in the combined organic acid production from each member of the lactic acid starter culture consortium. On the other hand, the commercial nono had a significantly higher viscosity (168 mpa.s) than that of the starter culture consortium (107.5 mpa.s). The possible reason behind this finding is related to the long hours of milk boiling before fermentation in commercial nono production, which possibly result in the loss of volume, coupled with the action of indigenous starter cultures that bind water through exopolysaccharide formation. Comparatively, the milk used for fermentation in this study was not boiled, but it was ultra-high-temperature (UHT)-pasteurized milk; thus, the difference in the heat treatment of both fermentation substrates possibly constituted the basic difference in their texture and viscosity. Lactalbumin denaturation and casein precipitation usually occur when milk is exposed to heating for a prolonged time, which impacts the gel structure of the milk [24] and could also be an additional reason for the difference in viscosity seen in both nono samples. However, the notable similarity in the viscosity of both the commercial nono and the test nono can be attributed to Bingham plastic behavior. It was observed that the viscosity value obtained from the starter culture consortium partly corresponds with the findings of Omola et al. [3], while the value of the commercial nono varied from it considerably.

Both products also had a notable difference in color, with the commercial product having a darker brown color than the starter culture consortium. This characteristic dark brown color of commercial nono could be a result of lactose caramelization and/or the production of humin and melanin commonly associated with protein breakdown in the presence of sugars at high temperatures [26]. This study used the CIE L*a*b* color system to calculate the lightness, redness, and yellowness, respectively, of both commercial nono and that made from the starter culture consortium. There was no significant difference in the lightness (L*) and redness (a*) of both products, but there was a significant difference in their yellowness (b*) ($p = 0.0017$), with the commercial nono having much higher b* value. This could possibly be a result of the long hours of boiling/caramelization it may have gone through before fermentation, coupled with the possible actions of some vitamins like riboflavin [27]. The change in color (ΔE) between both samples was 6.71, which implies that most people could easily tell the difference in color between the products. To our knowledge, no previous study has examined the chromatic components of nono, which is also another key finding in this study, particularly given that Milovanovic [28] opined that appearance attributes, like color, are very important in milk products because they often influence consumers' choices.

Sensory evaluation showed that nono made from the select starter culture consortium had an overall acceptance of 7.65 ± 0.47 on a nine-point hedonic scale. However, the untrained panelists chose the commercial nono over the product made from the constituted starter culture on the basis of taste, aroma and texture. It is assumed that the major difference between the experimentally fermented product and that available on the market stems from the nature of the starting material used for fermentation. This study used UHT-pasteurized milk, while the nono consumed in Nigeria is made from milk boiled for 10–12 h; this boiling is presumed to impact the color, aroma, and texture of the milk prior to fermentation. The sensory evaluation result obtained in this study partly corresponds with that of [12], who reported an overall acceptance of 7.01 ± 0.02. A key difference between this starter culture study and [12] is that the latter used isolates individually but did not proceed to check their fermentation capacities as a co-culture consortium, which was done in this study.

5. Conclusions

Synergistic interactions among lactic acid bacteria are required to produce nono with the desired characteristics. Starter culture formulation for nono provides a standardized product with known consistency and is beneficial for solving food safety problems caused by improper food sanitation processes. It ushers in an era of wholesome product formulation for the milkmaids who earn a living from this craft, as well as providing extra opportunity for people outside of Nigeria and West Africa to be able to make nono in their homes using the starter cultures. This study provides a conduit for introducing nono to the rest of the world, just like yoghurt, cheese, dahi, and kefir.

Author Contributions: Conceptualization, O.M.I. and G.A.C.; methodology, O.M.I., C.T.O., X.J. and G.A.C.; software, O.M.I. validation, X.J. and G.A.C.; formal analysis, O.M.I., C.T.O. and G.A.C.; investigation, C.T.O. and O.M.I.; resources, C.T.O., X.J. and G.A.C.; data curation, O.M.I. and C.T.O.; writing—original draft preparation, O.M.I.; writing—review and editing, G.A.C.; visualization, O.M.I. and G.A.C.; supervision, X.J., C.T.O. and G.A.C.; project administration, G.A.C.; funding acquisition, O.M.I. All authors have read and agreed to the published version of the manuscript.

Funding: This research received no external funding.

Institutional Review Board Statement: Institutional oversight was provided by Nnamdi Azikiwe University, under their existing approval for food tastings. Consent of participants was obtained via oral consent- the published oral consent guidelines provided by the Department of Applied Mi-crobiology and Brewing, Nnamdi Azikiwe University, Awka, Nigeria. Panelist who did not provide consent, or withdrew consent were excluded from the study and any data regarding their responses was destroyed.

Data Availability Statement: The raw data supporting the conclusions of this article will be made available by the authors on reasonable request.

Acknowledgments: The authors wish to express our heartfelt thanks to Ikechukwu Okoli, the Director of the TetFund Center of Excellence in Agriculture and Biomedical Research, Nnamdi Azikiwe University, Awka, Nigeria, for organizing administrative/funding assistance for the actualization of this project. We also appreciate the USDA-NRRL culture collection center for providing us with the fermented milk isolates we used for this starter culture formulation study. Lastly, the authors wish to thank Charles Santerre, Department of Food, Nutrition and Packaging Sciences, Clemson University, South Carolina, for approving the bench space and collaboration that led to this study during his tenure as department chair.

Conflicts of Interest: The authors declare no conflict of interest.

References

1. Abdulrahman, F.A.; Sanmi, E. Physicochemical Properties, Proximate Composition and Total Viable Counts of Staphylococcus aureus in 'Nono' and Yoghurt Samples in Kaduna, Nigeria. *Turk. J. Agric.-Food Sci. Technol.* **2021**, *9*, 15–20. [CrossRef]
2. Omola, E.M.; Kawo, A.H.; Bukar, A. Evaluation of nutritional quality and mineral composition of fermented milk (nono) sold in Kano, north-western Nigeria. *Bayero J. Pure Appl. Sci.* **2021**, *12*, 81–88. [CrossRef]
3. Omola, E.M.; Kawo, A.H.; Bukar, A. Microbiological Quality of Traditionally Fermented Fresh Cow Milk (Nono) Retailed in Selected Local Government Areas of Kano State, Nigeria. *UMYU J. Microbiol. Res.* **2019**, *4*, 45–52. [CrossRef]
4. Atanda, O.O.; Ikenebomeh, M.J. Microbiology quality of nono. *World J. Microbiol.* **2006**, *1*, 89–91.
5. Egwaikhide, P.A.; Malu, P.S.; Lawal, U.; Adelagun, R.O.; Andrew, C. Physicochemical and Microbiological analysis of Fermented cow milk (Nono) consumed within Kaduna town, North western, Nigeria. *Food Sci. Qual. Manag.* **2014**, *29*, 2225–2557.
6. Godwin, A.O.; Emmanuel, T.A. Extent of microbial contamination of nono, fresh cow milk and yoghurt sold in Makurdi, Benue State, Nigeria. *J. Microbiol. Biotechnol. Resour.* **2013**, *3*, 6–14.
7. Obi, C.N.; Ikenebomeh, M.J. Studies on the microbiological, nutritional quality of a Nigerian fermented milk products (nono). *Int. J. Dairy Sci.* **2007**, *2*, 95–99.
8. Ogbonna, I.O. Microbiological Analysis and Safety Evaluation of Nono: A Fermented milk product consumed in most part of Northern Nigeria. *Int. J. Dairy Sci.* **2011**, *6*, 181–189.

9. Uzeh, E.R.; Ohenhen, E.R.; Rojugbokan, K.A. Microbiological and Nutritional Qualities of Dairy Products: Nono and Wara. *J. Nat. Sci.* **2006**, *4*, 37–40.
10. Nebedum, J.O.; Obiakor, T. The effects of different preservation methods on the quality of nunu, a locally fermented Nigerian dairy product. *Afr. J. Biotechnol.* **2007**, *6*, 454–458.
11. Makut, M.D.; Nyam, M.A.; Amapu, T.Y.; Ahmed, A.M. Antibiogram of bacteria isolated from locally processed cow milk products sold in Keffi metropolis, Nasarawa state, Nigeria. *J. Biol. Agric. Healthc.* **2014**, *4*, 19–25.
12. Adesokan, I.; Odetoyinbo, B.; Ekanola, Y.; Avanrenren, R.; Fakorede, S. Production of nigerian nono lactic starter cultures. *Pak. J. Nutr.* **2011**, *10*, 2013–2017. [CrossRef]
13. Okiki, P.A.; Adeniji, C.A.; Oyetunji, O.A.; Omosolape, A.Y.; Omolara, A.P. Assessment of the Physicochemical and Bacteriological Qualities of Nono—A Fermented Cow Milk. *Potravin. Slovak J. Food Sci.* **2018**, *12*, 26–32. [CrossRef] [PubMed]
14. Obioha, P.I.; Anyogu, A.; Awamaria, B.; Ghoddusi, H.B.; Ouoba, L.I.I. Antimicrobial Resistance of Lactic Acid Bacteria from Nono, a Naturally Fermented Milk Product. *Antibiotics* **2023**, *12*, 843. [CrossRef]
15. Obioha, P.I.; Ouoba, L.I.I.; Anyogu, A.; Awamaria, B.; Atchia, S.; Ojimelukwe, P.C.; Sutherland, J.P.; Ghoddusi, H.B. Identification and characterisation of the lactic acid bacteria associated with the traditional fermentation of dairy fermented product. *Braz. J. Microbiol.* **2021**, *52*, 869–881. [CrossRef]
16. Oladapo, O.; Zakariya, M.; Olaniyi, A.J.; Daodu, O.; Olorunshola, I.; Akpabio, U. Prevalence of *Salmonella* Species in Locally Fermented Milk (Nono) in Gambari Market, Ilorin East Local Government, Kwara State, Nigeria. *Zagazig Vet. J.* **2023**, *51*, 92–100. [CrossRef]
17. Uzoaga, G.O.; Umeokonkwo, C.D.; Usman, A.B.; Kia, G.S.; Okolocha, E.C. Bacteriological quality of Nono, a milk product sold at retail outlets in Federal Capital Territory, Nigeria. *J. Interv. Epidemiol. Public Health* **2020**, 3. [CrossRef]
18. Oyedokun, N.O.; Olukotun, G.B.; Igwegbe, U.I.; Adamu, B.B.; Ideh, R.R.; Shekoni, O.C.; Igwegbe, A.O. Detection, Biochemical and Molecular Characterization of *Clostridium sporogens* in *Nono*: A Nigerian Traditionally Fermented Yoghurt Drink. *Open J. Med. Microbiol.* **2023**, *13*, 91–100. [CrossRef]
19. Dafur, G.S.; Iheukwumere, C.C.; Azua, E.T. Physicochemical and Sensory Quality Assessment of 'Nono' Sold In Mangu Local Government Area of Plateau State, Nigeria. *Int. J. Sci. Res. Publ.* **2018**, *8*, 594–601. [CrossRef]
20. Fagbemigun, O.; Cho, G.S.; Rösch, N.; Brinks, E.; Schrader, K.; Bockelmann, W.; Oguntoyinbo, F.A.; Franz, C.M.A.P. Isolation and characterization of potential starter cultures from the Nigerian fermented milk product nono. *Microorganisms* **2021**, *9*, 640. [CrossRef]
21. Ikele, M.O.; Umeoduagu, N.D.; Nwakoby, N.E.; Ogbukagu, C.M. Efficacy of Probiotic *Lactobacillus casei* in Bio-control of *Escherichia coli* O157:H7 in Nono. *Int. J. Innov. Res. Dev.* **2020**, *9*, 135–138.
22. Fabro, M.A.; Milanesio, H.V.; Robert, L.M.; Speranza, J.L.; Murphy, M.; Rodriguez, G.; Castenada, R. Technical Note: Determination of Acidity in Whole Raw Milk: Comparison of Results Obtained by Two Different Analytical Methods. *Am. Dairy Sci. Assoc.* **2006**, *89*, 859–861. [CrossRef] [PubMed]
23. Bachmann, H.; Huppertz, T.; Kok, J.; Tarazanova, M. Altering textural properties of fermented milk by using surface-engineered *Lactococcus lactis*. *Microb. Biotechnol.* **2018**, *11*, 770–780. [CrossRef]
24. Priyashantha, H.; Buldo, P.; Berg, T.; Gilleladen, C.; Ipsen, R. Understanding the fermentation factors affecting the separability of fermented milk: A model system study. *Food Struct.* **2021**, *30*, 100232. [CrossRef]
25. Li, H.; Gao, J.; Chen, W.; Qian, C.; Wang, Y.; Wang, J.; Chen, L. Lactic acid bacteria isolated from Kazakh traditional fermented milk products affect the fermentation characteristics and sensory qualities of yogurt. *Food Sci. Nutr.* **2022**, *10*, 1451–1460. [CrossRef]
26. Stojanovska, S.; Gruevska, N.; Tomovska, J.; Tasevska, J.; Krstanovski, A.; Menkovska, M. Maillard Reaction and Lactose Structural Changes during Milk Processing. *Chem. Res. J.* **2017**, *2*, 139–145.
27. Laverroux, S.; Picard, F.; Andueza, D.; Graulet, B. Vitamin B2 concentration in cow milk: Quantification by a new UHPLC method and prediction by visible and near-infrared spectral analysis. *Food Chem.* **2021**, *342*, 128310. [CrossRef]
28. Milovanovic, B.I.D.; Miocinovic, J.V.D.; Lorenzo, J.M.; Barba, F.J.; Daniel, M.; Igor, T. What Is the Color of Milk and Dairy Products and how is it measured? *Foods* **2020**, *9*, 1629. [CrossRef]

foods

MDPI

Article

Fatty Acids Profile and Consumers' Preferences of Pecorino Cheese Manufactured from Milk of Sheep Supplemented with Flaxseed and *Ascophyllum nodosum*

Antonella Santillo *, Maria Giovanna Ciliberti, Mariangela Caroprese, Agostino Sevi and Marzia Albenzio

Department of Agriculture, Food, Natural Resources, and Engineering (DAFNE), University of Foggia, Via Napoli, 25, 71122 Foggia, Italy
* Correspondence: antonella.santillo@unifg.it

Abstract: The impact of flaxseed and *Ascophyllum nodosum* supplementation in ewes during the summer season on the fatty acid and sensory profile and consumer preference for cheese was evaluated. Comisana ewes (n = 32) were divided into four groups: a control (CON) group fed (30 days) with pelleted concentrate, a flaxseed (FS) group fed with whole flaxseed supplementation (250 g/ewe per day), an *A. nodosum* (AN) group fed with 5% of *A. nodosum* (into 1 kg/ewe of pelleted concentrate), and an FS + AN group fed with a combination of algae and flaxseed. Pecorino cheeses were analysed after 1 day (curd) and after 45 days (cheese) of ripening. Curd from the FS and FS + AN groups registered higher contents of MUFA, n-3, and n-3/n-6, and lower levels of atherogenic and thrombogenic indexes than curd from the CON and AN groups, as well as a higher content of C18:3n-3, C18:2t9t12, and CLA9c11t, and n-3 and n-3/n-6 fatty acids. Consumers attributed the lowest scores for appearance attributes to AN Pecorino cheese; while Pecorino cheese from FS and FS + AN was judged to have a high-strength flavour attribute and a low rancid, mouldy, and piquant flavour, in comparison with cheese from AN. Flaxseed supplementation could be an effective strategy to improve the nutritional quality of the lipid fraction of cheese without having a detrimental impact on its sensory attributes, especially during the summer season.

Keywords: dairy products; small ruminants; fatty acids; human nutrition

Citation: Santillo, A.; Ciliberti, M.G.; Caroprese, M.; Sevi, A.; Albenzio, M. Fatty Acids Profile and Consumers' Preferences of Pecorino Cheese Manufactured from Milk of Sheep Supplemented with Flaxseed and *Ascophyllum nodosum*. *Foods* **2024**, *13*, 2165. https://doi.org/10.3390/foods13142165

Academic Editor: Rajka Božanić

Received: 29 May 2024
Revised: 1 July 2024
Accepted: 5 July 2024
Published: 9 July 2024

1. Introduction

Pecorino cheese is an Italian cheese made from ewe milk traditionally produced in Central and Southern Italy, contributing to the employment and income in the dairy sheep industry. Particularly, in Italy, several types of typical and traditional Pecorino cheeses are produced which are characterised by a short ripening time and semi-hard consistency [1], are made from raw, thermised or pasteurised milk [2], and are widely consumed as table cheese. However, very few studies describe the impact of different nutritional plans on Pecorino cheese quality and sensory attributes. In the last decades, there has been a re-discovery of natural and historical cheeses by the postmodern consumer which evaluates positively the place of origin by influencing their purchasing decisions towards a willingness to pay premium prices for traditional products [3]. Milk and dairy products are important components of the diet, playing an essential role in meeting nutritional requirements; in particular, cheese is rich in essential nutrients including fat, fatty acids, proteins, peptides, amino acids, vitamins, and minerals. However, cheese also contains relatively high levels of saturated fatty acids (SFAs) which are commonly perceived as negatively impacting the healthfulness of the diet and are associated with increased markers of cardiovascular risk and metabolic syndrome [4]. Dairy sheep farming systems in Mediterranean countries are based mainly on pasture, which is one of the major factors contributing to the enrichment of milk in beneficial fatty acids, especially conjugated linoleic acid (CLA). However, quantitative, and qualitative availability of pasture is greatly influenced by seasonality, and it is characterised by scarcity and poor quality during the summer season, in which dairy ewes are usually in late lactation. At this time of the year, dietary interventions based on different oils and oilseeds supplementation have been demonstrated to be beneficial for improving the fatty acid profile of milk [5]. Moreover, the manipulation of sheep diet with different plant oils has improved the fatty acid profile of cheese by raising the content of n-3 polyunsaturated fatty acid (PUFA) and CLA [6] and reducing the proportion of SFAs, thereby achieving a profile consistent with consumer perception and health recommendation. A limited number of experiments have been conducted on the evaluation of fat dietary supplementation during the summer season on the composition and quality of sheep dairy products, particularly, no previous studies have assessed the role of *A. nodosum* supplementation on dairy sheep products. Moreover, the novelty of the present study was the dietary intervention of sheep with *A. nodosum* supplementation and its combination with flaxseed, based on the hypothesis that PUFA supplementation during the summer season would improve the fatty acid profile of Pecorino-type cheese towards one with health-promoting features for human consumption. Therefore, the primary objective was to assess the effect of supplementation of ewes' diet with flaxseed, *Ascophyllum nodosum*, and their combination on the fatty acid profile of Pecorino curd and Pecorino cheese after 45 days of ripening. A further objective was to study the sensory profile and the consumer's preference for ripened Pecorino cheese.

2. Materials and Methods

2.1. Animals and Experimental Design

The experiment involved 32 late-lactating Comisana ewes and lasted for 5 weeks during the summer season on a dairy farm located in Foggia (Apulia region, Italy). The first 7 days of the trial were considered an adaptation period to the experimental diets. Ewes were balanced for days in milk (200 ± 2, DIM), milk yield (271 ± 8.42 g/d), body weight (55.15 ± 1.08 kg, BW), and body condition score (2.53 ± 0.1, BCS), and were divided into four experimental groups that were individually fed twice daily. The control group (CON, n = 8) was fed with 1 kg/ewe/d of pellet concentrate (Mangimificio Molino Gallo, Potenza, Italy); the FS group (n = 8) received a supplementation of 250 g/ewe/d of whole flaxseed (Lin Tech. Tecnozoo srl, Torreselle di Piombino Dese, Italy) which was substituted to the same amount of pellet concentrate; the AN group (n = 8) received supplementation of 5% *A. nodosum* directly incorporated into pellet concentrate (Tasco); and the FS + AN group (n = 8) received supplementation with 250 g/ewe/d of flaxseed and 750 g/ewe/d of pelleted concentrate with 5% *A. nodosum* incorporated in it. The experimental groups also individually received 1.8 kg/ewe/d of oat hay with water offered ad libitum. The EU Directive 2010/63/EU guidelines [7] on the protection of animals used for experimental and other scientific purposes were followed. Animals were carefully examined by veterinarians throughout the trial to monitor their health condition. Dry matter intake (DMI) was determined by weighing the refusals four times per day (0800, 1200, 1600, and 2000 h). No differences were found for DMI among groups, being 2.62 ± 0.04 kg/ewe/d (mean value ± SEM). Experimental diets were analysed for the fatty acid composition [8]. In brief, the flaxseed supplement was characterised by a total of 53.21% of C18:3n-3 (alpha-linolenic acid, ALA), while the *A. nodosum* supplement by a total of 37.03% C18:1 cis-9 (oleic acid) and 5.03% C20:5n3 (Eicosapentaenoic acid, EPA) calculated on fatty acids.

2.2. Sampling and Chemical Analyses of Milk and Pecorino Cheese

At the end of the dietary treatments, ewes' bulk milk from five consecutive milkings (morning and afternoon milking from 2.5 days) was collected to manufacture Pecorino cheese from each experimental group. Two cheese-making trials were performed. One fresh milk aliquot from each experimental group was collected for chemical composition determination, in terms of pH (GLP 21 Crison, Barcelona, Spain), fat, total protein, lactose, and casein content, by using MilkoScanTM FT120 (Foss Electric. DK-3400 Hillerød, Denmark) according to the International Dairy Federation standard [9]. Moreover, the evaluation of somatic cell count (SCC) by a Foss Electric Fossomatic 90 cell counter, and the renneting milk characteristics, (clotting time, rate of clot formation, and clot firmness after 30 min) using a Foss Electric formagraph, were performed. The traditional protocol of Pecorino cheese-making is shown in Figure 1.

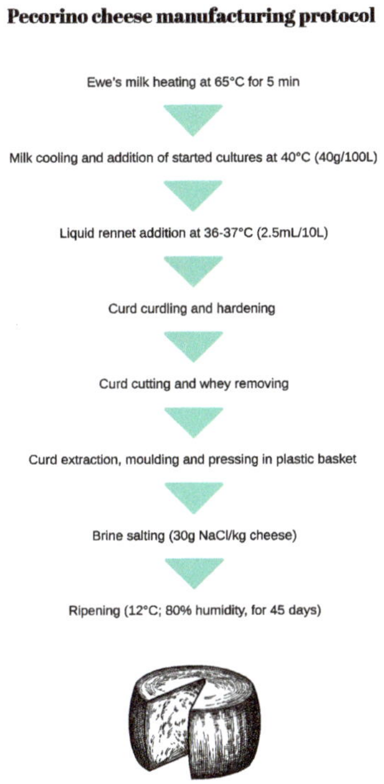

Figure 1. Pecorino cheese manufacturing protocol, adapted from [10].

The chemical composition of Pecorino curd (1 day of ripening) and Pecorino cheese (45 days of ripening) was determined in terms of pH, dry matter content, moisture, and NaCl. The total nitrogen (TN), non-casein nitrogen (NCN), and non-protein nitrogen (NPN) levels were determined using the standard Kjeldahl method procedures [11]. The casein nitrogen (CN) content in the curd and cheese samples was calculated by the formula CN = (TN − NCN) × 6.38 (conversion factor), the whey protein (WP) content by the formula WP = (NCN − NPN) × 6.38 (conversion factor), and the fat content by using the Soxhlet method.

2.3. Fatty Acids Profile of Pecorino Curd and Cheese

Extraction of fatty acids from Pecorino curd and cheese samples was performed following the O'Fallon et al. [8] procedure. In brief, 1.0 g of cheese sample was placed into a screw-capped Pyrex tube (16 × 125 mm) and treated with C13:0 internal standard (0.5 mg of C13:0/mL of methanol), 0.7 mL of 10 N KOH, and 5.3 mL of methanol. The mixture was incubated for 1.5 h at 55 °C in a water bath and shaken every 20 min. After cooling the tube to below room temperature, 0.58 mL of H_2SO_4 (24 N) was added and the sample was mixed by inversion and incubated for 1.5 h at 55 °C in a water bath and shaken every 20 min. Subsequent to the cooling step, which was performed in a cold tap water bath, 3 mL of hexane was added to the tube and it was vortexed for 5 min. The fatty acid methyl ester (FAME) was collected in the hexane layer obtained after centrifugation at 500× *g* for 5 min and then placed into a GC vial and stored at −20 °C until capillary gas chromatography (CG) analysis was performed. Specifically, a capillary column (Agilent Technologies Inc. Santa Clara, CA, HP-88, 100 m × 0.25 mm × 0.20 μm) installed on an Agilent Technologies 6890N GC equipped with a flame-ionisation detector and a split injection was used. The starting oven temperature was 70 °C (held for 4 min), it was increased to 175 °C (rate of 13 °C min^{-1}, held for 27 min), and finally raised to 215 °C (rate of 4 °C min^{-1}, held for 45 min). The carrier gas was represented by helium with a column head pressure of 175 kPa. The temperature of both the injector and the detector was 250 °C, and the split ratio was set at 20:1. Fatty acids were identified by comparing the retention times of the fatty acids of the samples with those of the fatty acid methyl standards (FIM-FAME-7-Mix, Matreya LLC, Pleasant Gap, PA, USA) and C18:1 trans-11, C18:2 cis-9,trans-11, C18:2 cis-9,cis-11, C18:2 trans-9,trans-11, and C18:2 trans-10,cis-12 (Matreya LLC, State College, PA, USA). The peak areas were quantified using Agilent Chemstation software (B.04.03). Short-chain fatty acids (SCFAs) represent the sum of C4:0, C6:0, C8:0, C10:0, and C12:0. Medium-chain fatty acids (MCFAs) represent the sum of C14:0, C14:1c, C15:0, C16:0, C16:1c, C17:0, and C17:1c. Long-chain fatty acids (LCFAs) represent the sum of C18:0, C18:1t11, C18:1c6, C18:1c9, C18:2t9t12, C19:1t10, C19:1t7, C18:2c9c12, C20:0, C18:3n3, CLA9c11t + C20:1c11, CLA 10t,12c, C22:0, C20:4n6, C20:5n3, and C22:5n3. Atherogenic and thrombogenic indexes were calculated using the Ulbricht and Southgate [12] formulas as follows: atherogenic index (AI) = (C12:0 + 4 × C14:0 + C16:0)/[Σ MUFA + Σ PUFA(n-6) and (n-3)]; thrombogenic index (TI) = (C14:0 + C16:0 + C18:0)/[0.5 × Σ MUFA + 0.5 × Σ PUFA(n-6) + 3 × Σ PUFA(n-3) + (n-3)/(n-6)].

2.4. Descriptive Sensory Analysis and Consumer Test

Staff and students of the DAFNE Department of the University of Foggia were recruited for Pecorino cheese (45 days of ripening) sensory analysis. Subject information and consent forms are reported in the Supplementary Materials (Figure S1). Firstly, consumers were asked to fill out a questionnaire (Figure S2) which included information about age, sex, and the frequency of consumption of Pecorino cheese. The enrolled panel consisted of a total of 58 consumers, characterised by men and women (46 and 54%, respectively), aged 21–43, who consumed cheese regularly (frequency of consumption of at least once a week). The panellists were trained for sensory evaluation procedures and shared a specific vocabulary describing the definitions of cheese attributes. Table 1 outlines the vocabulary used for cheese appearance, colour, odour, and flavour attribute characterisation.

Table 1. Descriptive vocabulary and their definition, adapted from [13,14] used by panellists to evaluate Pecorino cheese in the present study.

Attributes	Definition
Appeareance	
Chalky	Resembling chalk in appearance
Uniformity	Absence of cracks, pinholes, irregular-shaped holes
Grainy	The extent to which granular structures are formed as the sample breaks down (perceived in the second half of chewing)
Colour	
Mottling	The evenness of color shading within the cheese sample, with the most uniformly coloured cheese being free of mottling, marbling, or any other deficiencies in color
Colour Intensity	The color of cheese, ranging from pale yellow to orange, with the palest of yellow representing the start of the scale
Odour	
Strength	The overall intensity of aroma and flavor; the degree of mildness and maturity
Acidic	The smell associated with lactic and citric acids
Rancid	The smell associated with sour milk and oxidised fats, having the rank of an unpleasant aroma characteristic of oils and fats when no longer fresh
Flavour	
Strength	The overall intensity of aroma and flavor; the degree of mildness and maturity
Salty	The fundamental taste sensation of which sodium chloride is typical
Acidic	The fundamental taste sensation of which lactic and citric acids are typical
Piquant	The taste associated with an irritating or aggressive sensation perceived in the mouth or in the throat
Bitter	The fundamental taste sensation of which caffeine and quinine are typical
Sweet	The fundamental taste sensation of which sucrose is typical
Mouldy	The taste associated with moulds, usually earthy, dirty, stale, musty, and slightly sour
Rancid	The taste associated with sour milk and oxidised fats, having the rank of an unpleasant aroma or taste characteristic of oils and fats when no longer fresh

Pecorino cheeses that were ripened for 45 days were allowed to remain at room temperature (22 °C) for 1 h prior to the panel consumer test to achieve an optimum condition for sensory evaluation and obtain homogeneous cuts. The consumer test procedure was previously described in [13]. Briefly, each cheese sample was assigned a random number, and slices of 1.5 × 1.5 × 1.5 cm from all four experimental groups were randomly offered to the panellists. After each cheese tasting, the consumers were invited to take a small piece of unsalted crispy bread and drink a small quantity of water. Panellists used a 10-point intensity scale to sign the cheeses' perception. For the acceptance rating test, panellists were requested to express their overall liking on a 10-point hedonic scale from 0 (dislike extremely) to 10 (like extremely), with a neither like nor dislike neutral centre point. The order of cheese presentation between each consumer test session was performed to minimise any carryover effects [15]. At the end of the sessions, panellists were invited to express their preferred cheese among the ones tested.

2.5. Statistical Analysis

Data on the chemical composition of milk and the fatty acids profile of both the Pecorino curd and cheese were tested to determine a normal distribution. The differences between experimental groups were determined by ordinary one-way analysis of variance (ANOVA) with the Least Square Difference (LSD) post-hoc test for multiple pairwise comparisons. A value of $p < 0.10$ was considered as a tendency. The ability of the descriptive vocabulary to discriminate among cheeses was tested using one-way ANOVA and LSD for multiple pairwise comparison tests of the panel mean scores for each cheese. Principal component analysis was applied to a matrix of 16 sensory attributes (Appearance: chalky, uniformity, and grainy, Colour: colour uniformity, mottling, and intensity; Odour: strength, acidic, rancid; Flavour: strength, salty, acidic, piquant, bitter, sweet, mouldy, and rancid) using the PRINCOMP procedure of SAS, then the main significant two principal components were analysed using factorial analysis. All analyses were carried out in the statistical software SAS version 9.4, accessed through SAS University Edition [16].

3. Results and Discussion

3.1. Milk Chemical Composition

Recently, there has been a heightened interest in dairy production towards the enhancement of its human health-promoting properties, especially regarding the quality of the fat fraction which can be modified by proper fat dietary interventions. Notably, the response of dairy animals to fat supplementation may be related to fat source, level of supplementation, and stage of lactation [17]. In the present study, different sources of PUFA were added to the diet of sheep in late lactation during the summer season with the aim of studying their effect on milk and Pecorino cheese quality, with a focus on the fatty acids composition, sensory profile, and consumers' preference of Pecorino cheese. The effect of FS, AN, and the combination of FS and AN on bulk-milk composition used for Pecorino cheese-making is presented in Table 2.

No significant effect was registered in terms of milk yield (356.75 g/ewe/d $\pm$ 20.27) among experimental groups. Bulk milk significantly differed in fat (p = 0.03), with the AN and FS + AN groups showing higher fat content than the CON and FS groups. It was reported that the effects of marine algae supplementation in ewes' feeding are complex and not univocal, mainly due to different basal diet compositions and dosages. Accordingly, previous results on the milk fat content in ewes demonstrated its depression [18], no change [19], or increase [20]. In the present paper, AN alone or in combination with FS seemed to sustain the fat content, probably due to the low level of CLAt10c12 found which is responsible for milk fat depression syndrome [6]. Additionally, the literature on the effect of whole flaxseed on milk fat content is controversial; however, according to the results of the present study, most of the experimental trials on the administration of whole flaxseed report no effect on ewe milk fat content [21]. Protein content in milk tended to be higher ($p < 0.10$) in the AN and FS + AN groups than in the CON one, while casein content increased in all the supplemented groups ($p < 0.018$). The positive effect on milk protein secretion may be ascribed to the phlorotannins yielded by *A. nodosum* supplementation; such compounds are able to make complexes with proteins and carbohydrates thus reducing dietary protein degradation in the rumen environment and allowing the escaped amino acids to support milk protein synthesis in the mammary gland [22].

Table 2. Bulk-milk chemical composition for Pecorino cheese production as affected by the experimental diets.

	Experimental Diets [1]					
Items	**CON**	**AN**	**FS**	**FS + AN**	**SEM**	***p*-Value**
pH	6.57	6.60	6.64	6.57	0.029	NS
Fat, %	5.75 ^{b}	6.78 ^{a}	5.74 ^{b}	6.72 ^{a}	0.204	*
Protein, %	5.37 ^{b}	5.85 ^{a}	5.64 ^{ab}	5.83 ^{a}	0.084	<0.10
Lactose, %	4.26	4.64	4.56	4.66	0.074	NS
Casein, %	3.95 ^{b}	4.57 ^{a}	4.30 ^{a}	4.55 ^{a}	0.100	*
SCC [2], log_{10} n. cell/mL	2.81	3.10	3.04	2.89	2.31	NS
r [3], min	6.45	6.73	7.45	5.80	0.458	NS
a_{30} [4], mm	50.50	51.51	59.16	54.80	1.901	NS
k_{20} [5], min	1.30	1.23	1.30	1.23	0.032	NS

$^{a–b}$ Mean values in the same row with different superscripts differ ($p < 0.05$). * $p < 0.05$, <0.10 tendence, NS = $p >$ 0.10. [1] CON = sheep fed conventional diet; AN = sheep supplemented with 5% *Ascophyllum nodosum* directly incorporated into pellet concentrate (Tasco); FS = sheep supplemented with 250 g/ewe/d of whole flaxseed; FS + AN= sheep supplemented with 250 g/ewe/d of flaxseed and 750 g/ewe/d of pelleted concentrate with incorporated 5% *A. nodosum*. [2] SCC= Somatic Cell Count. [3] r = rennet coagulation time (min). [4] a_{30}= curd firmness 30 min after enzyme addition (mm). [5] k_{20} curd firmness traits (min) [time to curd firmness of 20 mm (k_{20})].

Moreover, in cows supplemented with *A. nodosum*, an improvement in macromineral content in milk, in particular, P and Ca, was observed due to the high mineral content supplied by seaweed [23]. Sheep milk contains higher Ca, P, and Mg levels than other common milk sources; these minerals are largely associated with the colloidal casein micelles [24]. Based on the previous statements, AN supplementation may contribute to improved Ca and P availability in the mammary gland, in turn enhancing the efficiency of casein micelle synthesis. Concerning whole flaxseed supplementation, it is widely reported that there is a suppressive effect of dietary fat on milk protein synthesis due to the reduced availability of amino acids to the mammary gland [21]. It was previously reported that flaxseed supplementation is able to sustain the immune response during heat stress exposition, mediated by C18:3 n-3 content [25]. Therefore, the increased level of casein observed in the FS group compared to the CON group could be an outcome of the improved health status of the late lactating ewes sustaining the secretory pattern of the mammary gland during the summer season. Finally, no significant differences were registered among experimental groups in terms of the pH, somatic cell count, and renneting properties of the bulk milk destined for Pecorino cheese-making.

3.2. Pecorino Curd Chemical and Fatty Acids Composition

The gross composition of Pecorino curds was not different across the experimental groups, denoting that the cheese-making technology of traditional cheese was able to standardise curd gross composition. Mean composition was 60.87 ± 0.6% (mean value ± SEM), 41.55 ± 1.38%, and 20.88 ± 1.15% for moisture, fat, and protein, respectively. The fatty acid composition of Pecorino curds obtained from the milk of ewes supplemented with different sources of PUFA is reported in Table 3.

Table 3. Fatty acids composition of Pecorino curd as affected by the experimental diets.

	Experimental Diets [1]					
Item	**CON**	**AN**	**FS**	**FS + AN**	**SEM**	***p*-Value**
FA, g/100 g of FA						
C4:0	3.38	4.28	3.58	3.99	0.233	NS
C6:0	1.69 ^{b}	2.56 ^{a}	1.42 ^{b}	1.79 ^{b}	0.144	*
C8:0	1.58 ^{b}	2.54 ^{a}	1.24 ^{b}	1.59 ^{b}	0.150	*
C10:0	4.54 ^{b}	7.14 ^{a}	3.47 ^{b}	4.38 ^{b}	0.401	**
C12:0	3.14 ^{b}	4.32 ^{a}	2.64 ^{b}	2.98 ^{b}	0.181	**
C14:0	10.52 ^{a}	11.15 ^{a}	8.37 ^{b}	8.64 ^{b}	0.327	*
C16:0	28.51 ^{a}	26.04 ^{b}	22.54 ^{c}	21.99 ^{c}	0.746	***
C16:1c	1.32 ^{a}	1.25 ^{a}	1.07 ^{b}	0.98 ^{b}	0.037	***
C18:0	8.59 ^{b}	7.26 ^{b}	9.71 ^{ab}	10.54 ^{a}	0.361	**
C18:1t11	2.11 ^{c}	3.28 ^{b}	6.21 ^{a}	3.79 ^{b}	0.411	***
C18:1c9	23.81 ^{b}	20.10 ^{c}	25.78 ^{ab}	26.90 ^{a}	0.733	***
C18:2t9t12	0.13 ^{b}	0.11 ^{b}	0.18 ^{a}	0.13 ^{b}	0.008	*
C18:2c9c12	2.85 ^{a}	2.73 ^{ab}	2.43 ^{c}	2.55 ^{bc}	0.050	**
C18:3n3	0.76 ^{b}	0.98 ^{b}	2.07 ^{a}	2.05 ^{a}	0.163	***
CLA9c11t	0.64 ^{c}	0.93 ^{bc}	1.93 ^{a}	1.31 ^{b}	0.139	***
CLAt10c12	0.04 ^{c}	0.05 ^{b}	0.11 ^{a}	0.04 ^{c}	0.007	***
C22:0	0.10 ^{a}	0.04 ^{b}	0.05 ^{ab}	0.04 ^{b}	0.009	*
C20:4n6	0.20 ^{ab}	0.23 ^{b}	0.19 ^{a}	0.17 ^{b}	0.008	<0.10
C20:5n3	0.08 ^{a}	0.05 ^{b}	0.07 ^{ab}	0.06 ^{ab}	0.005	NS
C22:5n3	0.08 ^{a}	0.06 ^{b}	0.09 ^{a}	0.09 ^{a}	0.003	**

$^{a–c}$ Mean values in the same row with different superscripts differ ($p < 0.05$). * $p < 0.05$, ** $p < 0.01$, *** $p < 0.001$, <0.10 tendence, NS = $p > 0.10$. [1] CON = sheep fed conventional diet; AN = sheep supplemented with 5% *Ascophyllum nodosum* directly incorporated into pellet concentrate (Tasco); FS = sheep supplemented with 250 g/ewe/d of whole flaxseed; FS + AN = sheep supplemented with 250 g/ewe/d of flaxseed and 750 g/ewe/d of pelleted concentrate with incorporated 5% *A. nodosum*.

Different sources of dietary PUFAs affected the fatty acid profile of cheese curd differently. The AN curd was characterised by the highest concentration of fatty acids, from C6:0 to C14:0. The dietary intervention with FS and FS + AN supplementation significantly decreased the level of C16:0, which was intermediate in the AN group and showed the highest values in the CON group. Both the levels of C18:0 and C18:1c9 were highest in the

FS and FS + AN groups. Moreover, plain flaxseed supplementation resulted in higher levels of C18:1t11, CLA 9c11t, and CLA t10c12 fatty acids than the other experimental groups.

The fatty acid indexes of Pecorino curd are presented in Table 4.

The SCFA level was the highest in the AN group ($p = 0.004$), and the MCFA level was higher in the AN and CON groups than the FS and FS + AN groups ($p < 0.001$). On the contrary, these last experimental groups registered higher levels of LCFAs. Moreover, the sole FS supplementation showed the lowest level of SFAs and the highest level of PUFA and n-6. The content of MUFA, n-3, and the index n-3/n-6 was higher in the FS and FS + AN groups than in the CON and AN groups. As expected, both the FS and FS + AN groups showed lower levels of AI and TI indexes than the CON and AN groups ($p < 0.001$). Within the medium chain FAs, flaxseed supplementation improved the fatty acid composition of Pecorino curd by reducing the level of C14:0 by about 19%, and C16:0 by 21%, with both fatty acids being mainly involved in human cardiovascular risk [26]. The decreased proportions of the mentioned fatty acids in cheese curd depend directly on the conditions of milk production; accordingly, in buffalo milk, a decreased proportion of medium-chain fatty acids was reported [27] due to the whole flaxseed which can alter the ruminal ability to produce fatty acids. Furthermore, the supplementation of flaxseed exerted a leading role in the enhancement of beneficial molecules for human consumption, including vaccenic acid (C18:1t11) and CLA. Interestingly, both flaxseed and the combination of flaxseed and *A. nodosum* contributed to obtaining a favourable distribution of fatty acids in terms of the degree of acidic carbon chain unsaturation, as demonstrated by the improvement of n-3, n-3/n-6, AI, and TI of the curd, indexes which are considered crucial for the prevention and management of chronic disease [28]. In accordance with our findings, sheep supplemented with flaxseed and a combination of flaxseed and *A. nodosum* resulted in reduced levels of AI and TI indexes in individual sheep milk during the summer season [5].

Table 4. Fatty acid indexes of Pecorino curd as affected by the experimental diets.

	Experimental Diets [1]					
Item	CON	AN	FS	FS + AN	SEM	p-Value
SCFA [2]	13.34 [b]	20.84 [a]	12.35 [b]	14.74 [b]	1.32	**
MCFA [3]	42.45 [a]	40.14 [a]	33.80 [b]	33.25 [b]	0.75	***
LCFA [4]	39.92 [b]	36.70 [b]	50.85 [a]	49.44 [a]	1.06	***
SFA [5]	64.57 [b]	67.48 [a]	55.47 [d]	58.15 [c]	1.289	***
MUFA [6]	30.30 [b]	27.12 [c]	37.07 [a]	35.16 [a]	1.059	***
PUFA [7]	5.13 [c]	5.41 [c]	7.45 [a]	6.70 [b]	0.260	***
P/S [8]	0.08 [c]	0.08 [c]	0.13 [a]	0.12 [b]	0.006	***
n-6	4.15 [b]	4.29 [b]	5.18 [a]	4.46 [b]	0.123	**
n-3	0.97 [b]	1.13 [b]	2.27 [a]	2.24 [a]	0.165	***
n-3/n-6	0.24 [b]	0.26 [b]	0.44 [a]	0.50 [a]	0.031	***
AI [9]	2.08 [b]	2.32 [a]	1.32 [c]	1.43 [c]	0.113	***
TI [10]	2.34 [a]	2.30 [a]	1.43 [b]	1.53 [b]	0.111	***

[a–c] Mean values in the same row with different superscripts differ ($p < 0.05$). ** $p < 0.01$, *** $p < 0.001$. [1] CON = sheep fed conventional diet; AN = sheep supplemented with 5% *Ascophyllum nodosum* directly incorporated into pellet concentrate (Tasco); FS = sheep supplemented with 250 g/ewe/d of whole flaxseed; FS + AN = sheep supplemented with 250 g/ewe/d of flaxseed and 750 g/ewe/d of pelleted concentrate with incorporated 5% *A. nodosum*. [2] SCFA= Short Chain Fatty Acids. [3] MCFA= Medium Chain Fatty Acids. [4] LCFA = Long Chain Fatty Acids. [5] SFA = Saturated Fatty Acids. [6] MUFA = Monounsaturated Fatty Acids. [7] PUFA = Polyunsaturated Fatty Acids. [8] P/S = Polyunsaturated Fatty Acids/Saturated Fatty Acids. [9] AI = Atherogenic Index. [10] TI = Thrombogenic index.

3.3. Pecorino Cheese Chemical and Fatty Acids Composition

As for the Pecorino curds, the chemical composition of Pecorino cheeses ripened for 45 d did not change across the experimental treatments; mean values for moisture of 36.39 ± 0.81%, for fat of 34.66 ± 2.01%, for protein of 35.31 ± 3.41%, and for casein of 30.51 ± 3.01% were found.

The fatty acid composition of Pecorino cheeses after 45 days of ripening is presented in Table 5.

Table 5. Fatty acid composition of Pecorino cheese as affected by the experimental diets.

	Experimental Diets [1]					
Item	**CON**	**AN**	**FS**	**FS + AN**	**SEM**	***p*-Value**
FA, g/100 g of FA						
C4:0	3.63	3.45	2.53	2.98	0.244	NS
C6:0	1.85 ^{ab}	2.13 ^{a}	1.19 ^{b}	1.41 ^{ab}	1.449	NS
C8:0	1.78 ^{ab}	2.2 ^{a}	1.18 ^{b}	1.4 ^{b}	0.137	*
C10:0	5.03 ^{ab}	6.26 ^{a}	3.47 ^{c}	3.9 ^{bc}	0.348	**
C12:0	3.62 ^{ab}	3.93 ^{a}	2.80 ^{b}	2.96 ^{b}	0.178	<0.10
C14:0	12.71 ^{a}	10.71 ^{ab}	9.49 ^{b}	9.46 ^{b}	0.549	<0.10
C16:0	28.11	26.75	27.96	26.64	2.458	NS
C16:1c	1.63 ^{a}	1.23 ^{ab}	1.26 ^{ab}	1.12 ^{b}	0.087	NS
C18:0	10.91 ^{b}	8.13 ^{c}	12.52 ^{b}	14.72 ^{a}	0.697	**
C18:1t11	3.36 ^{b}	3.79 ^{b}	9.21 ^{a}	5.73 ^{b}	1.05	**
C18:1c9	29.47	21.43	13.64	14.34	3.095	NS
C18:1c11	0.44	0.42	0.57	0.49	0.032	NS
C18:2t9t12	0.18 ^{b}	0.17 ^{b}	0.63 ^{a}	0.68 ^{a}	0.077	**
C18:2c9c12	3.67	2.92	3.07	3.23	0.176	NS
C18:3n3	0.79 ^{b}	0.80 ^{b}	1.93 ^{a}	2.16 ^{a}	0.182	**
CLA9c11t	0.93 ^{b}	0.89 ^{b}	1.97 ^{a}	1.77 ^{a}	0.165	*
CLAt10c12	0.05 ^{b}	0.06 ^{b}	0.11 ^{a}	0.06 ^{b}	0.007	**
C22:0	0.12 ^{a}	0.07 ^{b}	0.10 ^{ab}	0.08 ^{ab}	0.009	NS
C20:4n-6	0.21 ^{a}	0.18 ^{ab}	0.16 ^{b}	0.17 ^{ab}	0.008	NS
C20:5n-3	0.09	0.06	0.07	0.09	0.007	NS
C22:5n-3	0.09 ^{ab}	0.07 ^{b}	0.10 ^{a}	0.11 ^{a}	0.006	*

$^{a–c}$ Mean values in the same row with different superscripts differ ($p < 0.05$). * $p < 0.05$, ** $p < 0.01$, <0.10 tendence, NS = $p > 0.10$. [1] CON = sheep fed conventional diet; AN = sheep supplemented with 5% *Ascophyllum nodosum* directly incorporated into pellet concentrate (Tasco); FS = sheep supplemented with 250 g/ewe/d of whole flaxseed; FS + AN = sheep supplemented with 250 g/ewe/d of flaxseed and 750 g/ewe/d of pelleted concentrate with incorporated 5% *A. nodosum*.

Pecorino cheese produced with milk from the FS and FS + AN groups showed lower levels of C8:0 and C10:0 fatty acids than that of the AN group. The FS and FS + AN groups showed a tendency in the level of C12:0 and 14:0, which were lower than the CON and AN groups. The FS + AN group showed the highest level of C18:0, followed by the FS and CON groups ($p < 0.001$). The C18:1t11 content was the highest in the FS group ($p = 0.008$), while the contents of C18:3n3, C18:2t9t12, and CLA9c11t fatty acids were higher in the FS and FS + AN groups than in the CON and AN groups. The fatty acid indexes of Pecorino cheese are presented in Table 6.

Table 6. Fatty acid indexes of Pecorino cheese after 45 days of ripening as affected by the experimental diets.

	Experimental Diets [1]					
Item	CON	AN	FS	FS + AN	SEM	p-Value
SCFA [2]	15.91 [ab]	17.96 [a]	11.18 [b]	12.75 [ab]	1.67	*
MCFA [3]	30.93	40.50	41.07	39.28	3.96	NS
LCFA [4]	53.90	43.35	54.65	51.17	3.37	NS
SFA [5]	56.48	65.56	63.76	65.97	2.066	NS
MUFA [6]	37.05	28.95	28.17	25.27	2.376	NS
PUFA [7]	6.48 [ab]	5.49 [b]	8.59 [a]	8.76 [a]	0.531	*
P/S [8]	0.12 [ab]	0.08 [b]	0.14 [a]	0.13 [a]	0.008	*
n-6	5.43	4.52	6.43	6.34	0.380	NS
n-3	1.05 [b]	0.97 [b]	2.16 [a]	2.42 [a]	0.189	***
n-3/n-6	0.21 [b]	0.21 [b]	0.34 [a]	0.39 [a]	0.023	***
AI [9]	1.62	2.14	2.1	2.15	0.162	NS
TI [10]	1.62	2.3	2.21	2.22	0.160	NS

[a–b] Mean values in the same row with different superscripts differ ($p < 0.05$), * $p < 0.05$, *** $p < 0.001$, NS = $p > 0.10$. [1] CON = sheep fed conventional diet; AN = sheep supplemented with 5% *Ascophyllum nodosum* directly incorporated into pellet concentrate (Tasco); FS = sheep supplemented with 250 g/ewe/d of whole flaxseed; FS + AN = sheep supplemented with 250 g/ewe/d of flaxseed and 750 g/ewe/d of pelleted concentrate with incorporated 5% *A. nodosum*. [2] SCFA = Short Chain Fatty Acids. [3] MCFA = Medium Chain Fatty Acids. [4] LCFA = Long Chain Fatty Acids. [5] SFA = Saturated Fatty Acids. [6] MUFA = Monounsaturated Fatty Acids. [7] PUFA = Polyunsaturated Fatty Acids. [8] P/S = Polyunsaturated Fatty Acids/Saturated Fatty Acids. [9] AI = Atherogenic Index. [10] TI = Thrombogenic index.

The SCFA levels were higher in the AN group than in the FS group; the PUFA levels were higher in the FS and FS + AN groups than in the AN group. The n-3 and n-3/n-6 levels were higher in the FS and FS + AN groups than in the CON and AN groups. No significant differences emerged in the indexes AI and TI among the experimental groups. As for curd, the occurrence of the highest levels of both C18:1t11 and CLA in the FS group is attributed to the increase in biohydrogenation activity in the rumen supported by C18:3n-3 acid, which represented about 53% of the fatty acids yielded by flaxseed supplementation. Furthermore, vaccenic acid, which showed the highest content in the Pecorino curd and cheese, plays a key role in the synthesis of CLA cis-9, trans-11 [29] via both biohydrogenation conducted by ruminal bacteria and Δ9-desaturase activity in the udder. On the contrary, *A. nodosum* is mainly rich in eicosapentaenoic acid (representing about 5.03% of fatty acid), which was not found in different levels between the experimental groups. This result could be explained by the low transfer efficiency of EPA from the diet into milk, due to its extensive biohydrogenation at the rumen level. Moreover, the limited supply of EPA to the mammary gland, and therefore to milk, was also ascribed to the competition of its use between the mammary gland and adipose tissue [30].

The supplementation of flaxseed and the combination of flaxseed and *A. nodosum* resulted in an improved content of PUFA, n-3, and n-3/n-6 ratio also in Pecorino cheese ripened for up to 45 d, compared with plain *A. nodosum* supplementation. Therefore, especially in rearing systems where pasture is not available or during the time of the year characterised by scarcity and low-quality pasture, flaxseed supplementation can represent a feasible strategy capable of enhancing the nutritional value of the fat fraction of Pecorino cheese.

3.4. Descriptive Sensory Analysis and Consumer Test

The sensory attributes of Pecorino cheeses produced with milk from different dietary supplementations were categorised for appearance, colour, odour, and flavour (Table 7).

Table 7. Sensory attributes assessed by panellists after the consumer test of the Pecorino cheeses as affected by the experimental diets.

Attributes	Experimental Diets [1]					
	CON	AN	FS	FS + AN	SEM	*p*-Value
Appeareance						
Chalky	5.05	5.21	4.95	4.53	0.15	0.405
Uniformity	5.97 ^{a}	5.215 ^{b}	6.19 ^{a}	6.19 ^{a}	0.14	*
Grainy	5.19 ^{ab}	5.55 ^{a}	4.83 ^{b}	4.24 ^{b}	0.15	*
Colour						
Colour Uniformity	6.39	6.03	6.45	6.47	0.12	NS
Colour intensity	5.69 ^{a}	4.53 ^{b}	6.03 ^{a}	5.48 ^{a}	0.14	**
Odour						
Strength	5.92	5.66	5.53	5.91	0.12	NS
Acidic	3.28	3.64	3.00	3.07	0.15	NS
Rancid	2.14	2.19	2.05	2.02	0.15	NS
Flavour						
Strength	6.31 ^{ab}	6.60 ^{a}	5.72 ^{b}	6.36 ^{a}	0.12	<0.10
Salty	5.76	5.52	5.34	5.12	0.13	NS
Acidic	3.91	4.26	3.53	3.90	0.16	NS
Piquant	2.40	2.79	2.78	2.29	0.15	NS
Bitter	2.83	3.31	2.84	3.24	0.16	NS
Sweet	2.60	2.76	3.48	2.93	0.17	NS
Mould	1.41	1.55	1.17	1.21	0.13	NS
Rancid	2.19	2.24	1.62	1.86	0.16	NS

$^{a–b}$ Mean values in the same row with different superscripts differ ($p < 0.05$), * $p < 0.05$, ** $p < 0.01$, <0.10 tendency, NS = $p > 0.10$. [1] CON = sheep fed conventional diet; AN = sheep supplemented with 5% *Ascophyllum nodosum* directly incorporated into pellet concentrate (Tasco); FS = sheep supplemented with 250 g/ewe/d of whole flaxseed; FS + AN = sheep supplemented with 250 g/ewe/d of flaxseed and 750 g/ewe/d of pelleted concentrate with incorporated 5% *A. nodosum*.

Panellists scored the lowest appearance uniformity for AN Pecorino cheese ($p = 0.035$), while the grainy attribute was judged lower in the FS + AN and FS Pecorino cheeses than in the AN one ($p = 0.018$). Moreover, the perception of colour intensity showed the lowest value in the AN Pecorino cheese ($p = 0.001$). Odour attributes, in terms of strength, acidity, and rancidity were found comparable among Pecorino cheeses. As regards the flavour attributes, no significant differences emerged among experimental Pecorino cheeses, only the strength attribute tended to be lower ($p = 0.06$) in the FS cheese than in both the AN and FS + AN Pecorino cheeses.

For the PCA results on sensory attributes, the first two principal components (PCs) were chosen as the main representatives (Table S1). In particular, PC1 accounted for 24.93% of the total variance, with the main attributes that were positively correlated being rancidity, mould, and having a piquant flavour and a rancid and acidic odour. Furthermore, the second principal component (PC2 = 15.4% of total variance) was positively characterised by the following attributes: appearance uniformity, colour intensity, and uniformity. Figure 2 shows that the FS and FS + AN Pecorino cheeses were characterised by a negative PC1 score.

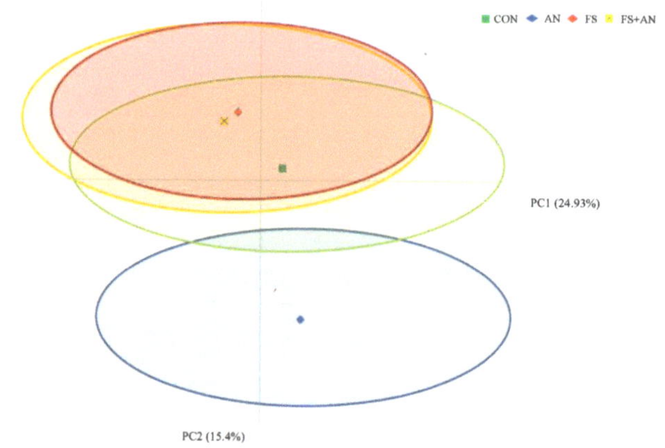

Figure 2. Principal Components Analysis of sensory parameters of Pecorino cheese produced from milk of sheep supplemented with flaxseed (FS, 250 g/ewe/d), *Ascophyllum nodosum* (AN, pelleted concentrate with incorporated 5% *A. nodosum*), the combination of flaxseed and *A. nodosum* (FS + AN, 250 g/ewe/d of flaxseed and 750 g/ewe/d of pelleted concentrate with incorporated 5% *A. nodosum*), or not supplemented (CON).

On the contrary, the AN cheese was in a well-defined zone of the plot with a negative PC2 score, categorising cheese with a lower appearance attribute. The CON cheese was located in the central part of the plot characterised by lower levels of both PC1 and PC2. The definition of a descriptive vocabulary is a key point to support the promotion and commercialisation of hard ovine cheese, especially to test the impact of managerial choices on the quality of typical and traditional dairy products. Consumers' attribute evaluations of Pecorino cheese from *A. nodosum* supplementation were characterised by reduced appearance, uniformity, and colour intensity. These attributes were perceived negatively by the panellists, influencing the position of the AN Pecorino cheese depicted in a defined zone of PCA which described cheese with a lower appearance attribute. In Torri et al. [31], the consumers' preferences for cheese enriched with grape skin powder (Barbera and Chardonnay) were positively influenced by the white colour, homogeneity, and elasticity of the paste. In the present experiment, the Pecorino cheese from sheep supplemented with flaxseed and the combination of flaxseed and *A. nodosum* was judged by the consumers to have a high-strength flavour attribute and a low rancid, mouldy, and piquant flavour, in comparison with the Pecorino cheese made from sheep milk supplemented with *A. nodosum*. Data on the effects of a linseed supplementation-based diet on the sensory properties of milk products are not consistent [32–34]. Polyunsaturated fatty acids are more prone to oxidation with the formation of unsaturated aldehydes resulting in the flavour defect referred to as oxidative rancidity; however, in cheese, the lipid oxidation is considered limited due to a low redox potential (−250 mV) [35]. Contrasting results were obtained when extruded flaxseed was integrated into the diet: production of off-flavours due to lipid oxidation was reported in cheese from cows [36] or goats [37] supplemented with flaxseed. Conversely, in sheep cheese, Caccamo et al. [38] found the absence of a detectable off-flavour in cheese matured for 40 days; this result agrees with the present research in which no differences emerged in the rancid perception among Pecorino cheese ripened for 45 days. In the present research, consumer preference was more oriented toward the conventional cheese, with the CON Pecorino cheese being preferred by 33.33% of the panel, followed by the FS group with a very close number of preferences of 28.07%, and then the FS + AN, which was preferred by 22.81%, and the AN Pecorino cheese, which was preferred by 17.54% of the panel. Consumers did not receive additional information about the different types of cheese tested, so it might be interesting to assess the impact of attributes related to the health qualities of the fatty acid profile of cheeses which cannot be ascertained from the direct tasting experience. Indeed, it was demonstrated that information about credence attributes of animal-based products is able to move the expectations towards an increase in quality perception and consumer liking [39]. A proper communication strategy would be expected to drive consumers' preferences towards products obtained from animals under specific dietary plans that are able to sustain human health upon consumption.

4. Conclusions

In the present study, different sources of dietary PUFA, namely, flaxseed, *Ascophyllum nodosum*, and a combination of the two, were administered to late-lactating sheep during the summer season. *A. nodosum* supplementation to dairy ewes led to a higher fat and protein content of bulk milk destined for cheese-making. The main focus of the experimental supplementation was the study of its effect on the fatty acid profile of ovine cheese ripened for up to 45 days, with the administration of flaxseed leading to the greatest impact in this regard. Dietary flaxseed led to an increase in the level of PUFA and CLA by about 25% and 50% compared to the control, respectively, enriching the cheese matrix with molecules beneficial for human consumption. Descriptive sensory analysis was applied to characterise the sensory attributes of the cheeses, and the flaxseed-supplemented cheeses showed a sensory pattern comparable to the control cheese, whereas cheeses from the *A. nodosum* supplementation were judged to be lower in appearance uniformity and colour intensity than the control.

Overall, data from the present study demonstrated that supplementation based on flaxseed, rather than on *A. nodosum*, could represent an effective strategy for improving the nutritional quality of the lipid fraction of cheeses, especially during seasons in which pasture is scarce and of poor quality, without exerting a negative impact on the sensory attributes of cheese.

Supplementary Materials: The following supporting information can be downloaded at: https://www.mdpi.com/article/10.3390/foods13142165/s1, Figure S1: Subject information and consent form, Figure S2: Questionnaire for Pecorino Cheese consumer test, Table S1: Results of Principal component analysis on sensory analysis consumer test Pecorino cheeses from the experimental diets, showing the loadings of the first five Principal Component.

Author Contributions: Conceptualisation, M.C. and A.S. (Antonella Santillo); methodology, M.G.C.; software, M.G.C. and A.S. (Antonella Santillo), formal analysis, M.G.C.; resources, A.S. (Agostino Sevi), M.C. and M.A.; data curation, M.G.C. and A.S. (Agostino Sevi); writing—original draft preparation, M.G.C. and A.S. (Antonella Santillo); writing—review and editing, M.G.C., A.S. (Agostino Sevi) and M.A., visualisation, A.S. (Agostino Sevi); supervision, M.A., M.C. and A.S. (Agostino Sevi); project administration, M.A. and M.C.; funding acquisition, M.A., M.C. and A.S. (Agostino Sevi). All authors have read and agreed to the published version of the manuscript.

Funding: This study was carried out within the Agritech National Research Center and received funding from the European Union Next-GenerationEU (PIANO NAZIONALE DI RIPRESA E RESILIENZA (PNRR)—MISSIONE 4 COMPONENTE 2, INVESTIMENTO 1.4—D.D. 1032 17/06/2022, CN00000022). This manuscript reflects only the authors' views and opinions; neither the European Union nor the European Commission can be considered responsible for them.

Institutional Review Board Statement: The experimental scheme involved in the sensory evaluation of Pecorino cheese did not need ethical approval. Informed consent was obtained from all participants involved in this study, ensuring their voluntary participation and understanding.

Informed Consent Statement: Informed consent was obtained from all subjects involved in this study.

Data Availability Statement: The original contributions presented in the study are included in the article and Supplementary Material, further inquiries can be directed to the corresponding author.

Conflicts of Interest: The authors declare no conflicts of interest.

References

1. Caridi, A.; Micari, P.; Caparra, P.; Cufari, A.; Sarullo, V. Ripening and seasonal changes in microbial groups and in physico-chemical properties of the ewes' cheese Pecorino del Poro. *Int. Dairy J.* **2003**, *13*, 191–200. [CrossRef]
2. Pirisi, A.; Comunian, R.; Urgeghe, P.P.; Scintu, M.F. Sheep's and goat's dairy products in Italy: Technological, chemical, microbiological, and sensory aspects. *Small Rumin. Res.* **2011**, *101*, 102–112. [CrossRef]
3. Braghieri, A.; Girolami, A.; Riviezzi, A.M.; Piazzolla, N.; Napolitano, F. Liking of traditional cheese and consumer willingness to pay. *Ital. J. Anim. Sci.* **2014**, *13*, 3029. [CrossRef]
4. Ference, B.A.; Kastelein, J.J.; Ray, K.K.; Ginsberg, H.N.; Chapman, M.J.; Packard, C.J.; Laufs, U.; Oliver-Williams, C.; Wood, A.M.; Butterworth, A.S. Association of triglyceride-lowering LPL variants and LDL-C–lowering LDLR variants with risk of coronary heart disease. *JAMA* **2019**, *321*, 364–373. [CrossRef] [PubMed]
5. Caroprese, M.; Ciliberti, M.G.; Marino, R.; Santillo, A.; Sevi, A.; Albenzio, M. Polyunsaturated fatty acid supplementation: Effects of seaweed *Ascophyllum nodosum* and flaxseed on milk production and fatty acid profile of lactating ewes during summer. *J. Dairy Res.* **2016**, *83*, 289–297. [CrossRef]
6. Bodas, R.; Manso, T.A.; Mantecon, R.; Juarez, M.; De la Fuente, M.A.; Gómez-Cortés, P. Comparison of the fatty acid profiles in cheeses from ewes fed diets supplemented with different plant oils. *J. Agric. Food Chem.* **2010**, *58*, 10493–10502. [CrossRef] [PubMed]
7. EU Directive 2010/63/EU 2010 on the Protection of Animals Used for Scientific Purposes, pp. 33–79. Official Journal L 276, European Communities Publication Office: Luxembourg. Available online: http://data.europa.eu/eli/dir/2010/63/oj (accessed on 12 May 2024).
8. O'Fallon, J.V.; Busboom, J.; Nelson, M.; Gaskins, C. A direct method for fatty acid methyl ester synthesis: Application to wet meat tissues, oils, and feedstuffs. *J. Anim. Sci.* **2007**, *85*, 1511–1521. [CrossRef]
9. *FIL-IDF Standard no. 141B*; International Dairy Federation. Determination of Milk Fat, Protein & Lactose Content—Guide for the Operation of Mid-Infrared Instruments. IDF: Brussels, Belgium, 1990.
10. Santillo, A.; Caroprese, M.; Marino, R.; Muscio, A.; Sevi, A.; Albenzio, M. Influence of lamb rennet paste on composition and proteolysis during ripening of Pecorino Foggiano cheese. *Int. Dairy J.* **2007**, *17*, 535–546. [CrossRef]
11. Gripon, J.; Desmazeaud, M.; Le Bars, D.; Bergere, J. Etude du rôle des micro-organismes et des enzymes au cours de la maturation des fromages. II.-Influence de la présure commerciale. *Le Lait* **1975**, *55*, 502–516. [CrossRef]
12. Ulbricht, T.L.V.; Southgate, D.A.T. Coronary heart disease: Seven dietary factors. *Lancet* **1991**, *338*, 985–992. [CrossRef]
13. Santillo, A.; Caroprese, M.; Ruggieri, D.; Marino, R.; Sevi, A.; Albenzio, M. Consumer acceptance and sensory evaluation of Monti Dauni Meridionali Caciocavallo cheese. *J. Dairy Sci.* **2012**, *95*, 4203–4208. [CrossRef]
14. Lawlor, J.B.; Delahunty, C.M. The sensory profile and consumer preference for ten speciality cheeses. *Int. J. Dairy Technol.* **2000**, *53*, 28–36. [CrossRef]
15. Muir, D.D.; Hunter, E.A. Sensory evaluation of Cheddar cheese: Order of tasting and carryover effects. *Food Qual. Prefer.* **1991**, *3*, 141–145. [CrossRef]
16. SAS Institute, I.N.C. 2013. SAS (University Edition). Available online: https://www.sas.com/it_it/software/on-demand-for-academics.html (accessed on 12 May 2024).
17. National Research Council; Committee on Animal Nutrition; Subcommittee on Dairy Cattle Nutrition. *Nutrient Requirements of Dairy Cattle: 2001*; National Academies Press: Washington, DC, USA, 2001.
18. Capper, J.; Wilkinson, R.; Mackenzie, A.; Sinclair, L. The effect of fish oil supplementation of pregnant and lactating ewes on milk production and lamb performance. *Animal* **2007**, *1*, 889–898. [CrossRef]
19. Reynolds, C.; Cannon, V.; Loerch, S. Effects of forage source and supplementation with soybean and marine algal oil on milk fatty acid composition of ewes. *Anim. Feed Sci. Technol.* **2006**, *131*, 333–357. [CrossRef]
20. Papadopoulos, G.; Goulas, C.; Apostolaki, E.; Abril, R. Effects of dietary supplements of algae, containing polyunsaturated fatty acids, on milk yield and the composition of milk products in dairy ewes. *J. Dairy Res.* **2002**, *69*, 357–365. [CrossRef]
21. Pulina, G.; Nudda, A.; Battacone, G.; Cannas, A. Effects of nutrition on the contents of fat, protein, somatic cells, aromatic compounds, and undesirable substances in sheep milk. *Anim. Feed Sci. Technol.* **2006**, *131*, 255–291. [CrossRef]
22. Brito, A. Effects of Seaweeds on Dairy Production. 2022. Available online: https://ecommons.cornell.edu/server/api/core/bitstreams/72e08088-d98c-447c-98af-b8058f5bd86a/content (accessed on 12 May 2024).
23. Newton, E.; Theodoridou, K.; Terré, M.; Huws, S.; Ray, P.; Reynolds, C.; Prat, N.; Sabrià, D.; Stergiadis, S. Effect of dietary seaweed (*Ascophyllum nodosum*) supplementation on milk mineral concentrations, transfer efficiency, and hematological parameters in lactating Holstein cows. *J. Dairy Sci.* **2023**, *106*, 6880–6893. [CrossRef] [PubMed]
24. Chia, J.; Burrow, K.; Carne, A.; McConnell, M.; Samuelsson, L.; Day, L.; Young, W.; Bekhit, A.E.-D.A. Minerals in sheep milk. In *Nutrients in Dairy and their Implications on Health and Disease*, 1st ed.; Watson, R.R., Collier, R.J., Preedy, V.R., Eds.; Elsevier: Amsterdam, The Netherlands, 2018; pp. 345–362.
25. Caroprese, M.; Ciliberti, M.G.; Annicchiarico, G.; Albenzio, M.; Muscio, A.; Sevi, A. Hypothalamic-pituitary-adrenal axis activation and immune regulation in heat-stressed sheep after supplementation with polyunsaturated fatty acids. *J. Dairy Sci.* **2014**, *97*, 4247–4258. [CrossRef]
26. Ohlsson, L. Dairy products and plasma cholesterol levels. *Food Nutr. Res.* **2010**, *54*, 5124. [CrossRef]

27. Santillo, A.; Caroprese, M.; Marino, R.; Sevi, A.; Albenzio, M. Quality of buffalo milk as affected by dietary protein l level and flaxseed supplementation. *J. Dairy Sci.* **2016**, *99*, 7725–7732. [CrossRef] [PubMed]
28. Simopoulos, A.P. Omega-3 fatty acids and antioxidants in edible wild plants. *Biol. Res.* **2004**, *37*, 263–277. [CrossRef] [PubMed]
29. Petit, H. Feed intake, milk production and milk composition of dairy cows fed flaxseed. *Can. J. Anim. Sci.* **2010**, *90*, 115–127. [CrossRef]
30. Rymer, C.; Givens, D.; Wahle, K. Dietary strategies for increasing docosahexaenoic acid (DHA) and eicosapentaenoic acid (EPA) concentrations in bovine milk: A review. *Nutr. Abstr. Rev. Ser. B Livest. Feed. Feed.* **2003**, *73*, 9R–25R.
31. Torri, L.; Piochi, M.; Marchiani, R.; Zeppa, G.; Dinnella, C.; Monteleone, E. A sensory-and consumer-based approach to optimize cheese enrichment with grape skin powders. *J. Dairy Sci.* **2016**, *99*, 194–204. [CrossRef] [PubMed]
32. Chilliard, Y.; Ferlay, A. Dietary lipids and forages interactions on cow and goat milk fatty acid composition and sensory properties. *Reprod. Nutr. Dev.* **2004**, *44*, 467–492. [CrossRef] [PubMed]
33. Branciari, R.; Valiani, A.; Trabalza-Marinucci, M.; Miraglia, D.; Ranucci, D.; Acuti, G.; Esposto, S.; Mughetti, L. Consumer acceptability of ovine cheese from ewes fed extruded linseed-enriched diets. *Small Rumin. Res.* **2012**, *106*, S43–S48. [CrossRef]
34. Mughetti, L.; Sinesio, F.; Acuti, G.; Antonini, C.; Moneta, E.; Peparaio, M.; Trabalza-Marinucci, M. Integration of extruded linseed into dairy sheep diets: Effects on milk composition and quality and sensorial properties of Pecorino cheese. *Anim. Feed Sci. Technol.* **2012**, *178*, 27–39. [CrossRef]
35. Fox, P.F.; Guinee, T.P.; Cogan, T.M.; McSweeney, P.L. Biochemistry of Cheese Ripening. In *Fundamentals of Cheese Science*, 2nd ed.; Fox, P.F., Guinee, T.P., Cogan, T.M., McSweeney, P.L., Eds.; Springer: New York, NY, USA, 2017.
36. Dubroeucq, H.; Martin, B.; Ferlay, A.; Pradel, P.; Verdier-Metz, I.; Chilliard, Y.; Agabriel, J.; Coulon, J. Cow's feeding may modify sensory properties of milk. In *9èmes Rencontres autour des Recherches sur les Ruminants*; Institut National de la Recherche Agronomique: Paris, France, 2002; Record Number: 20033018539.
37. Gaborit, P.; Raynal-Ljutovac, K.; Lauret, A.; Chabosseau, J.; Rouel, J.; Chilliard, Y. Flavour of goat milk and cheeses according to feeding: Alfalfa hay or maize silage with oleic sunflower or linseed oil supplementation. In *Multi-Function Grasslands: Quality Forages, Animal Products and Landscapes, Proceedings of the 19th General Meeting of the European Grassland Federation, La Rochelle, France, 27–30 May 2002*; Organizing Committee of the European Grassland Federation: Versailles, France, 2002; pp. 562–563.
38. Caccamo, M.; Valenti, B.; Luciano, G.; Priolo, A.; Rapisarda, T.; Belvedere, G.; Marino, V.M.; Esposto, S.; Taticchi, A.; Servili, M. Hazelnut as ingredient in dairy sheep diet: Effect on sensory and volatile profile of cheese. *Front. Nutr.* **2019**, *6*, 125. [CrossRef]
39. Napolitano, F.; Girolami, A.; Braghieri, A. Consumer liking and willingness to pay for high welfare animal-based products. *Trends Food Sci. Technol.* **2010**, *21*, 537–543. [CrossRef]

foods

MDPI

Article

Effect of Blueberry Pomace Addition on Quality Attributes of Buttermilk-Based Fermented Drinks during Cold Storage

Biljana Trajkovska [1,*], Gjore Nakov [2], Sari Thachappully Prabhat [3] and Prarabdh C. Badgujar [3]

1 Faculty of Biotechnical Sciences—Bitola, University "St. Kliment Ohridski"—Bitola, 7000 Bitola, North Macedonia
2 College of Sliven, Technical University of Sofia, 8800 Sliven, Bulgaria; gnakov@tu-sofia.bg
3 Department of Food Science and Technology, National Institute of Food Technology Entrepreneurship and Management, Kundli, Sonipat 131028, Haryana, India; sari.tp.tp@gmail.com (S.T.P.); prarabdh.badgujar@gmail.com (P.C.B.)
* Correspondence: biljana.trajkovska@uklo.edu.mk

Abstract: The fruit and beverage industry faces challenges related to waste management and environmental pollution due to rapid industrial expansion. Fruit industry waste, such as blueberry pomace, holds the promise of enhancing gut health and providing valuable antioxidants. Concurrently, buttermilk, a prominent dairy product, offers nutritional and technological benefits but remains underutilized. This study aimed to evaluate the incorporation of blueberry pomace (0%, 2%, 4%, 6%, 8%, and 10%) into buttermilk at varying levels and assess its impact on the physicochemical, antioxidant, microbiological, and sensory characteristics of the buttermilk. Buttermilk samples were supplemented with different concentrations of blueberry pomace and subjected to analysis over a two-week storage period (4 ± 1 °C). The addition of blueberry pomace led to alterations in the pH, dry matter, water holding capacity, color parameters, total phenolic content, and antioxidant activity. Microbiological analysis revealed the absence of *Enterobacteriaceae*, yeast, or molds. Sensory evaluation indicated significant differences among samples, with the highest scores observed for the buttermilk supplemented with 2% and 4% blueberry pomace. Incorporating blueberry pomace improved the overall acceptability and sensory properties. This research highlights the potential of fruit industry by-products to enhance the functionality and health benefits of dairy products, which is a promising way to effectively utilize waste.

Keywords: buttermilk; blueberry pomace; physicochemical properties; antioxidant activity; microbiological analysis; sensory evaluation

Citation: Trajkovska, B.; Nakov, G.; Prabhat, S.T.; Badgujar, P.C. Effect of Blueberry Pomace Addition on Quality Attributes of Buttermilk-Based Fermented Drinks during Cold Storage. *Foods* **2024**, *13*, 1770. https://doi.org/10.3390/foods13111770

Academic Editors: Michele Faccia and Giuseppe Natrella

Received: 2 May 2024
Revised: 3 June 2024
Accepted: 3 June 2024
Published: 5 June 2024

1. Introduction

The surge in global demand for dairy products has propelled the dairy sector's growth, transitioning from traditional to mechanized production and scaling up to meet consumer needs. However, this rapid industrial expansion not only leads to increased production but also raises concerns about higher concentrations of pollutants in water and land, contributing to environmental pollution and potential health risks [1]. By-products from the dairy industry pose a significant environmental threat due to their high organic compound content [2]. Buttermilk (BM) stands out as a major by-product in the dairy industry, formed in the serum phase during butter production [3]. Despite its nutritional and technological merits, BM remains underutilized [4]. Its composition typically includes lactose, proteins (casein and serum proteins), lipids, ash, and polar lipids (phospholipids and sphingolipids) originating from the milk fat globule membrane (MFGM). Notably, the concentration of polar lipids in BM is about five times higher than in whole milk [5,6]. Various types of BM are produced, including cultured buttermilk, sweet cream buttermilk, sour cream buttermilk, and commercial buttermilk [7]. On the other hand, dairy products are frequently enriched with a variety of ingredients such as fruit [8] and press cake flour [9,10] to amplify their positive health-promoting effects.

The agro-food industry has witnessed a notable increase in the production of by-products in recent years, which offer the potential for added value due to their functional and/or bioactive properties, promoting the concept of a circular economy [11]. Blueberry pomace (BP), a by-product of the juice industry, comprises seeds, skins, and pulp residue, constituting 20–30% of the fruit [12]. With its retained phenolic compounds and dietary fiber content, BP holds promise for enhancing gut health, potentially influencing gut microbiota composition [13]. Fermentation of BP by the probiotic *Lactobacillus casei* has been shown to enhance antioxidant activity and regulate fecal microbiota, offering potential health benefits [14]. Moreover, fruit by-products such as blueberry pomace are rich in bioactive compounds such as polyphenols, anthocyanins, phenolic acids, flavanols, and tannins, making them valuable sources of antioxidants [15,16]. These compounds exhibit significant antimicrobial activity, offering potential as innovative natural food additives [17].

The objective of this research was to evaluate the impact of incorporating BP, a fruit by-product, into BM at various levels (2%, 4%, 6%, 8%, and 10%) and to examine its effects on the physicochemical, technological, and sensory characteristics of the product. This study aimed at valorizing fruit by-products by incorporating them into BM to create a functional dairy product.

2. Materials and Methods

2.1. Materials and Buttermilk Production

Approximately 20 L of BM underwent chemical composition analysis (MilkoScanTM FT3, Foss, Hilleroed, Denmark). Before producing the various types of fortified BM-based fermented drinks, the chemical composition of the BM utilized was analyzed. The composition was as follows: 0.94% fat, 3.43% protein, 4.81% lactose, 9.18% dry matter, and a pH of 6.53. Additionally, we conducted a microbiological screening of the BM prior to fortification. The results showed that yeast and molds were not detected, *Escherichia coli* was absent (0 CFU/mL), the *Enterobacteriaceae* count was less than 10 CFU/mL, and the total bacterial/plate count was less than 100 CFU/mL. These analyses ensured the quality and safety of the BM before using it to produce fortified fermented drinks.

Following this, BM underwent thermal treatment at 72 ± 1 °C (Weck Inc., Luray, VA, USA) for 10 min, as described by Szkolnicka et al. [18]. After pasteurization, the BM was cooled to 35 ± 1 °C at room temperature. Mesophilic starter cultures (Selection TM Danica, CHR Hansen, Hørsholm, Denmark) were then added according to the manufacturer's guidelines (500 U/5000 L). The bacterial strains were *Lactococcus lactis* subsp. *cremoris*, *Lactococcus lactis* subsp. *lactis*, *Lactococcus lactis* subsp. *lactis biovar*, *Diacetylactis*, and *Leuconostoc*. These lactic acid bacteria are known for their ability to produce aroma and CO_2.

BP, obtained as a by-product from juice processing, was dried in a convection oven at 45 ± 2 °C for 20 h, ground, and passed through a 1 mm sieve [16]. The BP composition was as follows: 0.5 g/100 g fats, 87 g/100 g carbohydrates, 5 g/100 g fibers, and 0.2 g/100 g proteins. Various quantities of BP (2 g, 4 g, 6 g, 8 g, and 10 g per 100 g of BM) were added to sterile containers and the inoculated BM was divided among them, mixed, and incubated (yogurt maker Y 140, Elecrem, UK) at a constant temperature of 35 ± 1 °C. Incubation continued until the pH of the control BM samples without BP reached 4.6 ± 0.1 (9 h), at which point incubation was halted for all samples. Throughout the fermentation, the pH of the milk was monitored using a pH meter (Testo SE & Co., KGaA, Lenzkirch, Germany). After incubation, the BM samples were refrigerated (4 ± 1 °C) for subsequent analysis. Physicochemical and microbiological properties were evaluated over a 2-week period, with measurements taken weekly (on days 1, 7, and 14) during refrigerated storage at 4 ± 1 °C. Sensorial properties were evaluated only on the first day after production.

All BM-based fermented drinks were prepared in triplicate batches. Control samples without BP were referred to as plain samples, while those with BP were considered fortified samples (Figure 1).

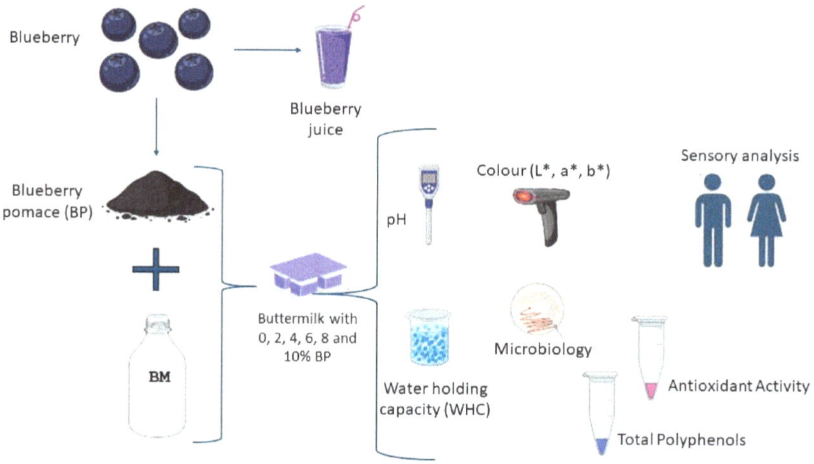

Figure 1. Experimental workflow for incorporating blueberry pomace into buttermilk and evaluating physicochemical, microbiological, and sensory properties.

2.2. Methods

2.2.1. pH Determination

The pH of the samples was measured at 20 °C using a pH meter (Testo SE & Co., KGaA, Lenzkirch, Germany) that was prior calibrated with pH 4.0 and 7.0 buffers.

2.2.2. Dry Matter

Two grams (2 g) of BM-based fermented drinks were weighed in an aluminum dish and dried at 102 ± 1 °C until a constant weight was achieved using a moisture analyzer MJ33 (Mettler Toledo, Greifensee, Switzerland) [19].

2.2.3. Water Holding Capacity and Color Determination

The water holding capacity (WHC) of the BM-based fermented samples was assessed using the centrifugation method, as per Grasso et al. [20]. The color of the BM-based fermented drink was evaluated using a colorimeter (Konica Minolta, Chroma Meter, CR400, Osaka, Japan), following the methodology outlined by Nakov et al. [9]. Parameters such as L* value (lightness), a* value (red-green intensity), and b* value (yellow-blue intensity) were measured for the BM samples. The total color change (ΔE) between the control sample and the BM-based fermented drink with different BP content was also determined: L_1, a_1, and b_1 are color parameters for the control sample and L_2, a_2, and b_2 are color parameters for the BM-based fermented drink with different BP content, respectively, after a specified period of time (weeks); (Equation (1)).

$$\Delta E = \sqrt{\left(L_1 - L_2\right)^2 + \left(a_1 - a_2\right)^2 + \left(b_1 - b_2\right)^2} \quad (1)$$

If ΔE < 1, the difference between colors is not visible to the human eye. When 1 < ΔE < 3, the color differences are not considered significant to the human eye. If ΔE > 3, the color differences are considered visible to the human eye [21].

2.2.4. UV-Vis Spectroscopy

UV-Vis experiments were conducted using a UV-1800 UV-VIS spectrophotometer (Shimadzu, Kyoto, Japan). To determine the total polyphenols (TPC) and antioxidant activity of BM-based fermented samples with different BP content, extraction was performed by mixing 10 g of the sample with 30 mL of an 80:20 methanol:water solution on a magnetic stirrer for 30 min. The mixture was then transferred to a centrifuge tube and centrifuged at 8000 rpm at 4 °C for 30 min [9].

Determination of Total Polyphenols

A volume of 0.3 mL of supernatant was transferred into a tube and mixed with 5 mL of Folin-Ciocalteu reagent (diluted 1:10). After 5 min, 1.5 mL of 6% Na_2CO_3 was added. The solutions were mixed with a vortex and left in the dark for 90 min. After the reaction time, the absorbance of the samples at 760 nm was measured using a UV-1800 UV-VIS spectrophotometer (Shimadzu, Japan). Results were expressed as µg gallic acid equivalent (GAE) per ml of fresh sample weight [9].

Determination of Antioxidant Activity

The free radical scavenging activity of BM-fortified samples was assessed following the method outlined by Barkallah et al. [22], employing 1,1-diphenyl-2-picrylhydrazyl (DPPH) as the indicator. Absorbance was measured at 517 nm using the same spectrophotometer. The results of all antioxidant activity determinations were expressed in mmol Trolox equivalents (TE), as Trolox is a stable antioxidant widely used as a standard for measuring antioxidative activity. A calibration curve ranging from 0.01 to 5.00 mmol of Trolox was used for the quantification of these activities.

2.2.5. Microbiological Analysis

To begin microbial count analysis, 10 g samples of each buttermilk-based fermented drink were refrigerated and then homogenized in 90 mL of a diluent solution comprising 0.85% sodium chloride and 0.1% tryptone. Following homogenization, serial 10-fold dilutions were meticulously prepared for further examination. Enumeration of *Enterobacteria* was performed on Violet Red Bile Glucose (VRBG) Agar (Condalab, Madrid, Spain, CAT 1092.00) according to ISO standard [23]. Yeast and mold analyses were conducted in accordance with ISO protocols [24].

2.2.6. Sensory Evaluation

The sensory evaluation of the BM-based fermented samples involved 20 semi-trained assessors who utilized a hedonic scale analysis. Approximately 10 g of homogenized samples were placed into glass cups labeled with randomized three-digit numbers and served chilled at 4 °C. Assessors rated the BM-based fermented drinks on a scale from 1 to 9, where 1 indicated "extremely dislike" and 9 indicated "extremely like", considering attributes such as flavor, mouth-feel, appearance, texture, and overall acceptance. The sensory analysis was conducted solely on the first day following the BM-based fermented drinks production.

2.2.7. Statistical Analysis

The results are presented as mean ± standard deviation. For physicochemical and microbiological parameters, $n = 3$ replicates were used, while, for the sensory evaluation, $n = 20$ assessors participated. Statistical analysis was conducted using factorial analysis of variance (2-way ANOVA), followed by Fisher's least significant difference (LSD) test, to assess the main effects (BP addition and storage) as well as their interaction effects on all the analyzed traits. XLSTAT software version 2019.2.2 (Addinsoft, New York, NY, USA) was employed for statistical analysis. The significance level was set at $p \leq 0.05$.

3. Results and Discussion

3.1. Chemical Characteristics

During production, pH value is a critical step in determining the safety and shelf life of a dairy product. On the first day, the pH of the plain BM (0%—without blueberry pomace) was 4.60. The addition of BP significantly decreased the pH value of the BM-based fermented samples ($p < 0.05$); namely, the BM with 10% BP had a pH of 4.47 on the first day of storage. Nevertheless, the pH value of both the plain and the fortified BM with BP decreased with the storage time ($p < 0.05$). Eventually, on Day 14 of the examination, for the plain BM, the pH was 4.57 and for the blueberry pomace-fortified BM, it was 4.39 (10% BP fortified BM) (Figure 2). This trend mirrors findings in other food applications, such as a blueberry flower fermented drink [25] and a berry fruit pomace-fortified mustard [26], where a similar inverse correlation with the inclusion of blueberry flowers or pomace was observed. Similarly, the addition of Spirulina platensis to BM samples led to a notable reduction in the time required to reach a pH of 4.5 [27].

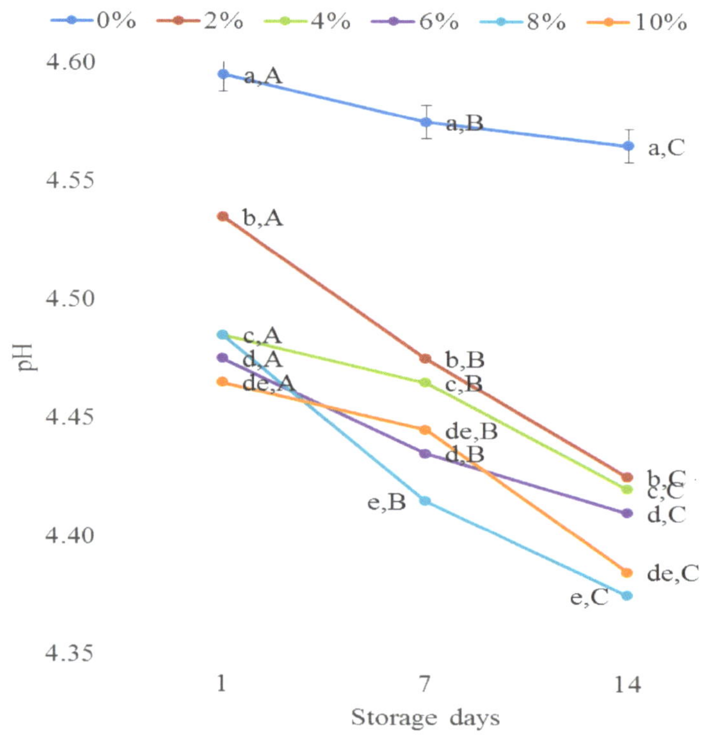

Figure 2. pH variation during storage of buttermilk fortified with different percentages of blueberry pomace. Note: Small letters refer to statistically significant differences ($p < 0.05$) between the samples with different quantities of BP; Capital letters refer to statistically significant differences ($p < 0.05$) between storage days.

Dry matter and WHC exhibited a significant increase with the incorporation of BP, both during storage ($p < 0.05$) (Table 1). Initially, the dry matter content in our research ranged from 11.6% ± 0.02 to 19.6% ± 0.01 on the first day of post-production, with a subsequent increase attributed to the addition of BP. This elevation in dry matter content is likely influenced by the composition of the raw materials [28]. Similarly, WHC initially ranged from 24.6% ± 0.07 to 34.9% ± 0.02 and showed a continuous increase during storage across all samples. This augmentation can be associated with the rise in total solids content, which enhances WHC and reduces syneresis [29].

Table 1. Mean values and standard deviation of dry matter—DM (%) and water holding capacity—WCH (%) on different percentages of BP in BM during storage days.

Parameters	BP (%)	Storage Period (Days)		
		1	7	14
Dry matter (DM %)	0	11.6 ± 0.02 ^{f,C}	12.4 ± 0.04 ^{f,B}	14.0 ± 0.02 ^{f,A}
	2	13.0 ± 0.04 ^{e,C}	14.8 ± 0.01 ^{e,B}	16.2 ± 0.02 ^{e,A}
	4	14.8 ± 0.04 ^{d,C}	15.3 ± 0.01 ^{d,B}	17.9 ± 0.04 ^{d,A}
	6	16.2 ± 0.06 ^{c,C}	17.4 ± 0.05 ^{c,B}	20.1 ± 0.04 ^{c,A}
	8	17.8 ± 0.06 ^{b,C}	19.6 ± 0.07 ^{b,B}	21.7 ± 0.07 ^{b,A}
	10	19.6 ± 0.01 ^{a,C}	20.1 ± 0.03 ^{a,B}	23.7 ± 0.05 ^{a,A}
WHC (%)	0	24.6 ± 0.07 ^{f,C}	27.3 ± 0.02 ^{f,B}	35.2 ± 0.04 ^{f,A}
	2	26.0 ± 0.11 ^{e,C}	28.9 ± 0.02 ^{e,B}	35.6 ± 0.07 ^{e,A}
	4	29.0 ± 0.04 ^{d,C}	31.2 ± 0.05 ^{d,B}	38.7 ± 0.02 ^{d,A}
	6	31.8 ± 0.01 ^{c,C}	34.2 ± 0.02 ^{c,B}	41.8 ± 0.18 ^{c,A}
	8	33.3 ± 0.12 ^{b,C}	37.4 ± 0.02 ^{b,B}	43.9 ± 0.04 ^{b,A}
	10	34.9 ± 0.02 ^{a,C}	36.7 ± 0.04 ^{a,B}	45.4 ± 0.04 ^{a,A}

Note: Small letters refer to statistically significant differences ($p < 0.05$) between the samples with different quantities of BP; Capital letters refer to statistically significant differences ($p < 0.05$) between storage days.

3.2. Color

Food selection is significantly influenced by its color and appearance, which serve as key indicators of food quality; often, color is the primary sensory attribute perceived by consumers, directly impacting their purchasing decisions [30]. Incorporating BP into the BM samples resulted in notable alterations in color parameters, including a decrease in the L* parameter value ($p < 0.05$). Furthermore, extended storage at 4 °C did not result in a significant decrease ($p > 0.05$) in the L* parameter values for the BP-fortified samples. Similar trends were observed with the yellowness coordinate (b*), whereby, compared to plain BM, the average b* value was 0.36 ± 7.27 on the first day after production, while, when fortified with 10% BP, the b* value increased to 9.69 ± 0.18. Additionally, both the blueberry concentration and storage duration significantly influenced the b* values, along with their interactions. Conversely, no statistically significant changes were detected in the redness coordinate (a*) for fortified BM samples, consistent with findings by Starkute et al. [31].

The color of fermented beverages often derives from pigments present in the raw materials used, as noted by Olukomaiya et al. [32]. Additionally, the lightness index of fermented milk products is affected by acidity, with a decrease in pH correlating with a decrease in lightness [33,34]. These findings align with our own results, which demonstrate that fortifying BM with BP leads to reduced acidity and a decrease in the L* parameter (Table 2), with more significant changes observed at 10% fortification with blueberry pomace. Furthermore, it is important to note that color parameters can vary over time, and storage duration can also influence these parameters [35]. Moreover, the size of fat globules and protein particles can significantly influence the brightness level [36].

The total color change (ΔE) between the control BM fermented drink sample and the BM fermented drink samples with varying BP content was also determined. According to the results presented in Table 2, it can be observed that these parameters changed during the storage period of the samples and with the fortification with BP. In both cases, the color differences are greater than 3, which are considered visible to the human eye [21].

Table 2. Mean values and standard deviation of color coordinates (L*, a*, and b*) on different percentages of blueberry pomace in buttermilk during storage days.

Parameters	BP (%)	Storage Period (Days)		
		1	7	14
L*	0	73.9 ± 7.60 ^{a,A}	79.9 ± 1.68 ^{a,A}	78.7 ± 2.83 ^{a,A}
	2	66.5 ± 0.60 ^{b,A}	66.9 ± 0.37 ^{b,A}	65.8 ± 0.35 ^{b,A}
	4	61.0 ± 0.36 ^{c,A}	60.2 ± 0.83 ^{c,A}	58.9 ± 0.00 ^{c,A}
	6	56.2 ± 0.08 ^{d,A}	55.1 ± 0.13 ^{d,A}	54.1 ± 0.30 ^{d,A}
	8	51.1 ± 0.03 ^{e,A}	49.9 ± 0.31 ^{e,A}	50.4 ± 0.17 ^{e,A}
	10	46.5 ± 0.34 ^{f,A}	47.0 ± 0.17 ^{f,A}	47.3 ± 0.55 ^{f,A}
a*	0	3.32 ± 7.87 ^{b,A}	−0.68 ± 3.56 ^{b,A}	−15.13 ± 15.29 ^{b,A}
	2	2.01 ± 0.08 ^{a,A}	2.31 ± 0.05 ^{a,A}	2.43 ± 0.11 ^{a,A}
	4	3.46 ± 0.11 ^{a,A}	3.45 ± 0.03 ^{a,A}	3.60 ± 0.05 ^{a,A}
	6	4.25 ± 0.02 ^{a,A}	4.73 ± 0.07 ^{a,A}	4.29 ± 0.18 ^{a,A}
	8	5.40 ± 0.06 ^{a,A}	5.83 ± 0.09 ^{a,A}	5.18 ± 0.13 ^{a,A}
	10	6.08 ± 0.04 ^{a,A}	6.22 ± 0.17 ^{a,A}	5.82 ± 0.11 ^{a,A}
b*	0	0.36 ± 7.27 ^{d,A}	−2.98 ± 2.68 ^{d,A}	−0.44 ± 3.25 ^{d,A}
	2	2.96 ± 0.05 ^{c,A}	2.44 ± 0.25 ^{c,A}	4.04 ± 0.21 ^{c,A}
	4	5.54 ± 0.53 ^{b,c,A}	5.83 ± 0.09 ^{b,c,A}	5.65 ± 0.08 ^{b,c,A}
	6	7.34 ± 0.03 ^{a,b,A}	7.95 ± 0.12 ^{a,b,A}	7.43 ± 0.44 ^{a,b,A}
	8	8.62 ± 0.07 ^{a,,A}	9.63 ± 0.20 ^{a,A}	8.36 ± 0.16 ^{a,A}
	10	9.69 ± 0.18 ^{a,A}	10.09 ± 0.03 ^{a,A}	9.11 ± 0.25 ^{a,A}
ΔE	2	11.4	14.7	23.8
	4	15.6	22.2	29.5
	6	25.2	27.9	33.8
	8	25.2	33.1	37.3
	10	29.3	36.12	40.3

Note: Small letters refer to statistically significant differences ($p < 0.05$) between the samples with different quantities of BP; Capital letters refer to statistically significant differences ($p < 0.05$) between storage days. ΔE—difference between control samples and samples with different quantities of BP.

3.3. *Total Polyphenols and Antioxidant Activity of Buttermilk with Different Amounts of Blueberry Pomace*

TPC without BP ranged from 4.9 ± 0.9 to 6.2 ± 0.04 mg GAE/g during the storage period (Figure 3), attributed to naturally occurring phenolic compounds in cultured BM-based fermented drinks. Similarly, Saberi et al. [37] observed significantly higher total phenolic content values in fortified yogurts with grape pomace and flaxseed oil compared to a control yogurt during storage. Similar tendencies were observed in the study of Starkute et al. [31], where the total phenolic content and antioxidant activity significantly increased in enriched samples, demonstrating the potential of raspberry, blueberry, and elderberry industry by-products for enhancing the nutritional and functional properties of unripened cow milk curd cheese. Various factors such as berry variety, cultivation region, and extraction and drying methods can influence phenolic compound levels [38]. The literature highlights blueberries as not only rich in fiber, minerals, and vitamins, but also a significant source of antioxidants, known for their documented health-promoting effects [13,14]. The addition of blueberry pomace (10%) significantly ($p < 0.05$) enhanced the total phenolic content to 16.6 ± 0.9 mg GAE/g, as compared to the plain BM-based fermented drink

(4.9 ± 0.9 mg GAE/g). Although TPC levels can increase with storage time within each sample, it is important to note that the Folin-Ciocalteau assay, commonly used to measure TPC, assesses the total reducing capacity of the sample, not just TPC [39]. Additionally, the temporary decrease in TPC observed in yogurt could be due to the decomposition of polymeric phenolics in the presence of lactic acid bacteria during refrigerated storage [40].

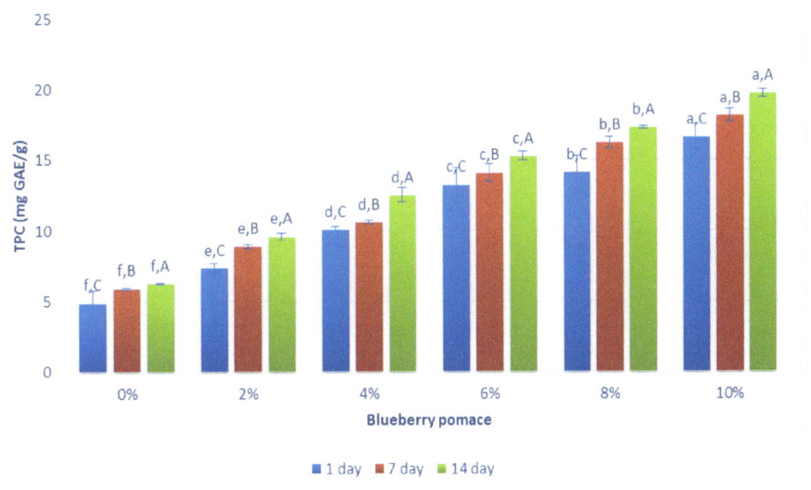

Figure 3. Total phenolic content (expressed as mg GAE/g) during storage days of buttermilk with different amounts of blueberry pomace. Small letters refer to statistically significant differences ($p < 0.05$) between the samples with different quantities of BP; Capital letters refer to statistically significant differences ($p < 0.05$) between storage days.

During the storage of both plain and fortified BM-based fermented samples, the antioxidant activity, expressed as the Trolox equivalent antioxidant capacity (TEAC), was estimated. With an increase in the amount of BP in the samples, the antioxidant activity also rose (from 0.04 ± 0.0% in plain BM-based fermented drinks to 0.25 ± 0.0% for 10% BP in BM-based fermented drinks) on the first day after production (Figure 4). Among six BM-based fermented samples, the radical scavenging activity (mmol of TA/l) was the highest in BM fortified with 10% BP. Plant polyphenols have been reported to effectively enhance the antioxidant activity of fermented dairy products [8,9]. It is worth mentioning that BP is rich in polyphenols with well-known antioxidant activity [41]. The greater the addition of BP, the higher the observed radical scavenging activity. This suggests that BP, containing numerous polyphenols, significantly boosts the antioxidant activity of BM. Consequently, BM containing BP exhibits substantial antioxidant activity, positioning it as a novel dairy product with notable health benefits. Our research agrees with those of previous reports, which described that the addition of berry industry by-products contained higher radical scavenging activity due to the higher amount of phenolic compounds [31,37]. Also, similar findings were presented by Najgebauer-Lejko and Sady [42] where BM flavored with strawberry and rhubarb exhibited higher radical scavenging activity (0.88 mmol of TE/kg) compared with the control sample. Additionally, antioxidant properties in fermented milk come from compounds such as casein, whey proteins, peptides, amino acids, coenzyme Q10, enzymatic systems, and lactic acid bacteria. In BM, lipophilic antioxidants are effective due to their variety, thermal stability, and synergistic interactions with hydrophilic antioxidants, protecting against undesirable compounds during processing [42]. On the other hand, Liu et al. [26] noticed that adding blueberry flower pulp to yogurt significantly increased the antioxidant properties of the yogurt. Throughout storage, these benefits remained relatively stable, with only minor fluctuations. However, according to their research, the addition of blueberry flower pulp led to a decrease in antioxidant activity. This could be attributed to the gradual decline of polyphenols, likely influenced by bacterial metabolic processes. Polyphenol oxidase might also play a role in this observed phenomenon [43].

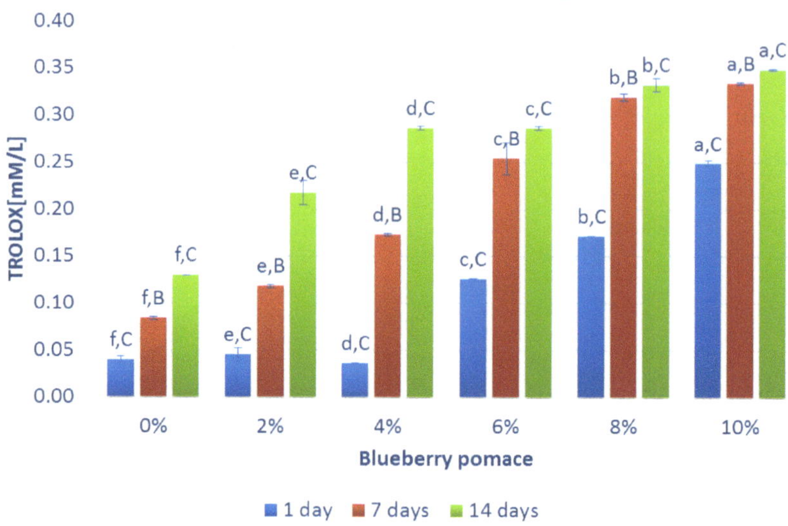

Figure 4. Antioxidant activity-AOA (expressed in Trolox mM/L) during storage days of buttermilk with different amounts of blueberry pomace. Small letters refer to statistically significant differences ($p < 0.05$) between the samples with different quantities of BP; Capital letters refer to statistically significant differences ($p < 0.05$) between storage days.

3.4. Microbiological Analysis

The perishability of dairy products is primarily governed by the microbiological quality of the product [44]. Throughout the 14-day storage period at 4 °C, none of the samples exhibited any signs of mold, yeast, or *Enterobacteriaceae*, which are considered indicative of hygiene standards. This absence of bacteria suggests that the BM-based fermented samples remained secure and uncontaminated, reflecting the adherence to clean and hygienic processing conditions. Similar findings were reported by Rose et al. [27], who studied cultured buttermilk fortified with Spirulina platensis, and by Parekh et al. [44], who investigated cultured buttermilk prepared by incorporating paneer whey.

3.5. Sensory Analysis of Buttermilk with Different Amounts of Blueberry Pomace

Various factors, such as the composition and quality of the product, influenced the sensory analysis [45]. The ANOVA (not presented) highlighted significant differences ($p < 0.05$) for all sensory parameters between the control sample (BM without BP) and other BMs with BP. (Figure 5a,b). The BM samples formulated with 2% and 4% BP received the highest sensory scores. Additionally, the highest overall acceptance score, averaging 8.3, was observed for the plain BM-based fermented drink and the 4% fortified BM-based fermented drink (Figure 5b). Significant differences were noted in the visual appearance, texture, and overall acceptance of BM-based fermented samples formulated with BP (6%, 8%, and 10%). A similar trend was observed in a previous study comparing Petit Suisse cheese made with and without blueberry bagasse powder [46]. While previous reports have highlighted the addition of BP to foods [47], there is limited research on its addition to dairy products. Liu and Lv [26] conducted a sensory analysis of yogurt containing blueberry flower pulp, finding that panelists preferred samples with 2–5 g of blueberry flower pulp over the control. These results align with our findings. Thus, incorporating blueberry-based ingredients improved the overall acceptability and sensory properties of the BM-based fermented drinks. Other findings suggest a decrease in sensory scores during storage, possibly due to the relationship between bacterial fermentation and the resulting acidity from oxidative decomposition of product ingredients [48].

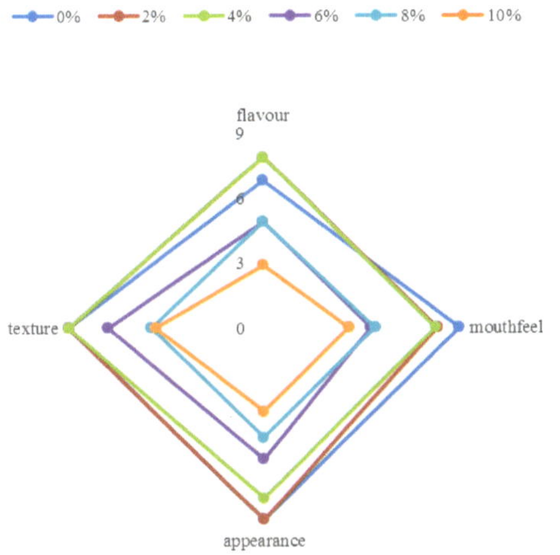

(a)

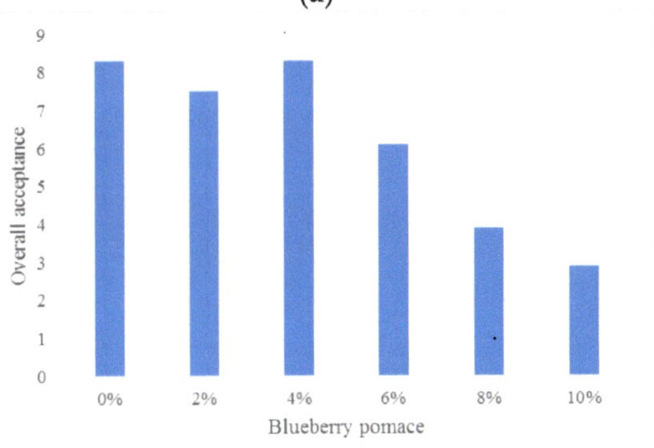

(b)

Figure 5. (**a**). Sensory evaluation of BM fortified with different concentrations of BP, assessing flavor, mouth-feel, appearance, and texture. (**b**). Overall acceptance of BM fortified with different concentrations of BP.

4. Conclusions

This study demonstrates the potential of incorporating blueberry pomace into BM to valorize fruit by-products and create functional dairy products with enhanced nutritional and sensory properties. Significant improvements were observed in the physicochemical parameters, including pH, dry matter content, water holding capacity, and color. The addition of blueberry pomace significantly increased the total phenolic content and antioxidant activity, highlighting its health-promoting benefits. Microbiological analysis confirmed the safety of all samples, with no presence of *Enterobacteriaceae*, yeast, or molds. Sensory evaluation revealed that BM fortified with 2% and 4% blueberry pomace received the highest scores, indicating favorable consumer acceptance. However, samples with 6%, 8%, and 10% pomace had lower sensory quality. These findings suggest that incorporating blueberry pomace can enhance the nutritional profile, functional properties, and overall appeal of BM, contributing to sustainable practices in the dairy industry. Further research is needed to optimize pomace incorporation, assess long-term stability, and evaluate broader consumer preferences.

Author Contributions: Conceptualization, B.T. and G.N.; methodology, G.N.; software, S.T.P. and P.C.B.; validation, B.T. and G.N.; formal analysis, B.T.; investigation, B.T.; resources, G.N.; data curation, B.T.; writing—original draft preparation, S.T.P.; writing—review and editing, G.N.; visualization, P.C.B.; supervision, B.T.; project administration, G.N.; funding acquisition, G.N. All authors have read and agreed to the published version of the manuscript.

Funding: This article/publication is based upon work from COST Action FULLRECO4US, CA20133, supported by COST (European Cooperation in Science and Technology).

Institutional Review Board Statement: The study was conducted in accordance with the Declaration of Helsinki, and approved by the Collage of Sliven, Technical University of Sofia, Resolution No. R1-1506/2023.

Informed Consent Statement: Informed Consent Statement: Informed consent was obtained from all subjects involved in the study.

Data Availability Statement: The original contributions presented in the study are included in the article, further inquiries can be directed to the corresponding author.

Acknowledgments: This article/publication is based upon work from COST Action FULLRECO4US, CA20133, supported by COST (European Cooperation in Science and Technology).

Conflicts of Interest: The authors declare no conflicts of interest.

References

1. Kyriakopoulou, K.; Keppler, J.K.; van der Goot, A.J.; Boom, R.M. Alternatives to Meat and Dairy. *Annu. Rev. Food Sci. Technol.* **2021**, *12*, 29–50. [CrossRef] [PubMed]
2. Ahmad, T.; Aadil, R.M.; Ahmed, H.; Ur Rahman, U.; Soares, B.C.; Souza Simone, L.Q.; Pimentel, T.C.; Scudino, H.; Guimarães, J.T.; Esmerino, E.A.; et al. Treatment and Utilization of Dairy Industrial Waste: A Review. *Trends Food Sci. Technol.* **2019**, *88*, 361–372. [CrossRef]
3. Morin, P.; Britten, M.; Jiménez-Flores, R.; Pouliot, Y. Microfiltration of buttermilk and washed cream buttermilk for concentration of milk fat globule membrane components. *J. Dairy Sci.* **2007**, *90*, 2132–2140. [CrossRef] [PubMed]
4. Krebs, L.; Bérubé, A.; Iung, J.; Marciniak, A.; Turgeon, S.L.; Brisson, G. Impact of ultra-high-pressure homogenization of buttermilk for the production of yogurt. *Foods* **2021**, *10*, 1757. [CrossRef] [PubMed]
5. Ali, A.H. Current knowledge of buttermilk: Composition, applications in the food industry, nutritional and beneficial health characteristics. *Int. J. Dairy Technol.* **2019**, *201972*, 169–182. [CrossRef]
6. Calvo, M.V.; Martín-Hernández, M.C.; García-Serrano, A.; Castro-Gómez, M.P.; Alonso-Miravalles, L.; García-Martín, R.; Megino-Tello, J.; Alonso, L.; Fontecha, J. Comprehensive characterization of neutral and polar lipids of buttermilk from different sources and its milk fat globule membrane isolates. *J. Food Compos. Anal.* **2020**, *86*, 103386. [CrossRef]
7. Hati, S.; Das, S.; Mandal, S. Technological advancement of functional fermented dairy beverages. In *Engineering Tools in the Beverage Industry*, 1st ed.; Grumezescu, A.M., Holban, A.M., Eds.; Woodhead Publishing: Sawston, UK, 2019; Volume 3, pp. 101–136. [CrossRef]
8. Feng, C.; Wang, B.; Zhao, A.; Wei, L.; Shao, Y.; Wang, Y.; Cao, B.; Zhang, F. Quality characteristics and antioxidant activities of goat milk yogurt with added jujube pulp. *Food Chem.* **2019**, *277*, 238–245. [CrossRef] [PubMed]
9. Nakov, G.; Trajkovska, B.; Atanasova-Pancevska, N.; Daniloski, D.; Ivanova, N.; Lučan Čolić, M.; Jukić, M.; Lukinac, J. The Influence of the Addition of Hemp Press Cake Flour on the Properties of Bovine and Ovine Yoghurts. *Foods* **2023**, *12*, 958. [CrossRef] [PubMed]
10. Nakov, G.; Trajkovska, B.; Zlatev, Z.; Jukić, M.; Lukinac, J. Quality characteristics of probiotic yoghurt enriched with honey and by-products left after the production of hemp oil by cold pressing the seeds of *Cannabis sativa* L. *Mljekarstvo* **2023**, *73*, 3–11. [CrossRef]
11. Faustino, M.; Veiga, M.; Sousa, P.; Costa, E.M.; Silva, S.; Pintado, M. Agro-food by-products as a new source of natural food additives. *Molecules* **2019**, *24*, 1056. [CrossRef] [PubMed]
12. Bener, M.; Shen, Y.; Apak, R.; Finley, J.W.; Xu, Z. Release and degradation of anthocyanins and phenolics from blueberry pomace during thermal acid hydrolysis and dry heating. *J. Agric. Food Chem.* **2013**, *61*, 6643–6649. [CrossRef] [PubMed]
13. Lee, S.; Keirsey, K.I.; Kirkland, R.; Grunewald, Z.I.; Fischer, J.G.; de La Serre, C.B. Blueberry supplementation influences the gut microbiota, inflammation, and insulin resistance in high-fat-diet–fed rats. *J Nutr.* **2018**, *148*, 209–219. [CrossRef] [PubMed]

14. Cheng, Y.; Wu, T.; Chu, X.; Tang, S.; Cao, W.; Liang, F.; Fang, Y.; Pan, S.; Xu, X. Fermented blueberry pomace with antioxidant properties improves fecal microbiota community structure and short chain fatty acids production in an in vitro mode. *LWT* **2020**, *125*, 109260. [CrossRef]
15. Ignat, I.; Volf, I.; Popa, V.I. A critical review of methods for characterisation of polyphenolic compounds in fruits and vegetables. *Food Chem.* **2011**, *126*, 1821–1835. [CrossRef] [PubMed]
16. Šarić, B.; Mišan, A.; Mandić, A.; Nedeljković, N.; Pojić, M.; Pestorić, M.; Đilas, S. Valorisation of raspberry and blueberry pomace through the formulation of value-added gluten-free cookies. *J. Food Sci. Technol.* **2016**, *53*, 1140–1150. [CrossRef] [PubMed]
17. Silva, S.; Costa, E.M.; Mendes, M.; Morais, R.; Calhau, C.; Pintado, M. Antimicrobial, antiadhesive and antibiofilm activity of an ethanolic, anthocyanin-rich blueberry extract purified by solid phase extraction. *J. Appl. Microbiol.* **2016**, *121*, 693–703. [CrossRef] [PubMed]
18. Szkolnicka, K.; Dmytrów, I.; Mituniewicz-Małek, A. Buttermilk ice cream—New method for buttermilk utilization. *Food Sci. Nutr.* **2020**, *8*, 1461–1470. [CrossRef] [PubMed]
19. *AOAC 925.10-1925*; Solids (Total) and Loss on Drying (Moisture) in Flour Air Oven Method. AOAC International: Gaithersburg, MD, USA, 2014.
20. Grasso, N.; Alonso-Miravalles, L.; O'Mahony, J.A. Composition, physicochemical and sensorial properties of commercial plant-based yoghurts. *Foods* **2020**, *9*, 252. [CrossRef] [PubMed]
21. Pojić, M.; Mišan, A.; Sakač, M.; Dapčević Hadnažev, T.; Šarkanj, B.; Milovanović, I.; Hadnadev, M. Characterization of byproducts originating from hemp oil processing. *J. Agric. Food Chem.* **2014**, *62*, 12436–12442. [CrossRef]
22. Barkallah, M.; Dammak, M.; Louati, I.; Hentati, F.; Hadrich, B.; Mechichi, T.; Ayadi, M.A.; Fendri, I.; Attia, H.; Abdelkafi, S. Effect of Spirulina Platensis Fortification on Physicochemical, Textural, Antioxidant and Sensory Properties of Yogurt during Fermentation and Storage. *LWT* **2017**, *84*, 323–330. [CrossRef]
23. *ISO 21528-2*; Microbiology of Food and Animal Feeding Stuffs—Horizontal Methods for the Detection and Enumeration of *Enterobacteriaceae*—Part 2: Colony-Count Method. ISO: Geneva, Switzerland, 2004.
24. *ISO 21527-1*; Microbiology of Food and Animal Feeding Stuffs—Horizontal Method for the Enumeration of Yeasts and Moulds—Part 1: Colony Count Technique in Products with Water Activity Greater than 0.95. ISO: Geneva, Switzerland, 2008.
25. Davis, L.; Jung, J.; Colonna, A.; Hasenbeck, A.; Gouw, V.; Zhao, Y. Quality and Consumer Acceptance of Berry Fruit Pomace–Fortified Specialty Mustard. *J. Food Sci.* **2018**, *83*, 1921–1932. [CrossRef] [PubMed]
26. Liu, D.; Lv, X.X. Effect of blueberry flower pulp on sensory, physicochemical properties, lactic acid bacteria, and antioxidant activity of set-type yogurt during refrigeration. *J. Food Process. Preserv.* **2019**, *43*, e13856. [CrossRef]
27. Rose, H.; Bakshi, S.; Kanetkar, P.; Lukose, S.J.; Felix, J.; Yadav, S.P.; Gupta, P.K.; Paswan, V.K. Development and Characterization of Cultured Buttermilk Fortified with Spirulina plantensis and Its Physico-Chemical and Functional Characteristics. *Dairy* **2023**, *4*, 271–284. [CrossRef]
28. Jeske, S.; Zannini, E.; Arendt, E.K. Evaluation of physicochemical and glycaemic properties of commercial plant-based milk substitutes. *Plant Foods Hum. Nutr.* **2017**, *72*, 26–33. [CrossRef]
29. Deshwal, G.K.; Tiwari, S.; Kumar, A.; Raman, R.K.; Kadyan, S. Review on factors affecting and control of post-acidification in yoghurt and related products. *Trends Food Sci. Technol.* **2021**, *109*, 499–512. [CrossRef]
30. Trajkovska, B.; Tobolková, B.; Kukurová, K.; Kubincová, J.; Sklářšová, B.; Koreňová, J. Evaluation of qualitative parameters of commercial fermented coconut plant-based yoghurt alternatives on the market in Slovakia. *J. Food Nutr. Res.* **2024**, *63*, 111–121.
31. Starkute, V.; Lukseviciute, J.; Klupsaite, D.; Mockus, E.; Klementaviciute, J.; Rocha, J.M.; Özogul, F.; Ruzauskas, M.; Viskelis, P.; Bartkiene, E. Characteristics of Unripened Cow Milk Curd Cheese Enriched with Raspberry (*Rubus idaeus*), Blueberry (*Vaccinium myrtillus*) and Elderberry (*Sambucus nigra*) Industry By-Products. *Foods* **2023**, *12*, 2860. [CrossRef] [PubMed]
32. Olukomaiya, O.O.; Fernando, W.C.; Mereddy, R.; Li, X.; Sultanbawa, Y. Physicochemical Microbiological and Functional Properties of Camelina Meal Fermented in Solid-State Using Food Grade Aspergillus Fungi. *Fermentation* **2020**, *6*, 44. [CrossRef]
33. García-Pérez, F.J.; Lario, Y.; Fernández-López, J.; Sayas, E.; PérezAlvarez, J.A.; Sendra, E. Effect of Orange Fiber Addition on Yogurt Color During Fermentation and Cold Storage. *Color Res. Appl. J.* **2005**, *30*, 457–463. [CrossRef]
34. Cais-Sokolińska, D.; Pikul, J. Use of colour measurement to evaluate yoghurt quality during storage. *IJFS Ital. J. Food Sci.* **2006**, *18*, 63–71.
35. Łopusiewicz, Ł.; Drozłowska, E.; Siedlecka, P.; Mężyńska, M.; Bartkowiak, A.; Sienkiewicz, M.; Zielińska-Bliźniewska, H.; Kwiatkowski, P. Development Characterization, and Bioactivity of Non-Dairy Kefir-Like Fermented Beverage Based on Flaxseed Oil Cake. *Foods* **2019**, *8*, 544. [CrossRef] [PubMed]
36. Qadi, W.S.; Mediani, A.; Benchoula, K.; Wong, E.H.; Misnan, N.M.; Sani, N.A. Characterization of physicochemical, biological, and chemical changes associated with coconut milk fermentation and correlation revealed by 1H NMR-based metabolomics. *Foods* **2023**, *12*, 1971. [CrossRef] [PubMed]
37. Saberi, M.; Saremnezhad, S.; Soltani, M.; Faraji, A. Functional stirred yogurt manufactured using co-microencapsulated or free forms of grape pomace and flaxseed oil as bioactive ingredients: Physicochemical, antioxidant, rheological, microstructural, and sensory properties. *Food Sci. Nutr.* **2023**, *11*, 3989–4001. [CrossRef] [PubMed]
38. Četojević-Simin, D.D.; Velićanski, A.S.; Cvetković, D.D.; Markov, S.L.; Ćetković, G.S.; Šaponjac, V.T.T.; Vulić, J.J.; Čanadanović-Brunet, J.M.; Djilas, S.M. Bioactivity of Meeker and Willamette raspberry (*Rubus idaeus* L.) pomace extracts. *Food Chem.* **2015**, *166*, 407–413. [CrossRef] [PubMed]

39. Muflihah, Y.M.; Gollavelli, G.; Ling, Y.C. Correlation study of antioxidant activity with phenolic and flavonoid compounds in 12 Indonesian indigenous herbs. *Antioxidants* **2021**, *10*, 1530. [CrossRef] [PubMed]
40. Cho, W.Y.; Yeon, S.J.; Hong, G.E.; Kim, J.H.; Tsend-Ayush, C.; Lee, C.H. Antioxidant activity and quality characteristics of yogurt added green olive powder during storage. *Korean J. Food Sci. Anim. Resour.* **2017**, *37*, 865. [CrossRef] [PubMed]
41. Zhang, M.Q.; Zhang, J.; Zhang, Y.T.; Sun, J.Y.; Prieto, M.A.; Simal-Gandara, J.; Putnik, P.; Li, N.Y.; Liu, C. The link between the phenolic composition and the antioxidant activity in different small berries: A metabolomic approach. *LWT* **2023**, *182*, 114853. [CrossRef]
42. Najgebauer-Lejko, D.; Sady, M. Estimation of the antioxidant activity of the commercially available fermented milks. *Acta Sci. Pol. Technol. Aliment.* **2015**, *14*, 387–396. [CrossRef]
43. Tseng, A.; Zhao, Y. Wine grape pomace as antioxidant dietary fibre for en-hancing nutritional value and improving storability of yogurt and salad dressing. *Food Chem.* **2013**, *138*, 356–365. [CrossRef]
44. Parekh, S.L.; Balakrishnan, S.; Jain, A.; Aparnathi, K.D. Storage stability and chemical constituents of cultured buttermilk prepared by incorporation of paneer whey. *J. Pure Appl. Microbiol.* **2016**, *10*, 3221–3230. [CrossRef]
45. Gupta, M.K.; Torrico, D.D.; Ong, L.; Gras, S.L.; Dunshea, F.R.; Cottrell, J.J. Plant and dairy-based yoghurts: A comparison of consumer sensory acceptability linked to textural analysis. *Foods* **2022**, *11*, 463. [CrossRef] [PubMed]
46. Hurtado-Romero, A.; Zepeda-Hernández, A.; Uribe-Velázquez, T.; Rosales-De la Cruz, M.F.; Raygoza-Murguía, L.V.; Garcia-Amezquita, L.E.; García-Cayuela, T. Utilization of blueberry-based ingredients for formulating a synbiotic Petit Suisse cheese: Physicochemical, microbiological, sensory, and functional characterization during cold storage. *LWT* **2023**, *183*, 114955. [CrossRef]
47. Curutchet, A.; Cozzano, S.; Tárrega, A.; Arcia, P. Blueberry pomace as a source of antioxidant fibre in cookies: Consumer's expectations and critical attributes for developing a new product. *Food Sci. Technol. Int.* **2019**, *25*, 642–648. [CrossRef] [PubMed]
48. Zolfaghari, A.; Ansari, S. Physicochemical and microbiological properties of Chaerophyllum, Oliveria and Zataria essential oils and their effects on the sensory properties of a fermented dairy drink,'doogh'. *Int. J. Food Prop.* **2020**, *23*, 1540–1555. [CrossRef]

foods

Article

The Influence of Whey Protein Isolate on the Quality Indicators of Acidophilic Ice Cream Based on Liquid Concentrates of Demineralized Whey

Artur Mykhalevych [1,*], Magdalena Buniowska-Olejnik [2], Galyna Polishchuk [1], Czesław Puchalski [3], Anna Kamińska-Dwórznicka [4] and Anna Berthold-Pluta [5,*]

[1] Department of Milk and Dairy Products Technology, Educational and Scientific Institute of Food Technologies, National University of Food Technologies, Volodymyrska 68 St., 01033 Kyiv, Ukraine; nuftmilk@i.ua
[2] Department of Dairy Technology, Institute of Food Technology and Nutrition, University of Rzeszow, Ćwiklinskiej 2D St., 35-601 Rzeszow, Poland; mbuniowska@ur.edu.pl
[3] Department of Bioenergetics, Food Analysis and Microbiology, University of Rzeszow, Ćwiklińskiej 2D, 35-601 Rzeszow, Poland; cpuchalski@ur.edu.pl
[4] Department of Food Engineering and Process Management, Institute of Food Sciences, Warsaw University of Life Sciences (WULS-SGGW), Nowoursynowska 159C, 02-776 Warsaw, Poland; anna_kaminska1@sggw.edu.pl
[5] Division of Milk Technology, Department of Food Technology and Assessment, Institute of Food Sciences, Warsaw University of Life Sciences—SGGW, Nowoursynowska 159c Street, 02-776 Warsaw, Poland
* Correspondence: artur0707@ukr.net (A.M.); anna_berthold@sggw.edu.pl (A.B.-P.)

Citation: Mykhalevych, A.; Buniowska-Olejnik, M.; Polishchuk, G.; Puchalski, C.; Kamińska-Dwórznicka, A.; Berthold-Pluta, A. The Influence of Whey Protein Isolate on the Quality Indicators of Acidophilic Ice Cream Based on Liquid Concentrates of Demineralized Whey. *Foods* **2024**, *13*, 170. https://doi.org/10.3390/foods13010170

Academic Editor: Wenjing Cui

Received: 6 December 2023
Revised: 19 December 2023
Accepted: 27 December 2023
Published: 3 January 2024

Abstract: The use of liquid whey concentrates in the composition of ice cream, especially in combination with other powdered whey proteins, is limited due to their understudied properties. This article shows the main rheological and thermophysical characteristics of ice cream mixes, as well as color parameters, microstructure, analysis of ice crystals and quality indicators of ice cream during storage. The most significant freezing of free water ($p \leq 0.05$) was observed in the temperature range from the cryoscopic temperature to −10 °C. The microscopy of experimental ice cream samples based on hydrolyzed whey concentrates indicates the formation of a homogeneous crystalline structure of ice crystals with an average diameter of 13.75–14.75 μm. Microstructural analysis confirms the expediency of using whey protein isolate in ice cream, which ensures uniform distribution of air bubbles in the product and sufficient overrun (71.98–76.55%). The combination of non-hydrolyzed whey concentrate and 3% whey protein isolate provides the highest stability to preserve the purity and color intensity of the ice cream during storage. The produced ice cream can be classified as probiotic (number of *Lactobacillus acidophilus* not lower than 6.2 log CFU/g) and protein-enriched (protein supply from 15.02–18.59%).

Keywords: liquid whey concentrates; protein; ice cream; microstructure; color; microscopy; cryoscopic temperature; probiotic

1. Introduction

The use of whey protein concentrates and isolates in dairy products, in particular ice cream, is becoming increasingly popular due to their functional and technological properties [1–3]. On the contrary, the use of liquid whey concentrates in the composition of food products is limited due to their understudied properties. The possibility of using liquid concentrates from whey of various origins in yogurts, cream, sour-milk desserts, sour cream and cheeses has been reported [3–8]. Whey ice cream based on liquid concentrates of demineralized whey with an increased solids content (39.61–41.61%, including 3.3% protein), in comparison with traditional ice cream recipes of this type [9], differs in improved quality indicators (original taste, absence of defects in consistency and structure), reduced sugar and lactose content in the case of using hydrolyzed concentrates [10–12]. An important aspect of whey ice cream production is the additional possibility of milk whey usage as a by-product of cheese manufacturing. The volume of whey processing in the world is still quite low [13], although in general, there is a trend of increasing interest in whey products [14].

At the previous stage of the research, it was shown that the introduction of protein ingredients into the composition of ice cream based on hydrolyzed whey concentrates significantly increases the thixotropy of ice cream mixes. Whey protein isolate (90%) had the most significant effect on the structural and mechanical properties of whey ice cream mixes among the studied protein additives (soy protein isolate, micellar casein, whey protein concentrate and isolate) [15]. The introduction of whey protein ingredients into the composition of ice cream is appropriate not only for the purpose of expanding the range of protein-enriched products [16,17] but also to prevent excessive freezing of free water in mixes and ice cream during low-temperature processing [18]; to ensure the formation of finely dispersed air bubbles in the thickness of the product during freezing [19,20]; and to provide the product with attractive consumer characteristics by improving resistance to melting, overrun and taste [21,22].

Studies on the effect of whey protein isolate on the quality parameters of ice cream have already reported its ability to increase the viscosity of ice cream mixes and resistance to melting and mask the absence or low content of milk fat in the product [23–25]. However, whey proteins could lead to the deterioration of the color of ice cream, especially during storage, and also form a bitter aftertaste, which is related to the specific sensory properties of whey [26–28]. Roy et al. [25] reported a reduction in ice cream overrun from 94.9% to 33.9% for increasing ice cream protein content from 4% to 10% due to the use of whey protein isolate. The use of hydrolyzed demineralized whey concentrates could be attractive both from the point of view of reducing the lactose content of the product and preventing the formation of sandy and snowy ice cream consistency [29]. However, monosaccharides in the composition of hydrolyzed whey can indirectly affect the formation of the ice cream structure by lowering the cryoscopic temperature [30]. A decrease in the sugar content in ice cream and the presence of lactose hydrolysis products, monosugars (glucose and galactose), which have a lower molecular weight and, accordingly, a higher concentration [31], are able to decrease the cryoscopic temperature and the amount of frozen free water during freezing and hardening [32]. Such changes can reduce the resistance to melting and overrun of ice cream [33,34]. The use of whey protein isolate in ice cream based on hydrolyzed whey concentrates could slow down the freezing process of free water due to its additional binding by whey proteins [35]. It was reported that due to the specificity of the amino acid composition, whey protein isolate belongs to the group of protein additives capable of inhibiting ice recrystallization [36–38]. Whey protein isolate could also be used as a functional and technological ingredient (emulsifier, thickener, gelling agent, foaming agent and water-binding agent) in the production of products with characteristics similar to those of classical formulations [39]. In addition, scientists have reported that whey protein isolate is able to support the activity of probiotic bacteria srtains *Lactobacillus* or *Bifidobacterium* in dairy products [40,41], which is an important aspect of this study.

Available scientific information on the influence of whey protein isolate on ice cream quality is characterized by certain contradictions, which are explained by the different degrees of its purification, the quality of the input raw materials and the chemical composition and technology of the selected type of ice cream [19–21,23–25,27,39].

To our knowledge, there are no publications on the use of whey protein isolate in the formulation of whey ice cream based on liquid concentrates of demineralized whey. The use of liquid hydrolyzed concentrates of demineralized whey in ice cream allows reducing the amount of sugar to 9% and lactose to 1% [9], which could be attractive to individual groups of consumers.

Thus, ***the purpose*** of this research was to study the influence of whey protein isolate on the physicochemical and sensory parameters of whey ice cream.

The following ***tasks*** of the research work were formulated:

- To study the influence of whey protein isolate on the physicochemical and rheological properties of mixes and ice cream;
- To compare the dynamics of changes in the freezing process in ice cream with whey protein isolate of free water during freezing and subsequent storage of the product at sub-zero temperatures;
- To study the microstructure of soft ice cream and determine the main physical and chemical parameters;
- To measure the microbiological indicators of ice cream during storage.

2. Materials and Methods

2.1. Raw Materials

Liquid concentrates of demineralized whey with a solids content of 40% were used to make experimental ice cream samples. Non-hydrolyzed concentrates of demineralized whey were produced by the reconstitution of whey powder with a degree of demineralization of 90% (Herkules, MLEKOVITA, High Mazovia, Poland) in water. Hydrolyzed concentrates of demineralized whey were obtained by lactose fermenting using the enzyme lactase (β-galactosidase) with an activity of 5200 NLU/g (SEROWAR s.c., Szczecin, Poland) and the starter preparation nu-trish® LA-5® containing L. acidophilus (Chr. Hansen A/S, Hoersholm, Denmark). Whey protein isolate 90% (SPOMLEK, Radzyń Podlaski, Poland) was used as a protein supplement. Water, white (regular) sugar, vanillin, activated starter and the stabilization system Cremodan SI 320 (Danisco A/S, Brabrand, Denmark) were used to prepare ice cream mixes. The content of whey protein isolate was explained by its influence on the thixotropic properties of ice cream mixes, which was explained at the previous stage of the experiment [15]. In ice cream based on non-hydrolyzed whey protein concentrate, its maximum effective amount is 3%, while in ice cream based on hydrolyzed whey concentrate, it is up to 5%. The difference in sugar content (11% and 9%) for ice cream samples with non-hydrolyzed and hydrolyzed concentrates of demineralized whey, respectively, is explained by the different degrees of sweetness, which increases during the hydrolysis of lactose into monosaccharides and, accordingly, allows reducing the amount of sugar to 9% in the case of using hydrolyzed concentrates [9]. The content of the stabilization system (0.6%) was chosen in accordance with the manufacturer's recommendations for the production of low-fat types of ice cream. Liquid concentrates of demineralized whey were made on the basis of water and demineralized whey powder 90% (Herkules, MLEKOVITA, High Mazovia, Poland). Enzymolysis of lactose in concentrates was carried out using the enzyme lactase (β-galactosidase) with an activity of 5200 NLU/g (SEROWAR s.c., Szczecin, Poland) and the starter nu-trish® LA-5® (Chr. Hansen A/S, Hoersholm, Denmark).

2.2. Ice Cream Production

2.2.1. Activated Starter

To obtain an activated starter, ultra-pasteurized skimmed milk was heated to an inoculation temperature of 38–42 °C, after which a pre-calculated amount of starter was added. Fermentation was carried out until pH 5.4–5.2.

2.2.2. Hydrolyzed Concentrates

Demineralized whey powder was reconstituted in water at a temperature of (40–45) °C to obtain concentrates with a solids content of 40%. The concentrates were filtered, pasteurized at (85–88) °C for 3–5 min and cooled to the storage temperature for non-hydrolyzed concentrates.

Hydrolyzed concentrates of demineralized whey were obtained according to the technology of Osmak et al. [42]. For the production of hydrolyzed whey concentrates, after pasteurization, they were cooled to (40–43) °C and simultaneously fermented with β-galactosidase preparation and starter nu-trish® LA-5®. With the simultaneous introduction of the β-galactosidase enzyme and the starter during the lag phase of L. acidophilus development (2–4 h), the enzyme has time to reveal hydrolytic activity at pH ≥ 5.7, which makes it possible to achieve maximum hydrolysis of lactose within 8 h of enzymolysis.

2.2.3. Ice Cream

Dry components (white sugar, stabilization system, vanilla and whey protein isolate) were mixed and added to preheated water (40–45 °C). Then, liquid concentrates of demineralized whey were added. The obtained mixes were filtered, pasteurized at 83–87 °C for 5 min and homogenized under a pressure of 12.0 ± 2.5 MPa using a laboratory homogenizer-disperser 15M-8TA "Lab Homogenizer & Sub-Micron Disperser" (GAULIN. CORPORATION, Boston, MA, USA). The homogenized mixes were cooled to 38–42 °C, and 3% activated starter was added. The fermentation process was carried out until pH 5.25–5.1, followed by cooling to 2–6 °C and maturation for 24 h. The matured mixes were frozen on a periodic freezer FPM-3.5/380-50 "Elbrus-400" (JSC "ROSS", Kharkiv, Ukraine). At the first stage of freezing, the mixes were cooled in a cooling cylinder (volume—7 L) to −1 °C at a rotation frequency of the scraper stirrer of 4.5 s^{-1} for 120 s. At the second stage, the mixes were frozen at a rotation frequency of 9 s^{-1} for 180 s to −5.0 ± 0.5 °C. The formulations of experimental samples of ice cream are given in Table 1.

Table 1. Formulations of experimental samples of whey ice cream.

Ingredients, %	Ice Cream Samples				
	NHC	NH3%	HC	H3%	H5%
Non-hydrolyzed concentrate of demineralized whey	75.0	75.0	–	–	–
Hydrolyzed concentrate of demineralized whey	–	–	75.0	75.0	75.0
White sugar	11.0	11.0	9.0	9.0	9.0
Whey protein isolate 90%	–	3.0	–	3.0	5.0
Stabilization system	0.6	0.6	0.6	0.6	0.6
Activated starter	3.0	3.0	3.0	3.0	3.0
Vanilla	0.1	0.1	0.1	0.1	0.1
Water	10.3	7.3	12.3	9.3	7.3
Total	100.0	100.0	100.0	100.0	100.0

Note. NHC—ice cream based on non-hydrolyzed whey concentrate; NH3%—ice cream based on non-hydrolyzed whey concentrate + 3% whey protein isolate; HC—ice cream based on hydrolyzed whey concentrate; H3%—ice cream based on hydrolyzed whey concentrate + 3% whey protein isolate; H5%—ice cream based on hydrolyzed whey concentrate + 5% whey protein isolate.

2.3. Methods

2.3.1. Chemical Composition

The total solids content in ice cream samples was determined by the arbitration method, the principle of which consists in drying the sample diluted with distilled water and mixed with sand at 102 °C to a constant mass, followed by weighing to determine the mass of the residue.

The protein content was determined using the Kjeldahl method, the fat content using the modified Gerber method [43], and the carbohydrate content using the Bertrand method [44]. The lactose content in whey concentrates and test samples of ice cream mixes was determined using iodometric and refractometric methods [45]. The degree of lactose hydrolysis was calculated based on the found lactose content [46]:

$$\text{Degree of lactose hydrolysis (\%)} = 100\% - \text{residual lactose content (\%)} \quad (1)$$

The level of protein supply (%) in the finished product was calculated as the ratio of the protein content to the sum of the protein, fat and carbohydrate content, multiplied by 100.

2.3.2. Rheological and Thermophysical Characteristics of Ice Cream Mixes

After cooling to 2–6 °C, the ice cream mixes were whipped for 5, 10 and 15 min with 5 min breaks according to the modified method of Lim et al. [47]. Foam overrun was determined as the ratio of the volume of the whipped mix to its initial volume, expressed as a percentage. The foam resistance of ice cream mixes was determined according to the modified method of Philips L., according to which a container with a hole in the bottom was used for the foam to drain after whipping [47]. The time during which 50% of the initial volume of the mix used for whipping is formed as a result of foam destruction was taken as an indicator of foam stability. The viscosity of ice cream mixes was determined from warmed samples using a viscometer IKA ROTAVISC lo-vi Complete (IKA, Staufen, Germany). A spindle SP-4 was used to measure viscosity, which was immersed in a prepared sample at 20 °C and a shear rate of 200 rpm. Viscosity values were read after 2 min [48]. Water activity was determined on a water activity analyzer "HygroLab 2" (Rotronic, Bassersdorf, Switzerland) at 20 °C in a measurement range of 0–1 Aw (0–100% rh) [49]. The cryoscopic temperature was set using a cryostat and Beckman thermometer (TL-1). Based on Raoult's law for non-dissociated molecular solutions, the amount of frozen water at different temperature stages was calculated using the following formula [50]:

$$\omega = (1 - t_{cr}/t) \times 100 \tag{2}$$

where ω is content of frozen water, %; t_{cr} is cryoscopic temperature, °C; and t is the temperature at each stage of technological processing, °C.

2.3.3. Color

Color parameters were determined using a colorimeter (Precision Colorimeter, Model NR 145, Shenzhen, China) and the CIE LAB system. The following parameters were defined: L* as whiteness (from 0—black to 100—white), a* as a color from red (+) to green (−), b* as a color from yellow (+) to blue (−), C* as the purity and intensity of color from gray (C* = 0) to the direction of pure colors (C* = 100) and h° as a color shade (within 0–360°). Before measurement, the device was calibrated against a white standard.

2.3.4. Microstructure

The ice cream sample was taken from the center of the sample in at least three different places and at a distance of 3 cm from the surface of the product, placed at 19 ± 1 °C in a Goryaev chamber covered with glass and immediately subjected to microscopy at a magnification of 160 times. At the same time, the ice crystals melted, but the foam remained, because under these conditions, the air bubble envelopes did not dehydrate. Photomicrographs were obtained using an Olympus CX41 light microscope (Olympus Corporation, Tokyo, Japan) and camera [19].

2.3.5. Analysis of Ice Crystals

Ice cream samples were taken from the center of the sample, from at least 3 different locations, at least 3 cm from the surface of the ice cream, and placed on a glass slide using a spatula and then covered with a glass. The samples were transferred to a microscope with a cooling system (Linkam LTS420, Tokyo, Japan). The recrystallization process was analyzed based on images of ice crystals taken after preparation, using an Olympus BX53 microscope (Olympus Corporation, Tokyo, Japan) with a Linkam LTS420 (Tokyo, Japan) cooling system (temperature range from −196 °C to −420 °C) and an Olympus SC50 camera (Olympus Corporation, Tokyo, Japan). The resulting images were then analyzed using software NIS Elements D (version 5.30.00, Nikon, Tokyo, Japan). Between 300 and 500 crystals were labeled for each sample, and area, equivalent diameter and standard deviation were then calculated using software NIS Elements D Imaging (version 5.30.00, Nikon). Crystal size frequency distributions were calculated using Microsoft Excel 2011 macro data analysis. The relative frequency of any class interval was calculated as the number of crystals in that class (class frequency) divided by the total number of crystals and expressed as a percentage. The X50 parameter was analyzed as the average diameter (DA) for 50% of the crystals in the sample. The mean diameter (DA) and standard deviation (SD) of each class were also calculated. The method of determination is given in scientific works [51–53].

2.3.6. Quality Indicators of Ice Cream

The overrun of the ice cream was determined using the weighing method, according to which the ice cream mix was weighed before freezing and the same volume of soft ice cream after freezing. Overrun (O, %) was calculated according to the following formula [54]:

$$O = (M_1 - M_2/M_2) \times 100 \quad (3)$$

where M_1 is the mass of the glass with the mix, g, and M_2 is the mass of a glass with ice cream, g.

Resistance to melting (the time of the first drop flowing out and the time of accumulation of 10 cm^3) was determined at an ambient temperature of 22 °C. The ice cream samples were placed on a grid (d = 95 mm, holes 5 × 5 mm, wire thickness 0.5 mm). Then, the time until complete melting was recorded for ice cream melting [51]. Hardness of ice cream samples was determined using a Brookfield CT-3 model texture analyzer. Hardness measurements were made between −6 °C and −2 °C, using texture software Pro CT V1.6 (Brookfield Engineering Laboratories Onc., ABD, Middleboro, MA, USA). A conical probe TA 15/1000 was used for analysis. Test speed was 2 mm/s, distance 15 mm, trigger load 6.8 g, length 40 mm and diameter 60 mm.

2.3.7. Microbiological Analysis

For the study, 5 g of each sample was diluted in 45 cm^3 of physiological saline. Microbiological analysis was performed under the following conditions: the total number of microorganisms on PCA (Plate Count Agar, Oxoid, Basingstoke, UK) at 30 °C for 48–72 h and the number of intestinal bacteria on VRBL agar (Violet Red Bile agar with lactose, Oxoid, Basingstoke, UK) at 37 °C for 24–48 h. Acidophilic bacilli counts on MRS (De Man Rogosa Sharpe, Oxoid, Basingstoke, UK) and MSE (Mayeux, Sandine & Elliker, Oxoid, Basingstoke, UK) were analyzed at 37 °C for 48–72 h and microscopic fungi and yeast on MEA (malt extract agar, Oxoid, Basingstoke, UK) at 25 °C for 5 days. For Lactobacillus acidophilus, the study was carried out using the plate method with MRS (Biocorp, Warsaw, Poland). Lactobacillus acidophilus was cultivated in microaerophilic conditions with 5% CO_2 at 37 °C.

2.3.8. Statistical Processing

Analysis of variance (ANOVA) was performed using STATISTICA 13 software. The significance of the test was set at $\alpha = 0.05$. Data are expressed as mean with standard deviations (±SD), and differences between groups were assessed using Tukey's HSD test. The study of physicochemical indicators of ice cream samples was conducted three times to ensure the reliability of the data obtained.

3. Results and Discussion

3.1. Chemical Composition

The chemical composition of liquid concentrates of demineralized whey (Table 2) differs slightly from the examples of liquid whey concentrates given in the literature, which usually have a higher protein content (5.09–11.93%) but a lower solids content (14.09–26.45%), while the content of fat varies from 0.43–0.78% for low-fat concentrates to 6.43–7.82% for full-fat concentrates [3,5,55]. The use of highly demineralized whey to obtain concentrates in this study significantly increases the lactose content to 30.5%, which could significantly affect the quality of ice cream during storage. The hydrolysis of lactose in whey concentrates allows reducing its content to 0.98%, which is lower than the value (4.95%) reported in another study [3]. At the same time, the high proportion of solids (39.92–40.01%) in these liquid concentrates could help to ensure the recommended level of solids in ice cream in the range of 25–35% [42], which is especially important for low-fat types of ice cream. The difference in the obtained values could be explained by the use of whey of different origins, as well as the use of special technologies for the production of concentrates.

Table 2. Chemical composition of whey concentrates and whey ice cream.

Sample	Total Solids, %	Protein, %	Fat, %	Carbohydrates, %	Lactose, %	The Degree of Lactose Hydrolysis, %	Level of Protein Supply, %	Ice Cream Category/ Nutritional Status
Whey concentrates								
NHCDW	39.92 ^{a} ± 0.30	4.45 ^{a} ± 0.12	0.43 ^{b} ± 0.01	30.71 ^{a} ± 0.68	30.50 ^{b} ± 1.19	–	–	–
HCDW	40.01 ^{a} ± 0.84	4.41 ^{a} ± 0.55	0.40 ^{a} ± 0.01	30.61 ^{a} ± 0.51	1.29 ^{a} ± 0.04	98.71 ^{a} ± 0.04	–	–
Whey ice cream								
NHC	41.61 ^{ab} ± 1.24	3.31 ^{a} ± 0.11	0.74 ^{abc} ± 0.02	33.05 ^{a} ± 1.52	22.90 ^{b} ± 0.53	–	8.92 ^{a} ± 0.52	premium
NH3%	44.55 ^{b} ± 0.87	6.10 ^{b} ± 0.25	0.72 ^{a} ± 0.01	33.46 ^{a} ± 0.79	22.87 ^{b} ± 0.12	–	15.11 ^{b} ± 0.35	super premium, enriched with protein
HC	39.58 ^{a} ± 1.53	3.30 ^{a} ± 0.14	0.72 ^{a} ± 0.03	33.10 ^{a} ± 0.65	0.97 ^{a} ± 0.01	99.03 ^{a} ± 0.01	8.89 ^{a} ± 0.27	super premium
H3%	42.54 ^{ab} ± 1.14	6.02 ^{b} ± 0.11	0.78 ^{bc} ± 0.01	33.28 ^{a} ± 0.47	0.98 ^{a} ± 0.03	99.02 ^{a} ± 0.03	15.02 ^{b} ± 0.19	super premium, enriched with protein
H5%	44.63 ^{b} ± 1.01	7.84 ^{c} ± 0.05	0.79 ^{b} ± 0.01	33.53 ^{a} ± 0.34	0.98 ^{a} ± 0.01	99.02 ^{a} ± 0.01	18.59 ^{c} ± 0.13	super premium, enriched with protein

Note. NHCDW—non-hydrolyzed concentrate of demineralized whey; HCDW—hydrolyzed concentrate of demineralized whey; NHC—ice cream based on non-hydrolyzed whey concentrate; NH3%—ice cream based on non-hydrolyzed whey concentrate + 3% whey protein isolate; HC—ice cream based on hydrolyzed whey concentrate; H3%—ice cream based on hydrolyzed whey concentrate + 3% whey protein isolate; H5%—ice cream based on hydrolyzed whey concentrate + 5% whey protein isolate. $^{a–c}$—different superscript letters in the columns represent significant differences in the mean values of the same parameter ($p \leq 0.05$).

Ice cream based on liquid whey concentrates in terms of solids content could be attributed to the full-fat analog (12–18% fat), namely the category of super-premium type of ice cream (40–42% solids) for all samples, except for NHC, which belongs to the premium category (38–40% solids) [56]. The content of high-value whey proteins in ice cream based on whey concentrates exceeds the average protein content (2.6–4.6%) in traditional types of ice cream (10–16% fat) [57]. However, it is insufficient for NHC and HC samples for their classification as ice cream enriched with protein (more than 12% protein supply according to EU regulation No. 1924/2006). Samples of ice cream with whey protein isolate are protein-enriched products, but their protein content does not allow them to be classified as a product source of protein (more than 20% protein supply in accordance with EU Regulation No. 1924/2006), which is a perspective for further research. In general, the data from Table 1 indicate a significant increase in the nutritional value of the developed ice cream compositions based on whey concentrates. However, in terms of the lactose content, the sample NH3% significantly exceeds the others, which could cause defects in consistency and taste during low-temperature storage of the product.

3.2. Physical Characteristics of Ice Cream Mixes

Table 3 shows the thermophysical characteristics of ice cream mixes based on liquid concentrates of demineralized whey. The addition of whey protein isolate (3–5%) increases foam overrun and foam resistance, and the maximum effect is observed in ice cream mixes based on hydrolyzed whey concentrates. The foam overrun of the mixes increases after 5 min of whipping and decreases after 10 min, except for samples with hydrolyzed whey concentrate and whey protein isolate, which could be explained by the lower viscosity of the mixes due to the presence of lactose hydrolysis products [58,59], as well as a slightly lower sugar content, which usually appears in ice cream mixes as a component that increases viscosity [56]. Lee and Duggan [60] found that the foam overrun of native mixes with whey protein isolate is significantly higher than microgels with WPI, but their foam stability is slightly lower. In this study, we observed a significant increase in both the foam overrun and foam resistance of whey ice cream mixes due to the fact that the ice cream mix is a multicomponent system in which the synergistic interactions of the ingredients could significantly influence its rheological properties. The combination of whey proteins with monosaccharides provides a significant increase in the foaming properties of mixes. Puangmanee et al. [61] compared the influence of ordinary whey protein isolate and its glycated species with various monosaccharides on the rheological properties of mixes. It was shown that the presence of d-glucose, d-fructose, d-allose and d-psicose additionally increases the foam resistance and foam overrun of the mixes, and this value only increases during the whipping interval from 15 to 30 min. This suggests that the presence of high amounts of monosaccharides in hydrolyzed whey concentrate ice cream mixes exhibits a synergistic effect with WPI. The obtained viscosity values are correlated with the indicators of foam overrun and foam resistance, confirming the regularity of the decrease in this indicator in mixes with a high content of monosaccharides (glucose, galactose). Akalin et al. [62] reported that the addition of 4% WPI resulted in the excessive thickening and gelation of ice cream with an increase in the viscosity of the ice cream mix. Another study reported that the combination of WPI with polysaccharides significantly increased the viscosity of ice cream mixes, which overall had a positive effect on its quality [23].

Table 3. Thermophysical characteristics of whey ice cream mixes.

Indicator		Sample				
		NHC	NH3%	HC	H3%	H5%
Foam overrun, %	5	144.58 ^{a} ± 1.67	158.74 ^{c} ± 2.10	150.08 ^{b} ± 1.5	203.33 ^{d} ± 1.7	192.14 ^{e} ± 2.0
	10	177.51 ^{a} ± 1.59	184.96 ^{a} ± 1.47	201.42 ^{b} ± 4.2	236.78 ^{d} ± 2.5	220.43 ^{c} ± 3.2
	15	164.23 ^{a} ± 2.54	179.7 ^{b} ± 1.89	187.54 ^{c} ± 1.6	246.75 ^{e} ± 4.7	221.59 ^{d} ± 1.2
Foam resistance, XB	5	31.57 ^{a} ± 0.87	33.89 ^{b} ± 0.27	44.82 ^{c} ± 0.67	48.41 ^{d} ± 0.74	45.90 ^{c} ± 0.81
	10	32.46 ^{a} ± 0.54	39.94 ^{b} ± 0.71	49.55 ^{c} ± 0.56	58.45 ^{d} ± 0.25	57.63 ^{d} ± 0.90
	15	32.11 ^{a} ± 0.68	38.43 ^{b} ± 0.44	48.13 ^{c} ± 1.36	61.94 ^{e} ± 0.49	59.33 ^{d} ± 1.01
Viscosity, Mpa × s		254.42 ^{b} ± 1.86	384.55 ^{d} ± 1.12	228.61 ^{a} ± 2.54	339.47 ^{c} ± 1.07	511.05 ^{e} ± 1.24
Cryoscopic temperature, °C		−2.88 ^{a} ± 0.01	−2.95 ^{a} ± 0.08	−2.39 ^{d} ± 0.01	−2.55 ^{c} ± 0.02	−2.71 ^{b} ± 0.05
Water activity, units		0.955 ^{e} ± 0.02	0.941 ^{d} ± 0.03	0.912 ^{c} ± 0.01	0.905 ^{b} ± 0.01	0.896 ^{a} ± 0.02

Note. NHC—ice cream based on non-hydrolyzed whey concentrate; NH3%—ice cream based on non-hydrolyzed whey concentrate + 3% whey protein isolate; HC—ice cream based on hydrolyzed whey concentrate; H3%—ice cream based on hydrolyzed whey concentrate + 3% whey protein isolate; H5%—ice cream based on hydrolyzed whey concentrate + 5% whey protein isolate. 5, 10, 15—time of ice cream mix whipping, min. $^{a–e}$—different superscript letters in the lines represent significant differences in the mean values of the same parameter ($p \leq 0.05$).

The cryoscopic temperature of ice cream mixes based on non-hydrolyzed concentrate is slightly higher than samples based on hydrolyzed concentrate, which is explained by the high content of monosaccharides in the latter, as a result of which a depression of the freezing point of ice cream is observed [63]. The high content of solids in all ice cream samples reduces the range of cryoscopic temperature depressions of ice cream mixes within the range of values from −2.39 to −2.95 °C, which is similar to the results of other scientists who studied low-fat or high-protein ice cream [19,64–66]. The study of osmotic pressure in the aqueous phase of whey ice cream mixes shows (Table 3) a decrease in water activity in mixes based on hydrolyzed whey concentrate. The addition of WPI reduces water activity, but the most significant effect on this indicator, in our opinion, is the presence of monosaccharides, which have a higher ability to bind free water than sucrose [67]. Thus, samples NHC, NH3%, HC and H3% could be attributed to systems with high water activity (aw = 1.0–0.9), while H5% could be attributed to a food system with intermediate moisture (aw = 0.9–0.6) [68]. A decrease in water activity will affect the quality of ice cream due to a decrease in the freezing point caused by the hydrolysis of lactose in the samples HC, H3% and H5%. This and other influences will be considered in the following sections of this article.

Water in ice cream is in a bound and free state [69]. The latter could freeze in the form of ice crystals at temperatures below the cryoscopic temperature [70]. During product storage, even with slight temperature fluctuations, migratory recrystallization of ice crystals and their fusion occur, which is accompanied by the disappearance of small crystals and the growth of large ones [71]. The recommended storage temperature of ice cream is not lower than −16 to −25 °C [19]; therefore, even with slight fluctuations in it, in ice cream with a high content of free water, the appearance of a coarse crystalline structure is possible. The change in the physical state of the aqueous phase of ice cream has been studied by many scientists [72–74], but the use of protein ingredients in its composition requires additional research. Proteins, as well as mono- and disaccharides, could, to some extent, affect the cryoscopic temperature of mixes and, accordingly, the proportion of frozen water and the structure of ice cream. Based on the values of the cryoscopic temperature of the studied mixes (Table 3), the content of frozen water in ice cream was calculated in the temperature range of technological processing from −5 to −40 °C, and it was shown that the proportion of frozen water reached values at the beginning and at the end of freezing from 42.40–52.20% to 92.80–94.03%, respectively (Figure 1).

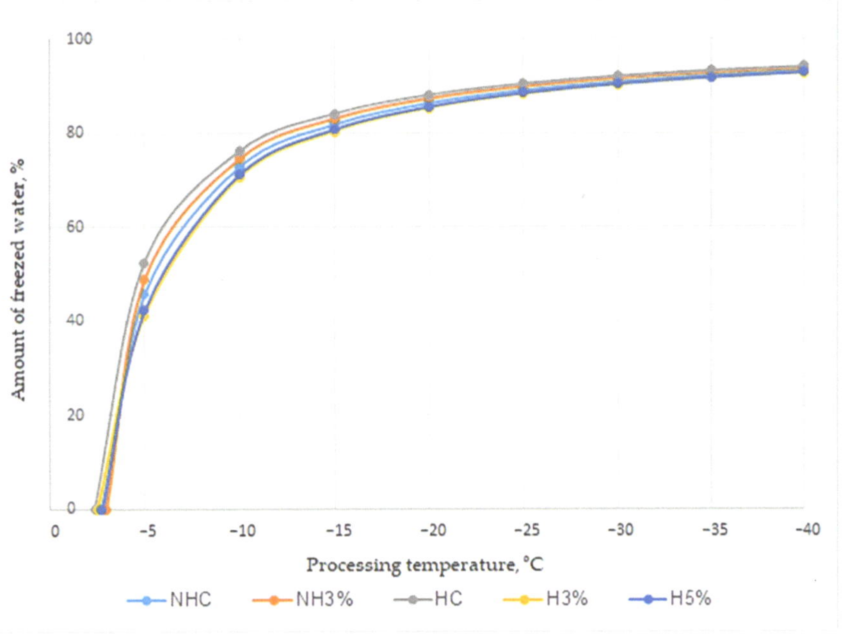

Figure 1. Free water freezing dynamics under different modes of low-temperature processing. **Note.** NHC—ice cream based on non-hydrolyzed whey concentrate; NH3%—ice cream based on non-hydrolyzed whey concentrate + 3% whey protein isolate; HC—ice cream based on hydrolyzed whey concentrate; H3%—ice cream based on hydrolyzed whey concentrate + 3% whey protein isolate; H5%—ice cream based on hydrolyzed whey concentrate + 5% whey protein isolate.

The maximum difference in the amount of frozen water in the studied samples (up to 9.8%) was found at −5 °C, but with a further decrease in temperature, the difference in the amount of frozen water in ice cream decreased to 1.4%. In general, the most significant freezing of free water was observed in the temperature range from the cryoscopic temperature to −10 °C. Under these conditions, 70.5–71.2% of water froze in the samples NHC and NH3% and 72.9–76.1% in the samples HC, H3% and H5%, which indicates the most significant cryoprotective activity of monosaccharides in the hydrolyzed whey concentrate, including combination with whey protein concentrate. A significant part of the water (up to 17.1–19.7%) continued to freeze when the mixes reached a temperature of −30 °C. The further change in this indicator in the temperature range from −30 to −40 °C was quite insignificant. The obtained data confirm the dynamics of freezing of water in ice cream based on whey concentrates, typical for classic types of ice cream, at low temperatures during the freezing of mixes and hardening and storage of the studied samples.

3.3. Color

The lowest value of whiteness L* ($p \leq 0.05$) was recorded for H5% (Table 4) which contained the largest amount of added whey protein isolate (5%), which gives the product a yellow color [24]. Barros et al. [75] also reported a decrease in whiteness in ice cream samples from 87.65 to 82.45 units depending on the content of concentrated whey (70–280 g per ice cream sample). The obtained L* values for all ice cream samples are lower than the reported results [76], which is explained by the higher content of whey ingredients than in the recipes of known analogs, as well as the absence in these samples of pasteurized or dry milk as raw components that give whiteness to dairy products due to the presence of colloidal particles, such as milk fat globules and casein micelles, capable of scattering light in the visible spectrum [77]. During 14 days of storage, the value of whiteness (L*) in ice cream decreases; however, for samples with hydrolyzed whey concentrate, this process occurs less intensively, which is associated with the effect of lactose hydrolysis, which increases the whiteness of the food system [78]. The values of parameters a* and b* indicate the predominance of green and yellow colors in the product, respectively, which is due to the color characteristics of whey and its processing products [79,80], which were the main ingredients in ice cream production for this study.

Table 4. Color parameters of ice cream mixes.

Sample			Color Parameters				
			L*	a*	b*	C*	h°
NHC	mix		$72.80^{a} \pm 1.08$	$-4.38^{b} \pm 0.12$	$20.50^{a} \pm 0.88$	$20.96^{a} \pm 0.66$	$102.07^{a} \pm 2.28$
	ice cream	1st day	$81.66^{a} \pm 2.54$	$-4.09^{b} \pm 0.04$	$20.16^{a} \pm 0.50$	$14.73^{a} \pm 0.05$	$101.50^{a} \pm 2.05$
		14th day	$76.81^{a} \pm 1.27$	$-4.11^{b} \pm 0.17$	$21.84^{ab} \pm 0.17$	$13.52^{a} \pm 0.64$	$103.17^{b} \pm 1.84$
NH3%	mix		$69.56^{a} \pm 1.59$	$-3.61^{a} \pm 0.09$	$20.18^{a} \pm 0.01$	$20.37^{a} \pm 0.95$	$100.24^{a} \pm 1.10$
	ice cream	1st day	$82.78^{a} \pm 2.22$	$-2.98^{a} \pm 0.13$	$21.03^{a} \pm 0.67$	$21.26^{b} \pm 0.54$	$98.08^{a} \pm 2.89$
		14th day	$76.07^{a} \pm 2.18$	$-3.05^{a} \pm 0.01$	$21.56^{ab} \pm 0.94$	$20.90^{b} \pm 0.96$	$98.99^{a} \pm 2.17$
HC	mix		$79.03^{a} \pm 2.01$	$-3.94^{a} \pm 0.11$	$24.07^{a} \pm 0.58$	$24.41^{a} \pm 1.14$	$99.19^{a} \pm 2.74$
	ice cream	1st day	$82.31^{a} \pm 0.87$	$-2.42^{a} \pm 0.10$	$14.55^{b} \pm 0.69$	$20.57^{a} \pm 0.71$	$99.11^{a} \pm 2.54$
		14th day	$80.98^{a} \pm 1.85$	$-2.55^{a} \pm 0.02$	$14.69^{a} \pm 0.54$	$18.09^{b} \pm 0.05$	$99.54^{a} \pm 1.71$
H3%	mix		$81.36^{a} \pm 1.54$	$-3.17^{a} \pm 0.13$	$24.43^{a} \pm 0.47$	$24.66^{a} \pm 1.19$	$97.46^{a} \pm 0.86$
	ice cream	1st day	$83.25^{a} \pm 1.73$	$-2.84^{a} \pm 0.08$	$18.84^{a} \pm 0.54$	$18.57^{a} \pm 0.55$	$98.23^{a} \pm 3.57$
		14th day	$78.56^{a} \pm 1.25$	$-2.91^{a} \pm 0.11$	$19.63^{ab} \pm 0.07$	$18.08^{b} \pm 0.76$	$99.07^{a} \pm 2.60$
H5%	mix		$77.20^{a} \pm 1.89$	$-2.71^{a} \pm 0.01$	$23.39^{a} \pm 1.05$	$23.08^{a} \pm 1.00$	$96.54^{a} \pm 0.97$
	ice cream	1st day	$80.01^{a} \pm 2.12$	$-3.30^{a} \pm 0.13$	$21.58^{a} \pm 0.68$	$21.84^{a} \pm 0.58$	$98.33^{a} \pm 1.18$
		14th day	$77.88^{a} \pm 2.57$	$-3.35^{a} \pm 0.05$	$22.42^{b} \pm 0.84$	$20.05^{b} \pm 1.13$	$98.74^{a} \pm 1.36$

Note. NHC—ice cream based on non-hydrolyzed whey concentrate; NH3%—ice cream based on non-hydrolyzed whey concentrate + 3% whey protein isolate; HC—ice cream based on hydrolyzed whey concentrate; H3%—ice cream based on hydrolyzed whey concentrate + 3% whey protein isolate; H5%—ice cream based on hydrolyzed whey concentrate + 5% whey protein isolate. ^{a,b}—different superscript letters in the columns represent significant differences in the mean values of the same parameter ($p \leq 0.05$).

A decrease in the parameter b* was observed for HC which means a decrease in the intensity of the yellow color and correlates with the L* indicator, which, on the contrary, increases for this ice cream sample. The addition of whey protein isolate to samples based on hydrolyzed concentrate leads to an increase in b* (increased yellowness). Barros et al. [75] also reported an increase in the b* indicator from 16.27 to 18.02% with an increase in the content of concentrated whey in the composition of ice cream. However, Meneses et al. [80] showed that the b* indicator increased from 57.64 to 60.92 units when adding whey at different mass fractions. A significant difference in the results obtained in this study may be related to the hydrolysis of lactose, as well as the type of whey used. During storage, the degree of green and yellow coloring increases ($p \leq 0.05$) for all samples. Color intensity and purity (C*) increase for HC, but further addition of whey protein isolate (H3% and H5%) shifts this parameter toward gray colors (less pure and intense color). During storage, the largest decrease in the C* index occurs for NHC, which emphasizes the instability of its color at low temperatures. The NH3% sample shows greater stability to maintain purity and color intensity. The value of the hue (h°) by the position in the spectrum makes it possible to conclude that the ice cream samples are between yellow and green with a predominance toward the first, which is correlated with indicators a* and b*.

3.4. Microstructure

Analysis of the microstructure indicates that whey protein isolate (3–5%) contributes to the increase in the size of air bubbles in ice cream, as well as non-hydrolyzed whey concentrate as an ice cream base (Figure 2). The sample HC is characterized by a finer dispersion (average diameter of air bubbles—6.4 μm) and uniform distribution of the air phase in the thickness of the product, compared to NHC (average diameter of air bubbles—8.8 μm), which is due to the higher content of solids in the latter, as well as slightly higher viscosity as a result of significant lactose content. At the same time, the addition of 3% whey protein isolate in combination with non-hydrolyzed concentrate (NH3%) leads to the formation of large air bubbles (up to 12.5–24.1 μm in diameter) due to the increased viscosity of the ice cream mix, which makes it difficult to saturate mixes with air during the freezing process. Whey protein isolate in the samples NH3% and NH5% also leads to the formation of larger air bubbles than in the control samples (NHC, HC); however, their number is significantly lower than for NH3%. In addition, for the above-mentioned samples, a significantly larger number of finely dispersed air bubbles are observed, which show the ability to aggregate among themselves and concentrate around larger air bubbles.

The effect of sticking air bubbles and their uniform distribution in ice cream was observed by scientists when using polysaccharides, protein concentrates and isolates [64,72,81]. In whey ice cream, the size of air bubbles depends not only on the type of raw materials and components but also on the solid content. The small number of solids in MSNF (milk solids non-fat) leads to a structure with inclusions of large air cells [82] and, accordingly, a decrease in the melting rate [25]. The presence of monosaccharides in samples HC, H3% and H5% indirectly affects the distribution of the air phase in the product thickness [83], primarily due to a decrease in the viscosity of ice cream mixes, which increases their saturation by air during freezing. El-Hadad et al. [84] reported that emulsifiers and stabilizers can also affect the even distribution of air bubbles in ice cream. At the same time, the homogenization under pressure of ice cream mixes could have an indirect effect [65], since finely dispersed fat takes an active part in the stabilization of air bubbles [85]. In low-fat ice cream, as in our case, the stabilizing role is performed by whey proteins [86], the high content of which forms a coagulation-type structure, which subsequently leads to a high overrun of ice cream after freezing.

NHC

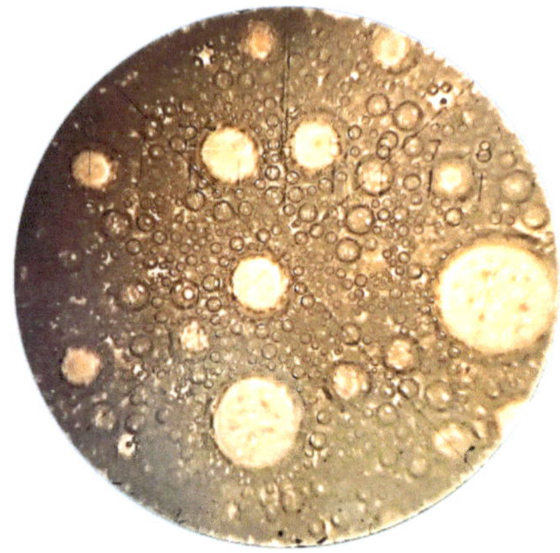

NH3%

HC

H3%

H5%

Figure 2. Microstructure of soft ice cream at a magnification of 160 times. NHC—ice cream based on non-hydrolyzed whey concentrate; NH3%—ice cream based on non-hydrolyzed whey concentrate + 3% whey protein isolate; HC—ice cream based on hydrolyzed whey concentrate; H3%—ice cream based on hydrolyzed whey concentrate + 3% whey protein isolate; H5%—ice cream based on hydrolyzed whey concentrate + 5% whey protein isolate.

3.5. Microscopy Analysis

The recommended size of ice crystals in ice cream should not exceed 50 μm [87]; however, the formation of ice crystals with a size of 10 to 20 μm, according to some scientists, gives ice cream the proper smoothness and creaminess [53,88], while ice crystals larger than 50 μm give the product an undesirable texture [89,90]. Crystal diameters smaller than 20 μm were found in ice cream after 24 h in the freezer storage and may contribute to the stabilization of the crystal structure during longer storage of the product, which needs to be verified experimentally in further studies. Table 5 clearly demonstrates changes in the equivalent diameter of ice crystals for the studied samples. Samples H3% and H5% are characterized by crystals with the smallest sizes at the level of 12.23–13.18 μm (Table 5, Figure 3), which corresponds to the average diameter of ice crystals in milk ice cream stabilized with special cryoprotectants [53].

Table 5. Comparison of ice crystal sizes after 24 h of storage at −18°C.

Sample	Minimum Diameter of Ice Crystals (μm)	Maximum Diameter of Ice Crystals (μm)	The Average Value of the Diameter of Ice Crystals (μm)
NHC	$7.55^{d} \pm 0.12$	$44.36^{d} \pm 2.03$	$25.96^{c} \pm 1.04$
NH3%	$6.85^{c} \pm 0.09$	$26.00^{c} \pm 0.52$	$16.43^{b} \pm 1.17$
HC	$4.68^{a} \pm 0.10$	$22.68^{b} \pm 0.16$	$13.68^{a} \pm 0.02$
H3%	$5.41^{b} \pm 0.24$	$19.04^{a} \pm 0.12$	$12.23^{a} \pm 0.18$
H5%	$5.18^{a} \pm 0.02$	$21.17^{ab} \pm 0.95$	$13.18^{a} \pm 0.56$

Note. NHC—ice cream based on non-hydrolyzed whey concentrate; NH3%—ice cream based on non-hydrolyzed whey concentrate + 3% whey protein isolate; HC—ice cream based on hydrolyzed whey concentrate; H3%—ice cream based on hydrolyzed whey concentrate + 3% whey protein isolate; H5%—ice cream based on hydrolyzed whey concentrate + 5% whey protein isolate. $^{a–d}$—different superscript letters in the columns represent significant differences in the mean values of the same parameter ($p \leq 0.05$).

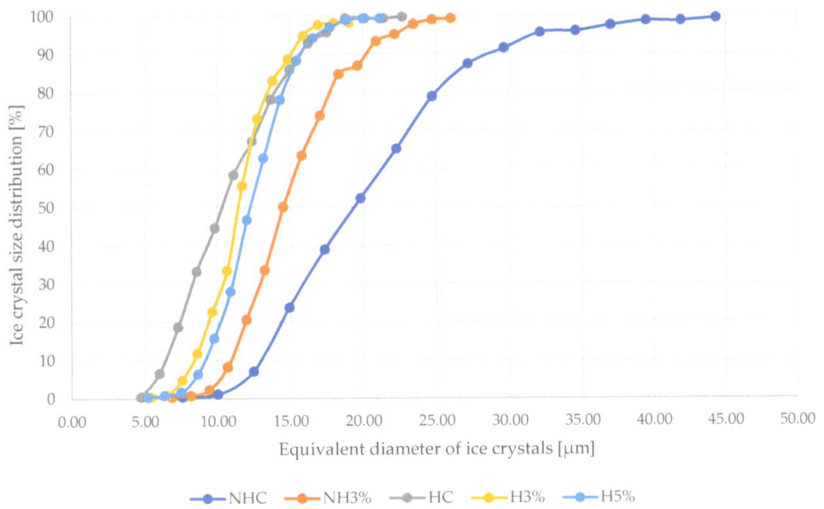

Figure 3. Distribution of ice crystals in whey ice cream samples. **Note.** NHC—ice cream based on non-hydrolyzed whey concentrate; NH3%—ice cream based on non-hydrolyzed whey concentrate + 3% whey protein isolate; HC—ice cream based on hydrolyzed whey concentrate; H3%—ice cream based on hydrolyzed whey concentrate + 3% whey protein isolate; H5%—ice cream based on hydrolyzed whey concentrate + 5% whey protein isolate.

In ice cream samples (HC, H3% and H5%), 50% of the tested crystal diameters (parameter X50) did not exceed 12.23–13.68 μm (Figure 3). This crystal size after 24 h of ice cream production will ensure stable storage of this product at the specified temperature (−18 °C). Even if temperature fluctuations occur and crystals grow as a result of the recrystallization process, they will not exceed the recommended size (25–50 μm). It has been reported that the growth of ice crystals to diameters greater than 50 μm after one month of storage in dairy ice cream is possible if stabilizers are not used [51,91,92]. Analyzing the diameter of ice crystals according to the studied parameter X50, it is clearly seen that for the samples NHC and NH3%, the largest ice crystals were formed at the level of 16.43–25.96 μm (Table 5, Figure 3). Whey protein isolate contributes to the formation of a structure with increased water-holding capacity, which leads to the formation of a strong three-dimensional network in the NH3%, HC, H3% and H5%. The resulting structure can even mechanically counteract the growth of ice crystals.

Scientists have proven that adding proteins to ice cream can lead to the growth of non-hexagonal crystals, the structure of which is more favorable for binding numerous water molecules [93]. It was also shown that the shape of ice crystals strictly depends on the type of added stabilizers, while their diameter is also affected by the ice cream composition [53,89,90]. This result was also obtained for proteins binding free water in ice cream [94].

Based on the observations (Figure 4), it could be assumed that the mechanisms of retention of water molecules by the whey protein isolate added to the studied samples (NH3%, H3% and H5%) may be similar. Changes in the shape of the ice crystals, especially visible in NHC and NH3%, may indicate that recrystallization processes have occurred. In contrast, HC, H3% and H5% showed smaller crystal diameters, which moreover have regular shapes, compared to samples based on non-hydrolyzed concentrate (NHC and NH3%). A noticeable feature of NHC and NH3% is also a different appearance of the crystal structure, in which the crystals are arranged quite densely, and clearly defined edges create a three-dimensional effect. The shape of the crystals in NH3% indicates that coalescence and migration processes have taken place.

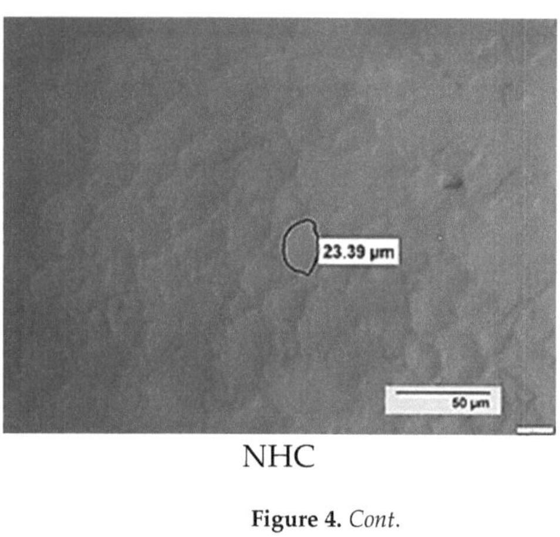

NHC

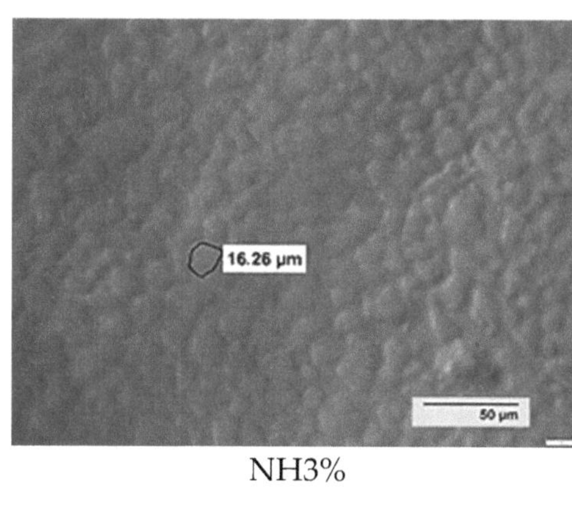

NH3%

Figure 4. *Cont.*

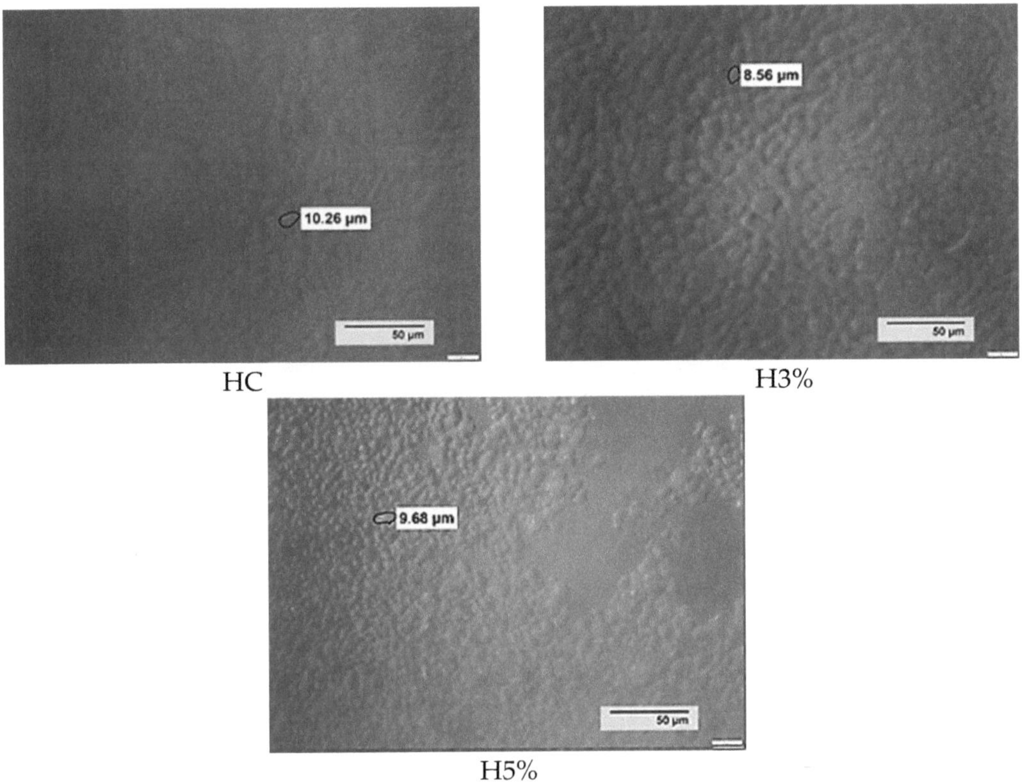

Figure 4. Photographs of ice crystals after 24 h of storage at −18 °C. Note. NHC—ice cream based on non-hydrolyzed whey concentrate; NH3%—ice cream based on non-hydrolyzed whey concentrate + 3% whey protein isolate; HC—ice cream based on hydrolyzed whey concentrate; H3%—ice cream based on hydrolyzed whey concentrate + 3% whey protein isolate; H5%—ice cream based on hydrolyzed whey concentrate + 5% whey protein isolate.

3.6. Quality Indicators

The overrun of ice cream for HC, H3% and H5% is the highest (71.98–79.18%) (Table 6), which is associated with the presence of monosaccharides (glucose, galactose) and their influence on the viscosity of ice cream mixes.

At the same time, even with the addition of whey protein isolate, the H3% and H5% have higher overrun than NHC and NH3%, although it is still slightly lower than HC. Lee and Duggan [60] reported on the foaming properties of whey protein isolate microgels. Due to the formation of elastic bonds, whey protein isolate contributes to the formation of more intense intermolecular interactions [95], as well as the uniform distribution of air bubbles, which ensure the production of ice cream with a high overrun. On the other hand, the addition of whey protein isolate, even in a smaller amount (3%), leads to a decrease in overrun, compared to the control sample (HC), because it increases the viscosity of the mixes, which limits the saturation of the mix with air during freezing [96]. Ice cream with non-hydrolyzed whey concentrates has a higher resistance to melting than samples based on hydrolyzed concentrates. The high content of monosaccharides, which are more effective cryoprotectants, leads to a decrease in the resistance to melting of HC, H3% and H5%. Whey protein isolate increases resistance to melting, but in samples with hydrolyzed concentrates, this indicator is still lower than in samples NHC and NH3%. A correlation was observed between the values of overrun and resistance to melting: the higher the air saturation of the ice cream, the higher the speed of its melting, respectively. This trend is comparable to the data of other scientists who determined overrun and resistance to melting in ice cream and frozen desserts [96–98]. However, Warren and Hartel [99] found that ice cream with a low overrun melts quickly, while ice cream with a high overrun slows down the melting rate.

Table 6. Physicochemical indicators of ice cream samples.

Indicator			Ice Cream Samples				
			NHC	NH3%	HC	H3%	H5%
Overrun, %			71.84 ^{b} ± 1.45	59.3 ^{a} ± 0.86	79.18 ^{c} ± 2.55	76.55 ^{bc} ± 3.08	71.98 ^{b} ± 2.72
pH		1st day	5.25 ^{a} ± 0.01	5.22 ^{a} ± 0.05	5.23 ^{a} ± 0.08	5.19 ^{a} ± 0.03	5.17 ^{a} ± 0.01
		14th day	5.20 ^{b} ± 0.10	5.13 ^{ab} ± 0.04	5.12 ^{ab} ± 0.01	5.09 ^{ab} ± 0.01	5.05 ^{a} ± 0.04
Resistance to melting	1 drop	1st day	29.81 ^{b} ± 0.53	34.58 ^{c} ± 1.04	24.85 ^{a} ± 0.95	26.07 ^{a} ± 1.17	29.11 ^{b} ± 1.50
		14th day	30.11 ^{bc} ± 1.08	35.87 ^{d} ± 1.55	25.04 ^{a} ± 0.37	28.54 ^{b} ± 1.20	32.42 ^{c} ± 1.16
	10 cm^{3}	1st day	44.08 ^{b} ± 1.62	49.76 ^{c} ± 1.89	33.69 ^{a} ± 1.36	36.25 ^{a} ± 1.24	40.67 ^{b} ± 1.33
		14 th day	46.11 ^{c} ± 1.91	52.32 ^{d} ± 1.01	34.78 ^{a} ± 0.88	39.82 ^{b} ± 1.45	43.83 ^{c} ± 1.57
Hardness, g/cm^{3}	1st day		1734.88 ^{b} ± 44.37	2512.46 ^{d} ± 50.27	1580.27 ^{a} ± 41.86	1941.43 ^{c} ± 51.87	2409.74 ^{d} ± 47.08
	14th day		1808.51 ^{b} ± 30.27	2567.09 ^{d} ± 14.90	1602.82 ^{a} ± 55.68	2154.37 ^{c} ± 44.20	2618.74 ^{d} ± 36.17

Note. NHC—ice cream based on non-hydrolyzed whey concentrate; NH3%—ice cream based on non-hydrolyzed whey concentrate + 3% whey protein isolate; HC—ice cream based on hydrolyzed whey concentrate; H3%—ice cream based on hydrolyzed whey concentrate + 3% whey protein isolate; H5%—ice cream based on hydrolyzed whey concentrate + 5% whey protein isolate. $^{a–d}$—different superscript letters in the lines represent significant differences in the mean values of the same parameter ($p \leq 0.05$).

Other scientists reported that there is no interdependence between the overrun and resistance to melting [54,100]. The speed of ice cream melting could also be affected by the size of ice crystals [54], and the larger they are, the lower the resistance to melting, which is consistent with our data on the size of ice crystals in experimental samples based on non-hydrolyzed whey concentrates (NHC, NH3%). The presence of hydrolysis products in the samples HC, H3% and H5% increases the melting rate, despite the formed small-size ice crystals in these samples. Lindamo et al. [101] reported a similar trend of decreasing melting resistance as the degree of lactose hydrolysis increased in ice cream samples. Due to its high moisture-binding capacity, whey protein isolate increases the viscosity of ice cream mixes [51], which limits the excessive growth of ice crystals, ensures the formation of a creamy consistency of ice cream and increases the resistance to melting of ice cream due to the structural and mechanical factor of foam stabilization as a dispersed system. For the purpose of a deeper study of the regularities of the formation of physicochemical parameters of ice cream, their hardness was determined. The hardness of HC with hydrolyzed concentrate ($p < 0.05$) is less than that of NHC, but the further addition of whey protein isolate to the composition of ice cream increases the hardness in samples H3% and H5%, although it is slightly less, compared to NH3%. During 14 days of storage, the hardness of all ice cream samples increased, and this was most noticeable for H3% and H5%. Patel, Baer and Acharya [102] reported that increasing the protein content of vanilla ice cream resulted in excessive firmness with lower overrun. The authors explained this by a hard three-dimensional gel formed by proteins, which led to an increase in viscosity. The obtained data coincide with the studied dynamics of crystal formation in ice cream, confirming that the larger the ice crystals, the higher the hardness of the ice cream, but this correlation is also not always confirmed by scientists [12,54]. Alfaifi and Stathopoulos [103] reported that the addition of whey protein isolate increased ice cream hardness, and Danesh et al. [104], on the contrary, found that hardness decreases when using whey protein isolate. From the data in Table 6, there is no direct dependence of the hardness on the resistance to melting or overrun, which is related to the peculiarities of the chemical composition of whey ice cream samples as a food system, in particular, the presence of lactose or monosaccharides, a reduced MSNF content and an increased protein content. The existing contradictions in the scientific literature regarding the interdependence of whipping, resistance to melting and hardness, as well as their correlation with the processes of ice crystal formation and saturation of ice cream with air bubbles are due to the difficulty of obtaining data on ice cream samples with an individual recipe composition. These dynamics may also depend on many factors and require a complex approach to analysis.

3.7. Microbiological Analysis

The number of *Lactobacillus acidophilus* probiotic cells for all samples was not lower than 6.2 log CFU/g during two weeks of storage (Table 7), which allows the developed ice cream compositions to be classified as probiotic, which means it contains at least 6 log CFU/g of probiotic cultures [105].

Table 7. Microbiological indicators of ice cream samples.

Sample	Storage Time (Days)	Counting (log CFU/g)			
		Coliform	LA-5®	Yeasts	Fungi
NHC	1	ND	6.6	5.1	6.0
	14	ND	6.6	5.5	6.2
NH3%	1	ND	6.3	5.4	ND
	14	ND	6.2	5.7	5.1
HC	1	ND	>7.7	5.3	5.3
	14	ND	>7.7	5.5	5.5
H3%	1	ND	7.6	5.0	ND
	14	ND	7.6	5.3	ND
H5%	1	ND	7.4	5.0	ND
	14	ND	7.3	5.2	ND

Note. ND—not detected. NHC—ice cream based on non-hydrolyzed whey concentrate; NH3%—ice cream based on non-hydrolyzed whey concentrate + 3% whey protein isolate; HC—ice cream based on hydrolyzed whey concentrate; H3%—ice cream based on hydrolyzed whey concentrate + 3% whey protein isolate; H5%—ice cream based on hydrolyzed whey concentrate + 5% whey protein isolate.

The decrease in *Lactobacillus acidophilus* from 6.6 log CFU/g (sample 1) to 6.2–6.3 log CFU/g (sample 2) is likely due to a high solid content, resulting in a decrease in water activity and an increase in osmotic pressure, which negatively affects the vital activity of starter microorganisms. The increase in the number of *Lactobacillus acidophilus* bacterial cells in samples with hydrolyzed whey concentrate, despite an even greater increase in the solid content, is due to the presence of glucose and galactose, which are a nutrient medium and stimulate the development process of probiotic cultures [42]. The presence of whey protein isolate could also contribute to the development of *Lactobacillus acidophilus* to some extent. Afzaal et al. [106] showed that whey protein isolates were more effective due to their amino acid composition as a protective medium for probiotic cell strains. Burgain et al. [107] reported that some of the molecules present in the cells of probiotic bacteria are involved in adhesion with polysaccharides, acids, proteins and lipids. Milk protein ingredients, as representatives of biopolymers, are common components of bioactive agents (encapsulants, protectors) used to protect probiotic bacteria [108,109]. The combination of probiotics with whey protein isolate can add more value to processed foods, but the main disadvantage associated with this combination is the instability of bacteria, since WPI is an ideal food source for the growth and reproduction of microorganisms at high moisture content [40]. The results of counting yeasts and fungi in the ice cream samples indicate that H3% and H5% have a lower content of them compared to other samples. Due to the active binding of free water and increased osmotic pressure in the product, whey protein isolate creates unfavorable conditions for the development of fungi and yeast, which correlates with the obtained data on the activity of water in ice cream mixes.

4. Conclusions

Whey protein isolate has a significant effect on the rheological properties of ice cream mixes based on hydrolyzed whey concentrates, namely, an increase in viscosity indicators. Substantial freezing of free water in ice cream occurs in the temperature range from the cryoscopic temperature to −10 °C, which ensures the freezing of 70.5–71.2% of water and, in samples HC, H3% and H5% up to 72.9–76.1%. Whey protein isolate contributes to the formation of a structure with increased water-holding capacity, which leads to the formation of a more uniform crystal structure (average diameter of ice crystals from 13.75 to 14.75 μm) in HC, H3% and H5% samples. Analysis of the microstructure confirms the feasibility of using whey protein isolate (3–5%), which ensures an even distribution of air bubbles in the ice cream and contributes to obtaining a product with a high whipping index. The combination of non-hydrolyzed whey concentrate and 3% whey protein isolate (NH3%) provides the greatest stability to preserve the purity and color intensity of the ice cream during storage. According to the microbiological analysis, the developed types of ice cream can be classified as probiotic based on the amount of Lactobacillus acidophilus in all samples not being lower than 6.2 log CFU/g. According to the results of the research, H3% and H5% were selected as the best ice cream recipes, which ensure the proper formation of quality indicators of whey ice cream. The prospect of further research is the study of a complex of quality indicators for acidophilic ice cream based on hydrolyzed whey concentrates during storage.

Author Contributions: Conceptualization, A.M. and G.P.; methodology, A.M.; software, M.B.-O.; validation, A.M., A.K.-D. and A.B-P.; formal analysis, M.B.-O.; investigation, M.B.-O. and A.M.; resources, A.B.-P.; data curation, A.K.-D.; writing—original draft preparation, A.M.; writing—review and editing, M.B.-O.; visualization, A.B.-P.; supervision, G.P.; project administration, C.P.; funding acquisition, C.P. All authors have read and agreed to the published version of the manuscript.

Funding: The project is financed by the program of the Minister of Education and Science named "Regional Initiative of Excellence" in the years 2019–2023, project number 026/RID/2018/19, the amount of financing PLN 9 542 500.00.

Institutional Review Board Statement: Not applicable.

Informed Consent Statement: Not applicable.

Data Availability Statement: Data is contained within the article.

Conflicts of Interest: The authors declare no conflicts of interest.

References

1. Yadav, J.S.S.; Yan, S.; Pilli, S.; Kumar, L.; Tyagi, R.D.; Surampalli, R.Y. Cheese whey: A potential resource to transform into bioprotein, functional/nutritional proteins and bioactive peptides. *Biotechnol. Adv.* **2015**, *33*, 756–774. [CrossRef] [PubMed]
2. Patel, S. Functional food relevance of whey protein: A review of recent findings and scopes ahead. *J. Funct. Foods* **2015**, *19*, 308–319. [CrossRef]
3. Henriques, M.H.F.; Gomes, D.M.G.S.; Borges, A.R.; Pereira, C.J.D. Liquid whey protein concentrates as primary raw material for acid dairy gels. *Food Sci. Technol.* **2019**, *40*, 361–369. [CrossRef]
4. Henriques, M.H.F.; Gomes, D.M.G.S.; Pereira, C.J.D.; Gil, M.H.M. Effects of Liquid Whey Protein Concentrate on Functional and Sensorial Properties of Set Yogurts and Fresh Cheese. *Food Bioprocess Technol.* **2013**, *6*, 952–963. [CrossRef]
5. Henriques, M.; Gomes, D.; Rodrigues, D.; Pereira, C.; Gil, M. Performance of bovine and ovine liquid whey protein concentrate on functional properties of set yoghurts. *Procedia Food Sci.* **2011**, *1*, 2007–2014. [CrossRef]
6. Pereira, C.; Henriques, M.; Gomes, D.; Gomez-Zavaglia, A.; de Antoni, G. Novel functional whey-based drinks with great potential in the dairy industry. *Food Technol. Biotechnol.* **2015**, *53*, 307–314. [CrossRef] [PubMed]
7. Mykhalevych, A.; Kostenko, O.; Polishchuk, G.; Bandura, U. Application of milk protein concentrates in preparation of reduced fat sour cream. *Ukr. Food J.* **2022**, *11*, 429–447. [CrossRef]
8. Polishchuk, G.; Sharakhmatova, T.; Shevchenko, I.; Manduk, O.; Mykhalevych, A.; Pukhlyak, A. Scientific substantiation of cream heating duration in the technology of sour cream, enriched with protein. *Food Sci. Technol.* **2023**, *17*, 3–10. [CrossRef]
9. Shevchenko, O.; Mykhalevych, A.; Polischuk, G.; Buniowska-Olejnik, M.; Bass, O.; Bandura, U. Technological functions of hydrolyzed whey concentrate in ice cream. *Ukr. Food J.* **2022**, *11*, 498–517. [CrossRef]

10. Dekker, P.; Koenders, D.; Bruins, M. Lactose-Free Dairy Products: Market Developments, Production, Nutrition and Health Benefits. *Nutrients* **2019**, *11*, 551. [CrossRef]
11. Wang, G.; Guo, M. Manufacturing Technologies of Whey Protein Products. In *Whey Protein Production, Chemistry, Functionality, and Applications*; Guo, M., Ed.; John Wiley & Sons Ltd.: Chichester, UK, 2019; pp. 13–37. [CrossRef]
12. Sofjan, R.P.; Hartel, R.W. Effects of overrun on structural and physical characteristics of ice cream. *Int. Dairy J.* **2004**, *14*, 255–262. [CrossRef]
13. Panghal, A.; Patidar, R.; Jaglan, S.; Chhikara, N.; Khatkar, S.K.; Gat, Y.; Sindhu, N. Whey valorization: Current options and future scenario–a critical review. *Nutr. Food Sci.* **2018**, *48*, 520–535. [CrossRef]
14. Pires, A.F.; Marnotes, N.G.; Rubio, O.D.; Garcia, A.C.; Pereira, C.D. Dairy By-Products: A Review on the Valorization of Whey and Second Cheese Whey. *Foods* **2021**, *10*, 1067. [CrossRef] [PubMed]
15. Mykhalevych, A.; Polishchuk, G.; Buniowska-Olejnik, M.; Tomczyńska-Mleko, M.; Mleko, S. Functional and technological properties of protein ingredients in whey ice cream. *Ukr. J. Food Sci.* **2022**, *10*, 125–135. [CrossRef]
16. Axentii, M.; Stroe, S.-G.; Codină, G.G. Development and Quality Evaluation of Rigatoni Pasta Enriched with Hemp Seed Meal. *Foods* **2023**, *12*, 1774. [CrossRef] [PubMed]
17. Arranz, E.; Segat, A.; Velayos, G.; Flynn, C.; Brodkorb, A.; Giblin, L. Dairy and plant based protein beverages: In vitro digestion behaviour and effect on intestinal barrier biomarkers. *Food Res. Int.* **2023**, *169*, 112815. [CrossRef] [PubMed]
18. Liu, Y.; Liu, A.; Liu, L.; Kan, Z.; Wang, W. The relationship between water-holding capacities of soybean–whey mixed protein and ice crystal size for ice cream. *J. Food Process Eng.* **2021**, *44*, e13723. [CrossRef]
19. Tvorogova, A.A.; Gurskiy, I.A.; Shobanova, T.V.; Smykov, I.T. Effect of Protein Concentrates and Isolates on the Rheological, Structural, Thermal and Sensory Properties of Ice Cream. *Curr. Res. Nutr. Food Sci. J.* **2023**, *11*, 294–306. [CrossRef]
20. El-Zeini, H.M.; El-Abd, M.M.; Mostafa, A.Z.; El-Ghany, F.H.Y. Effect of incorporating whey protein concentrate on chemical, rheological and textural properties of ice cream. *J. Food Process. Technol.* **2016**, *7*, 1000546.
21. Hossain, M.K.; Petrov, M.; Hensel, O.; Diakité, M. Microstructure and Physicochemical Properties of Light Ice Cream: Effects of Extruded Microparticulated Whey Proteins and Process Design. *Foods* **2021**, *10*, 1433. [CrossRef]
22. Das, N.; Hooda, A. Chemistry and Different Aspects of Ice Cream. In *The Chemistry of Milk and Milk Products*; Apple Academic Press: New York, NY, USA, 2023; pp. 65–86.
23. Saentaweesuk, S.; Chaikham, P. Effect of whey protein isolate incorporated with various carbohydratebased fat replacers on physicochemical and sensorial properties of low-fat chocolate ice cream. *Food Res.* **2023**, *7*, 167–176. [CrossRef] [PubMed]
24. Saentaweesuk, S.; Aukkanit, N. Effects of Whey Protein Isolate and Soy Protein Isolate as Fat Replacers on the Physicochemical and Sensory Properties of Low-Fat Chocolate Ice Cream. *Burapha Sci. J.* **2022**, *27*, 686–701.
25. Roy, S.; Hussain, S.A.; Prasad, W.G.; Khetra, Y. Quality attributes of high protein ice cream prepared by incorporation of whey protein isolate. *Appl. Food Res.* **2022**, *2*, 100029. [CrossRef]
26. Song, X.; Perez-Cueto, F.; Bredie, W. Sensory-Driven Development of Protein-Enriched Rye Bread and Cream Cheese for the Nutritional Demands of Older Adults. *Nutrients* **2018**, *10*, 1006. [CrossRef] [PubMed]
27. Salem, S.A.; Hamad, E.M.; Ashoush, I.S. Effect of partial fat replacement by whey protein, oat, wheat germ and modified starch on sensory properties, viscosity and antioxidant activity of reduced fat ice cream. *Food Nutr. Sci.* **2016**, *7*, 397–404. [CrossRef]
28. Zhang, A.Q.; Xu, D.; Liu, B.H.; Shi, B.M.; Zhang, Y.H. Low-fat ice cream model system: Impact of incorporation of alcalase hydrolyzed zein. *Food Funct.* **2023**, *14*, 4430–4439. [CrossRef]
29. Dhingra, M.; Singh, J. Enzymes in Food Industry and Their Regulatory Oversight. In *Microbes Food Ind.*; Scrivener Publishing LLC: Beverly, MA, USA, 2023; pp. 249–274. [CrossRef]
30. Dadan, M.; Nowacka, M.; Czyzewski, J.; Witrowa-Rajchert, D. Modification of food structure and improvement of freezing processes by pulsed electric field treatment. In *Pulsed Electric Fields to Obtain Healthier and Sustainable Food for Tomorrow*; Academic Press: Cambridge, MA, USA, 2020; pp. 203–226. [CrossRef]
31. Queiroz, E.S.; Rezende, A.L.L.; Perrone, Í.T.; Francisquini, J.D.A.; de Carvalho, A.F.; Alves, N.M.G.; de Oliveira, L.F.C.; Stephani, R. Spray drying and characterization of lactose-free goat milk. *LWT* **2021**, *147*, 111516. [CrossRef]
32. Pertsevoy, F.; Ladyka, V.; Smetanska, I.; Bienias, D.; Ianchyk, M.; Grynchenko, N.; Omelchenko, S.; Hrynchenko, O. *Technology of Thermostable and Frozen Fillings Using Dairy Raw Materials and Sesame Seeds Concentrate*; Dissa+: Kharkiv, Ukraine, 2022; 192p.
33. Arellano, M.; Benkhelifa, H.; Flick, D.; Alvarez, G. Online ice crystal size measurements during sorbet freezing by means of the focused beam reflectance measurement (FBRM) technology. Influence of operating conditions. *J. Food Eng.* **2012**, *113*, 351–359. [CrossRef]
34. Buyck, J.R.; Baer, R.J.; Choi, J. Effect of storage temperature on quality of light and full-fat ice cream. *J. Dairy Sci.* **2011**, *94*, 2213–2219. [CrossRef]
35. Tay, R.R.E.; Agatha, T.; Somang, G.; Yuliarti, O.; Tan, E.L.L. Structuring wheat flour-based crackers using whey protein isolate. *Int. Dairy J.* **2022**, *128*, 105314. [CrossRef]
36. Attia, Y.A.; Al-Harthi, M.A.; Korish, M.A.; Shiboob, M.H. Protein and Amino Acid Content in Four Brands of Commercial Table Eggs in Retail Markets in Relation to Human Requirements. *Animals* **2020**, *10*, 406. [CrossRef] [PubMed]
37. Loveday, S.M. Food proteins: Technological, nutritional, and sustainability attributes of traditional and emerging proteins. *Annu. Rev. Food Sci. Technol.* **2019**, *10*, 311–339. [CrossRef] [PubMed]

38. Van Vlierberghe, S.; Graulus, G.J.; Keshari Samal, S.; Van Nieuwenhove, I.; Dubruel, P. Porous hydrogel biomedical foam scaffolds for tissue repair. In *Biomedical Foams for Tissue Engineering Applications*; Netti, P.A., Ed.; Woodhead Publishing: Cambridge, UK, 2014; pp. 335–390. [CrossRef]
39. de Castro, R.J.S.; Domingues, M.A.F.; Ohara, A.; Okuro, P.K.; dos Santos, J.G.; Brexó, R.P.; Sato, H.H. Whey protein as a key component in food systems: Physicochemical properties, production technologies and applications. *Food Struct.* **2017**, *14*, 17–29. [CrossRef]
40. Khem, S.; Small, D.M.; May, B.K. The behaviour of whey protein isolate in protecting Lactobacillus plantarum. *Food Chem.* **2016**, *190*, 717–723. [CrossRef] [PubMed]
41. Maleki, O.; Khosrowshahi Asl, A.; Alizadeh Khaledabad, M.; Amiri, S. Production and characterization of synbiotic ice cream using microencapsulation and cryopreservation of Lactobacillus rhamnosus in whey protein/bio-cellulose/inulin composite microcapsules. *J. Food Meas. Charact.* **2023**, *17*, 3909–3917. [CrossRef]
42. Osmak, T.; Mleko, S.; Bass, O.; Mykhalevych, A.; Kuzmyk, U. Enzymatic hydrolysis of lactose in concentrates of reconstituted demineralized whey, intended for ice cream production. *Ukr. Food J.* **2021**, *10*, 277–288. [CrossRef]
43. Instituto Português da Qualidade. *Milk—Fat Content Determination, Gerber Method (NP Standard No. 469 in Portuguese)*; IPQ: Monte de Caparica, Portugal, 2002.
44. Slashcheva, A.; Nykyforov, R.; Popova, S.; Korenets, Y. Rationale for the use of protein-carbohydrate mix in the technology of disperse products. *East.-Eur. J. Enterp. Technol.* **2016**, *2*, 64–71. [CrossRef]
45. Romanchuk, I.; Minorova, A.; Krushelnytska, N. Physical-chemical composition and technological properties of demineralized milk whey received by membrane methods. *Agric. Sci. Pract.* **2018**, *5*, 33–39. [CrossRef]
46. Frenzel, M.; Zerge, K.; Clawin-Rädecker, I.; Lorenzen, P.C. Comparison of the galacto-oligosaccharide forming activity of different β-galactosidases. *LWT-Food Sci. Technol.* **2015**, *60*, 1068–1071. [CrossRef]
47. Lim, S.Y.; Swanson, B.G.; Clark, S. High hydrostatic pressure modification of whey protein concentrate for improved functional properties. *J. Dairy Sci.* **2008**, *91*, 1299–1307. [CrossRef]
48. Nazarewicz, S.; Kozłowicz, K.; Kobus, Z.; Gładyszewska, B.; Matwijczuk, A.; Ślusarczyk, L.; Skrzypek, T.; Sujka, M.; Kozłowicz, N. The Use of Ultrasound in Shaping the Properties of Ice Cream with Oleogel Based on Oil Extracted from Tomato Seeds. *Appl. Sci.* **2022**, *12*, 9165. [CrossRef]
49. Kuzmyk, U.; Marynin, A.; Svyatnenko, R.; Zheludenko, Y.; Kurmach, M.; Shvaiko, R. Prospects of use of vegetable raw materials in the technology of sour-milk dessert. *EUREKA Life Sci.* **2021**, *3*, 29–35. [CrossRef]
50. Polischuk, G.; Sharahmatova, T.; Breus, N.; Bass, O.; Shevchenko, I. Studies of water freezing features in ice cream with starch syrup. *Food Sci. Technol.* **2019**, *13*, 71–77. [CrossRef]
51. Goff, H.D.; Hartel, R.W. Ice cream structure. In *Ice Cream*; Springer: Boston, MA, USA, 2013; pp. 313–352.
52. Herrera, M.L.; M'Cann, J.I.; Ferrero, C.; Hagiwara, T.; Zaritzky, N.E.; Hartel, R.W. Thermal, mechanical, and molecular relaxation properties of frozen sucrose and fructose solutions containing hydrocolloids. *Food Biophys.* **2007**, *2*, 20–28. [CrossRef]
53. Kamińska-Dwórznicka, A.; Łaba, S.; Jakubczyk, E. The effects of selected stabilizers addition on physical properties and changes in crystal structure of whey ice cream. *LWT* **2022**, *154*, 112841. [CrossRef]
54. Muse, M.R.; Hartel, R.W. Ice cream structural elements that affect melting rate and hardness. *J. Dairy Sci.* **2004**, *87*, 1–10. [CrossRef] [PubMed]
55. Henriques, M.; Gomes, D.; Pereira, C. Liquid whey protein concentrates produced by ultrafiltration as primary raw materials for thermal dairy gels. *Food Technol. Biotechnol.* **2017**, *55*, 454. [CrossRef] [PubMed]
56. Romulo, A.; Meindrawan, B. Effect of Dairy and Non-Dairy Ingredients on the Physical Characteristic of Ice Cream. *IOP Conf. Ser. Earth Environ. Sci.* **2021**, *794*, 012145. [CrossRef]
57. Arbuckle, W.S. *Ice Cream*; Springer: New York, NY, USA, 2013. [CrossRef]
58. Özdemir, C.; Dağdemir, E.U.; Özdemir, S.; Sağdıç, O. The Effects of Using Alternative Sweeteners to Sucrose on Ice Cream Quality. *J. Food Qual.* **2008**, *31*, 415–428. [CrossRef]
59. Özdemir, C.; Arslaner, A.; Özdemir, S.; Özdemir, G.U.C. Ice-Cream Production from Lactose-Free UHT Milk. *J. Food Sci. Eng.* **2018**, *8*, 210–214. [CrossRef]
60. Lee, J.; Duggan, E. Whey protein microgels for stabilisation of foams. *Int. Dairy J.* **2022**, *132*, 105399. [CrossRef]
61. Puangmanee, S.; Hayakawa, S.; Sun, Y.; Ogawa, M. Application of whey protein isolate glycated with rare sugars to ice cream. *Food Sci. Technol. Res.* **2008**, *14*, 457. [CrossRef]
62. Akalın, A.S.; Karagözlü, C.; Ünal, G. Rheological properties of reduced-fat and low-fat ice cream containing whey protein isolate and inulin. *Eur. Food Res. Technol.* **2008**, *227*, 889–895. [CrossRef]
63. Koretska, I.; Polyovyk, V.; Maslikov, M.; Kuzmin, O. Thermophysical characteristics of frozen semifinished products for restaurant technology. *Ukr. J. Food Sci.* **2020**, *8*, 231–240. [CrossRef]
64. Kozłowicz, K.; Nazarewicz, S.; Różyło, R.; Nastaj, M.; Parafiniuk, S.; Szmigielski, M.; Bieńczak, A.; Kozłowicz, N. The Use of Moldavian Dragonhead Bagasse in Shaping the Thermophysical and Physicochemical Properties of Ice Cream. *Appl. Sci.* **2021**, *11*, 8598. [CrossRef]
65. Kot, A.; Jakubczyk, E.; Kamińska-Dwórznicka, A. The Effectiveness of Combination Stabilizers and Ultrasound Homogenization in Milk Ice Cream Production. *Appl. Sci.* **2023**, *13*, 7561. [CrossRef]

66. Landikhovskaya, A.V.; Tvorogova, A.A. Ice cream and frozen desserts nutrient compositions: Current trends of researches. *Food Syst.* **2021**, *4*, 74–81. [CrossRef]
67. Fernández-Garía, E.; McGregor, J.U.; Traylor, S. The addition of oat fiber and natural alternative sweeteners in the manufacture of plain yogurt. *J. Dairy Sci.* **1998**, *81*, 655–663. [CrossRef]
68. Belokurova, E.S.; Pankina, I.A.; Sevastianova, A.D.; Asfondiarova, I.V.; Katkova, N.M. The effect of functional additives on the indicator "water activity" of biscuit semi-finished products. *IOP Conf. Ser. Earth Environ. Sci.* **2021**, *640*, 022022. [CrossRef]
69. de Souza Fernandes, D.; Leonel, M.; Del Bem, M.S.; Mischan, M.M.; Garcia, É.L.; Dos Santos, T.P.R. Cassava derivatives in ice cream formulations: Effects on physicochemical, physical and sensory properties. *J. Food Sci. Technol.* **2017**, *54*, 1357–1367. [CrossRef]
70. Cook, K.L.K.; Hartel, R.W. Mechanisms of ice crystallization in ice cream production. *Compr. Rev. Food Sci. Food Saf.* **2010**, *9*, 213–222. [CrossRef]
71. Kumar, P.K.; Rasco, B.A.; Tang, J.; Sablani, S.S. State/phase transitions, ice recrystallization, and quality changes in frozen foods subjected to temperature fluctuations. *Food Eng. Rev.* **2020**, *12*, 421–451. [CrossRef]
72. Goff, H.D. Ice cream. In *Advanced Dairy Chemistry*; Springer: Boston, MA, USA, 2013; pp. 441–450.
73. Syed, Q.A.; Anwar, S.; Shukat, R.; Zahoor, T. Effects of different ingredients on texture of ice cream. *J. Nutr. Health Food Eng.* **2018**, *8*, 422–435.
74. Leducq, D.; Ndoye, F.T.; Alvarez, G. Phase change material for the thermal protection of ice cream during storage and transportation. *Int. J. Refrig.* **2015**, *52*, 133–139. [CrossRef]
75. Barros, E.L.D.S.; Silva, C.C.; Canella, M.H.M.; Verruck, S.; Prestes, A.A.; Vargas, M.O.; Maran, B.M.; Esmerino, E.A.; Silva, R.; Balthazar, C.F.; et al. Effect of replacement of milk by block freeze concentrated whey in physicochemical and rheological properties of ice cream. *Food Sci. Technol.* **2021**, *42*, 1–9. [CrossRef]
76. de Meneses, R.B.; Moura, D.C.C.; de Almeida, D.T.; da Silva Bispo, E.; Maciel, L.F.; da Rocha-Leão, M.H.M.; Conte-Junior, C.A. Impact of different dairy wheys on quality parameters of ice cream. *Rev. Bras. Ciências Agrárias* **2021**, *16*, 1–10. [CrossRef]
77. García-Pérez, F.J.; Lario, Y.; Fernández-López, J.; Sayas, E.; Pérez-Alvarez, J.A.; Sendra, E. Effect of orange fiber addition on yogurt color during fermentation and cold storage. *Color Res. Appl.* **2005**, *30*, 457–463. [CrossRef]
78. Skryplonek, K.; Gomes, D.; Viegas, J.; Pereira, C.; Henriques, M. Lactose-free frozen yogurt: Production and characteristics. *Acta Sci. Pol. Technol. Aliment.* **2017**, *16*, 171–179. [CrossRef]
79. Meneses, R.B.; Silva, M.S.; Monteiro, M.L.G.; Rocha-Leão, M.H.M.; Conte-Junior, C.A. Effect of dairy by-products as milk replacers on quality attributes of ice cream. *J. Dairy Sci.* **2020**, *103*, 10022–10035. [CrossRef]
80. Moschopoulou, E.; Dernikos, D.; Zoidou, E. Ovine ice cream made with addition of whey protein concentrates of ovine-caprine origin. *Int. Dairy J.* **2021**, *122*, 105146. [CrossRef]
81. Buniowska-Olejnik, M.; Mykhalevych, A.; Polishchuk, G.; Sapiga, V.; Znamirowska-Piotrowska, A.; Kot, A.; Kamińska-Dwórznicka, A. Study of Water Freezing in Low-Fat Milky Ice Cream with Oat β-Glucan and Its Influence on Quality Indicators. *Molecules* **2023**, *28*, 2924. [CrossRef] [PubMed]
82. Mykhalevych, A.; Sapiga, V.; Polischuk, G.; Osmak, T. Functional and technological properties of oat beta-glucan in acidophilic-whey ice cream. *Food Environ. Saf. J.* **2022**, *21*, 116–128. [CrossRef]
83. Zhang, H.; Chen, J.; Li, J.; Wei, C.; Ye, X.; Shi, J.; Chen, S. Pectin from citrus canning wastewater as potential fat replacer in ice cream. *Molecules* **2018**, *23*, 925. [CrossRef] [PubMed]
84. El-Hadad, S.S.; Tikhomirova, N.A.; Tvorogova, A.A.; Shobanova, T.V.; El-Aziz, M.A. Physical properties and microstructure of ice cream supplemented with minor components of wheat germ oil. *Int. J. Dairy Sci.* **2020**, *15*, 189–199. [CrossRef]
85. Arief, M.F.; Andini, R.D.; Rosyidi, D.; Radiati, L.E. Effect of Goat Kefir Utilization on Physicochemical Quality and Sensory Attributes of Ice Cream Probiotic. In *Proceedings of the 3rd International Conference on Environmentally Sustainable Animal Industry 2022 (ICESAI 2022)*; Atlantis Press: Dordrecht, The Netherlands, 2022.
86. Liu, X.; Sala, G.; Scholten, E. Structural and functional differences between ice crystal-dominated and fat network-dominated ice cream. *Food Hydrocoll.* **2023**, *138*, 108466. [CrossRef]
87. Drewett, E.M.; Hartel, R.W. Ice crystallization in a scraped surface freezer. *J. Food Eng.* **2007**, *78*, 1060–1066. [CrossRef]
88. Kamińska-Dwórznicka, A. Wpływ stabilizatorów na ograniczenie rekrystalizacji w lodach typu sorbet. *Przemysł Spożywczy* **2016**, *1*, 34–37. [CrossRef]
89. Arellano, M.; Gonzalez, J.E.; Alvarez, G.; Benkhelifa, H.; Flick, D.; Leducq, D. Online ice crystal size measurements by the focused beam reflectance method (FBRM) during sorbet freezing. *Procedia Food Sci.* **2011**, *1*, 1256–1264. [CrossRef]
90. Kamińska-Dwórznicka, A.; Matusiak, M.; Samborska, K.; Witrowa-Rajchert, D.; Gondek, E.; Jakubczyk, E.; Antczak, A. The influence of kappa carrageenan and its hydrolysates on the recrystallization process in sorbet. *J. Food Eng.* **2015**, *167*, 162–165. [CrossRef]
91. Lomolino, G.; Zannoni, S.; Zabara, A.; Da Lio, M.; De Iseppi, A. Ice recrystallisation and melting in ice cream with different proteins levels and subjected to thermal fluctuation. *Int. Dairy J.* **2020**, *100*, 104557. [CrossRef]
92. Regand, A.; Goff, H.D. Structure and ice recrystallization in frozen stabilized ice cream model systems. *Food Hydrocoll.* **2003**, *17*, 95–102. [CrossRef]
93. Gruneberg, A.K.; Graham, L.A.; Eves, R.; Agrawal, P.; Oleschuk, R.D.; Davies, P.L. Ice recrystallization inhibition activity varies with ice-binding protein type and does not correlate with thermal hysteresis. *Cryobiology* **2021**, *99*, 28–39. [CrossRef] [PubMed]

94. Kaleda, A.; Tsanev, R.; Klesment, T.; Vilu, R.; Laos, K. Ice cream structure modification by ice-binding proteins. *Food Chem.* **2018**, *246*, 164–171. [CrossRef]
95. Keim, S.; Hinrichs, J. Influence of stabilizing bonds on the texture properties of high-pressure-induced whey protein gels. *Int. Dairy J.* **2004**, *14*, 355–363. [CrossRef]
96. O'Chiu, E.; Vardhanabhuti, B. Utilizing whey protein isolate and polysaccha-ride complexes to stabilize aerated dairy gels. *J. Dairy Sci.* **2017**, *100*, 3404–3412. [CrossRef] [PubMed]
97. Liu, X.; Sala, G.; Scholten, E. Effect of fat aggregate size and percentage on the melting properties of ice cream. *Food Res. Int.* **2022**, *160*, 111709. [CrossRef] [PubMed]
98. Azari-Anpar, M.; Khomeiri, M.; Daraei Garmakhany, A.; Lotfi-Shirazi, S. Development of camel and cow's milk, low-fat frozen yoghurt incorporated with Qodume Shahri (*Lepidium perfoliatum*) and cress seeds (*Lepidium sativum*) gum: Flow behavior, textural, and sensory attributes' assessment. *Food Sci. Nutr.* **2021**, *9*, 1640–1650. [CrossRef]
99. Warren, M.M.; Hartel, R.W. Effects of emulsifier, overrun and dasher speed on ice cream microstructure and melting properties. *J. Food Sci.* **2018**, *83*, 639–647. [CrossRef]
100. Wu, B.; Freire, D.O.; Hartel, R.W. The effect of overrun, fat destabilization, and ice cream mix viscosity on entire meltdown behavior. *J. Food Sci.* **2019**, *84*, 2562–2571. [CrossRef]
101. Lindamood, J.B.; Grooms, D.J.; Hansen, P.M.T. Effect of hydrolysis of lac-tose and sucrose on firmness of ice cream. *Food Hydrocoll.* **1989**, *3*, 379–388. [CrossRef]
102. Patel, M.R.; Baer, R.J.; Acharya, M.R. Increasing the protein content of ice cream. *J. Dairy Sci.* **2006**, *89*, 1400–1406. [CrossRef] [PubMed]
103. Alfaifi, M.S.; Stathopoulos, C.E. Effect of egg yolk substitution by sweet whey protein isolate on texture, stability and colour of Gelato-style vanilla ice cream. *Int. J. Dairy Technol.* **2010**, *63*, 593–598. [CrossRef]
104. Danesh, E.; Goudarzi, M.; Jooyandeh, H. Effect of whey protein addition and transglutaminase treatment on the physical and sensory properties of reduced-fat ice cream. *J. Dairy Sci.* **2017**, *100*, 5206–5211. [CrossRef] [PubMed]
105. Pimentel, T.C.; de Oliveira, L.I.G.; de Souza, R.C.; Magnani, M. Probiotic ice cream: A literature overview of the technological and sensory aspects and health properties. *Int. J. Dairy Technol.* **2022**, *75*, 59–76. [CrossRef]
106. Afzaal, M.; Saeed, F.; Hussain, M.; Ismail, Z.; Siddeeg, A.; Ammar, A.F.; Aljobair, M.O. Influence of encapsulation on the survival of probiotics in food matrix under simulated stress conditions. *Saudi J. Biol. Sci.* **2022**, *29*, 103394. [CrossRef]
107. Burgain, J.; Scher, J.; Lebeer, S.; Vanderleyden, J.; Cailliez-Grimal, C.; Corgneau, M.; Francius, G.; Gaiani, C. Significance of bacterial surface molecules interactions with milk proteins to enhance microencapsulation of Lactobacillus rhamnosus GG. *Food Hydrocoll.* **2014**, *41*, 60–70. [CrossRef]
108. Heidebach, T.; Först, P.; Kulozik, U. Microencapsulation of probiotic cells for food applications. *Crit. Rev. Food Sci. Nutr.* **2012**, *52*, 291–311. [CrossRef]
109. Tavares, G.M.; Croguennec, T.; Carvalho, A.F.; Bouhallab, S. Milk proteins as encapsulation devices and delivery vehicles: Applications and trends. *Trends Food Sci. Technol.* **2014**, *37*, 5–20. [CrossRef]

foods

MDPI

Article

Characterization of Physicochemical and Sensory Properties of Cheeses Added with Bovine Colostrum

Idiana de Macêdo Barbosa [1,†], Katya Anaya [2], Cláudia Souza Macêdo [1], Robson Rogério Pessoa Coelho [1], Claudio Cipolat-Gotet [3], Emerson Gabriel dos Santos Oliveira Silva [1], Nkarthe Guerra Araújo [1], Bruna Maria Emerenciano das Chagas [4], Juliana Paula Felipe de Oliveira [5], Cleube Andrade Boari [6], Danielle Cavalcanti Sales [1], Emmanuella de Oliveira Moura Araújo [1,*], Josemir Araújo Neves [7] and Adriano Henrique do Nascimento Rangel [1]

1 Academic Unit Specialized in Agricultural, Federal University of Rio Grande do Norte (UFRN), Macaíba 59280-000, RN, Brazil; idiana_corregо@yahoo.com.br (I.d.M.B.); adrianohrangel@yahoo.com.br (A.H.d.N.R.)
2 Health Sciences College of Trairi, Federal University of Rio Grande do Norte, Santa Cruz 59200-000, RN, Brazil; katya.anaya@ufrn.br
3 Department of Veterinary Science, University of Parma, 43121 Parma, Italy
4 Infrastructure Superintendence, Federal University of Rio Grande do Norte, Natal 59078-970, RN, Brazil
5 Rural Health and Technology Center, Federal University of Campina Grande, Patos 58708-110, PB, Brazil
6 Department of Animal Science, Federal University of the Jequitinhonha and Mucuri Valleys, Diamantina 39100-000, MG, Brazil
7 Agricultural Research Company of Rio Grande do Norte, Natal 59062-500, RN, Brazil
* Correspondence: manu_moura9@yahoo.com.br
† This study is a part of the master's thesis of Idiana de Macêdo Barbosa.

Citation: Barbosa, I.d.M.; Anaya, K.; Macêdo, C.S.; Coelho, R.R.P.; Cipolat-Gotet, C.; Silva, E.G.d.S.O.; Araújo, N.G.; Chagas, B.M.E.d.; Oliveira, J.P.F.d.; Boari, C.A.; et al. Characterization of Physicochemical and Sensory Properties of Cheeses Added with Bovine Colostrum. *Foods* **2023**, *12*, 4474. https://doi.org/10.3390/foods12244474

Academic Editors: Michele Faccia and Giuseppe Natrella

Received: 19 August 2023
Revised: 25 September 2023
Accepted: 8 October 2023
Published: 14 December 2023

Abstract: The objective of this study was to develop fresh and matured cheeses with different bovine colostrum levels, aiming to promote the consumption of dairy products with the addition of colostrum. Four different cheese formulations were produced with a mixture of 0:100, 15:85, 20:80, and 25:75, bovine colostrum:milk (*v*:*v*), and aged for 0, 10, 20, and 40 days. Milk, colostrum, and fresh and matured cheeses were submitted to physicochemical characterization. Moreover, microbiological quality, yield, texture profile, color, and sensory acceptance of cheese samples were evaluated. Colostrum supplementation favored low acidity, high moisture, a pH range of 5.0–6.2, and water activity of 0.94–99. Sensory attributes and overall evaluation of all cheese formulations achieved an Acceptability Index above 70, indicating good acceptability. Since cheese with colostrum presented the potential to be used as human food, assessing the presence of colostrum bioactive components in those dairy products is a promising goal for further research.

Keywords: sensory evaluation; dairy product; cheese maturation; food composition

1. Introduction

The colostrum is the secretion produced by the mammary gland immediately after birth [1]. It represents a rich source of vitamin A with relatively high concentrations of caseins and albumin, containing many essential nutrients in higher concentrations than those usually found in mature milk [1,2]. For instance, Jersey cows' colostrum 24 h after birth presents 4% fat, 15% proteins (12% caseins), 1,5% lactose and 489 IU/dL vitamin A [3]. Compared to milk, colostrum generally contains less lactose and has a higher content of other components such as fat, protein, ash, vitamins, hormones, and immunoglobulins. After three days, the lactose content increases, while the percentage of other components gradually decreases [1,4]. In addition, bovine colostrum (BC) contains bioactive components in relevant amounts, including growth factors, immunoglobulins, lactoperoxidase, lysozyme, lactoferrin, nucleosides, vitamins, peptides, and oligosaccharides, which are extremely relevant to health [5]. Immunoglobulins can account for more than 80% of total proteins in BC; therefore, its commercial value is currently quoted by the protein concentration and immunoglobulin levels, especially IgG [6,7]. Recent research has shown that immune defense proteins in bovine colostrum may be related to protective effects against respiratory diseases such as those caused by SARS-CoV-2 infection (COVID-19) [8]. Bioactive molecules are allegedly responsible for the effect of passive immunity when BC is consumed by humans [9]. Despite considerable differences between BC and human colostrum compositions, BC has good tolerability in the human organism [6] and it is well accepted by most consumers [7].

The demand for innovative foods grows every day as consumers consider nutrition to address a wide range of health conditions. In this way, industries and research centers promote initiatives to support the production of nutraceuticals and functional foods. Food companies are developing their product lines with a primary focus on promoting people's health [10]. In this sense, some products based on bovine colostrum have been developed and are already commercialized as food supplements in countries such as New Zealand, United States, Europe, and China [11]. Nevertheless, a wide variation in the bioactivity was detected in commercially available colostrum products such as powders and capsules [7]. A crescent number of patents has been registered across the globe claiming health benefits to colostrum products, with emerging processing technologies being developed to preserve the therapeutic properties of BC bioactive compounds [12].

Regarding food products, yogurts [13,14], mixtures for beverages [15], cheeses [16–18], ice cream [19], nutritional bars, jellies, and ready-to-drink beverages have used or might use BC as a functional ingredient [20]. Dairy products, such as cheeses, are a valuable source of proteins, lipids, vitamins, and minerals, and their global consumption is expected to increase by around 13.8% between 2019 and 2029 [21]. Thus, the production of mixed cheese with milk and colostrum has emerged as an alternative way of utilizing surplus colostrum and an innovation in the dairy industry for providing possible benefits because of the presence of immunoglobulins, which adhere to the intestinal mucosa, functioning as a protective layer and preventing pathogenic microorganisms from colonizing [22].

Considering the worldwide consumption of cheese, developing this type of product with the inclusion of colostrum could be an innovative alternative to utilizing the surplus production of this raw material. Furthermore, once the health benefits of colostrum and its potential for new products are understood, there is an opportunity for industries to cater to different consumer profiles [18].

In this sense, this study aimed at (i) developing fresh and mature cheeses added with different bovine colostrum levels and (ii) characterizing the physical-chemical and sensory properties of these cheeses.

2. Materials and Methods

2.1. Milk and Colostrum Sampling and Thermal Processing

The milk and colostrum were collected on a commercial farm located in the municipality of São Gonçalo do Amarante, Rio Grande do Norte, Brazil, between January and November 2019. The herd consisted of 62 Jersey cows (30 primiparous and 32 multiparous), managed under a Compost Barn system. Colostrum samples were collected at second and third milking after calving (up to 24 h) in plastic bottles with a capacity of one liter, labeled and packed in isothermal boxes at a temperature of 4 to 5 °C, transported to the Milk Quality Laboratory (LABOLEITE) of the Federal University of Rio Grande do Norte at the Macaíba Campus, and then stored at −18 °C. The colostrum was subsequently thawed in a cold chamber at a refrigeration temperature of 5 °C to perform thermal processing, physical-chemical analysis, and cheese-making. Subsequently, milk and colostrum mixtures in the proportions 100:0, 85:15, 80:20, and 75:25 (*v*:*v*; milk:colostrum, respectively) were subjected to heat treatment (60 °C for 45 min) according to Das and Seth [23], aiming at ensuring the microbial quality of the raw material to be used in cheese-making, while preserving the IgG and IgA immunoglobulins, with no visual protein coagulation.

The preliminary pasteurization tests showed high thermal instability of the second milking colostrum, which led us to follow the cheese production stage with only the third milking colostrum.

2.2. Physical-Chemical Analysis of Milk and Colostrum

Milk and colostrum were analyzed for fat, total protein, lactose, casein, total solids, and milk solids non-fat using an infrared spectrophotometer (Dairy Spect®, Bentley Instruments Inc., Chaska, MI, USA). The pH of milk and colostrum samples was measured using a digital pH meter (Lucadema, São José do Rio Preto-SP, Brazil). Acidity was determined according to the standards of the Instituto Adolfo Lutz [24] using a Dornic Acidimeter (CAP-LAB, Ipiranga-SP, Brazil), and the result was expressed in g of lactic acid/100 mL. All these measurements were performed in triplicate.

2.3. Experimental Treatments and Cheese-Making

This study tested four treatments characterized by a different colostrum addition (0, 15, 20, 25 mL 100 mL^{-1}, respectively named 0, 1, 2, 3). Cheese produced using 100 mL of milk was defined as the control treatment. The colostrum proportions were chosen according to preliminary tests on the thermal stability of the mixture of colostrum and milk. Increasing proportions of colostrum promoted the coagulation of the mixture during the pasteurization. Furthermore, higher amounts of colostrum could potentially lead to remarkable sensory changes, negatively influencing the final acceptance of the products.

Different ripening periods were assessed for each treatment, from 0 days (fresh; F) to 10, 20, and 40 days, respectively named A, B, and C. For each treatment, cheese-making was performed following several procedures: after pasteurization, milk with or without colostrum was cooled to 37 °C, inoculated with a mixed starter culture (*Lactococcus lactis subsp. Lactococcus lactis subsp. cremoris, Streptococcus salivarius subsp. thermophilus*) (Vilac Foods®, Macaíba, Brazil), in the proportion of 0.6 mL/10 L of milk, then allowed to rest for 30 min. Following the manufacturer's instructions, commercial 0.04 g 100 g^{-1} calcium chloride and 0.01 g 100 g^{-1} calf rennet (LacRen 1000 IMCU/mL; Vilac Foods®, Brazil) were homogenized and added to the vat, and milk was left for coagulation (40 to 60 min). The final curd pH was monitored to be no lower than 6.4. The curd was cut into about 1.5 to 2.0 cm^3 cubes and scalded to 45 °C (rising by 1 °C every 3 min). Then, draining, salting (0.9 g 100 g^{-1} sodium chloride), molding, and pressing in a manual press were performed. During pressing, each cheese was turned three times (each turn lasted 1 h), and the total pressing time was 4 h. The resulting cheese yield was measured by the ratio of liters of milk used to produce one kilogram of cheese (L/kg) [25].

The F cheeses were vacuum-packed the next day after manufacture. During the ripening stage, the cheeses were unmolded and placed in plastic trays, then stored in a cold chamber at 5 °C with a relative humidity of 75%. The cheeses were vacuum-packed after completing the ripening phase. The cheese production steps followed the norms of Regulation No. 146 [26].

2.4. Chemical Composition of Cheese Samples

The chemical composition of cheese samples was determined in triplicate according to the official methods of AOAC International [27], analyzing pH, total titratable acidity, moisture, and total solids and ash. Water activity was measured on an a_w meter (LabSwift-Novasina, Piracicaba-SP, Brazil). A conversion factor of 6.38 was used to determine proteins after total nitrogen analysis following the Kjeldahl method (AOAC, 1970). The fat content was determined by the Gerber methodology, described in Regulation No. 68 [26].

2.5. Colorimetric Analysis

The color of cheese samples was analyzed by an instrumental method using an ACR-1023® colorimeter (Instrutherm, São Paulo, Brazil) in the RGB system and converted into CIELAB by the OpenRGB® program (Logicol, Italy). The L* coordinate readings were taken to assess the luminosity ($L^* = 0$, black; $L^* = 100$, white), while a^* and b^* refer to the chromaticity coordinates: a^* represents the variation degree between green and red (a^* negative = green; a^* positive = red), and b^* expresses a hue between blue and yellow (b^* negative = blue; b^* positive = yellow) [28]. The device was previously calibrated, and the colorimetry readings were performed in the geometric center of each cheese in triplicate. The Yellowness Index (YI) was calculated from the average of the values of L^* a^* b^*, according to the following equation [29]:

$$YI = 142.86\ (b^*/L^*)$$

2.6. *Texture Profile Analysis*

The Texture Profile Analysis (TPA) was performed in a universal texturometer (TA-XT plus, Stable MicroSystems, Godalming, UK) equipped with integrated Exponent Stable MicroSystems 32 software, 36 mm AACC cylindrical probe (model P-36R), and a 30 kg load cell. The double compression method was used, simulating chewing, with two cycles (20% compression; 5 s between cycles; constant speed of 2 mm/s). The cheese samples were 20 mm in diameter and were kept at room temperature (30 °C) for around 4 h before testing [30]. The firmness, cohesiveness, chewability, and resilience parameters were analyzed. A total of four repetitions were performed for each cheese sample.

2.7. *Microbiological Analyses*

The different formulations (milk: colostrum) used to produce the cheeses were subjected to the analysis of total and thermotolerant coliforms (MPN/g) after the pasteurization process to verify their hygienic quality [31].

All the different formulations of fresh cheeses were subjected to the analysis of thermotolerant coliforms (MPN/g), molds and yeasts, coagulase-positive staphylococci (*Staphylococcus aureus*), *Listeria monocytogenes*, and *Salmonella* spp. [32]. The results were evaluated based on the specific quality parameters for cheeses established by the Brazilian health agency Agência Nacional de Vigilância Sanitária [26,32–34].

2.8. *Sensory Analysis*

Hedonic sensory tests were carried out by a panel of 93 untrained volunteer tasters (ranging from 18 to 51 years), recruited in November 2019, among students and workers from the Federal University of Rio Grande do Norte (Escola Agrícola de Jundiaí, Macaíba, RN, Brazil).

For conducting sensory tests, this work was approved by the Research Ethics Committee of the Health Sciences College of Trairi, Universidade Federal of Rio Grande do Norte (UFRN), under approval number 3.696.904. The written informed consent was properly collected. The authors had no access to information that could identify individual participants during or after data collection.

The tests occurred in individual sensory analysis booths, with temperature controlled to 22 °C and artificial white light. Water and salty crackers were offered to the volunteers between the samples. The evaluated parameters were appearance, color, aroma, texture, flavor, and overall acceptance using a hedonic scale of 9 points, ranging from disliked very much (1) to liked very much (9). Cheeses were also evaluated for purchase intent by a numerical and nominal five-point scale (1 = certainly would not buy, to 5 = would certainly buy).

The Acceptability Index (AI) was calculated according to Dutcosky [35], which classifies a product with good acceptability when the AI is greater than 70, using the equation:

$$AI\ (\%) = M/h \times 100$$

in which:

M = arithmetic mean of the scores assigned to the parameter, and
h = highest score given by the tasters to the parameter under analysis.

2.9. Statistical Analyses

The results were expressed as means and standard deviations. Differences between treatments were determined through an analysis of variance (ANOVA) with the R version 3.5.0 software program using the "Agricolae" package, complemented by the Tukey test with a 0.05-significance level. The results of the acceptability tests for sensory attributes and purchase intention were subjected to the Dunnet test by comparing cheeses with the colostrum addition to the control sample.

3. Results and Discussion

The characteristics of the milk and colostrum used to manufacture the cheeses can be seen in Table 1. The milk composition is within the legal standards of Regulation No. 76 (IN 76) of the Ministry of Agriculture, Livestock and Supply [28].

Table 1. Composition, pH, and acidity of milk and colostrum used in cheese production (mean ± standard deviation).

Composition	Milk ± SD	Colostrum ± SD
Fat (g 100 g^{-1})	5.40 ± 0.01 ^{a}	4.05 ± 0.01 ^{b}
Total protein (g 100 g^{-1})	3.87 ± 0.01 ^{b}	15.20 ± 0.01 ^{a}
Lactose (g 100 g^{-1})	5.34 ± 0.01 ^{a}	1.48 ± 0.01 ^{b}
Casein (g 100 g^{-1})	3.02 ± 0.01 ^{b}	12.07 ± 0.01 ^{a}
SNF (g 100 g^{-1})	10.02 ± 0.01 ^{b}	17.46 ± 0.14 ^{a}
TS (g 100 g^{-1})	15.42 ± 0.01 ^{a}	21.52 ± 0.01 ^{b}
Acidity (g lactic acid 100 mL^{-1})	0.18 ± 0.01 ^{b}	0.25 ± 0.01 ^{a}
pH	6.80 ± 0.01 ^{a}	6.40 ± 0.01 ^{b}

SNF: Solids-not-fat; TS: Total solids. SD: Standard Deviation. Means with different letters on the same line differ from each other by the Tukey test ($p < 0.05$).

The average colostrum composition values are higher when compared with milk, except for lactose and fat, which had a lower concentration. Conte and Scarantino [36] report that a gradual increase in lactose values during lactation is expected, particularly in the first days after calving. Sobczuk-Szul et al. [37] show 23.34 g 100 g^{-1} for solids-not-fat in colostrum from Jersey cows, which is higher than the value found in this study (17.46 g 100 g^{-1}). The 23 g 100 g^{-1} total solids value colostrum from Jersey dairy cows is reported by Morrill et al. [38], being slightly higher than that found in the present study (21.52 g 100 g^{-1}).

Oliveira et al. [39] report that the high protein concentration in colostrum is related to the greater amount of casein and immunoglobulins. These immunoglobulins have the function of protecting the calf from various diseases until its body is able to develop its own defense cells. Clinical trials have proven that bovine colostrum immunoglobulins may also benefit human health. Among them, Potiroglu and Kondolot [40] conducted a study on using bovine colostrum in treating children with upper respiratory tract infections due to IgA deficiency. They observed a reduction in the severity of the infections. Those researchers emphasize that no adverse effects are followed by the patients' mothers. The use of colostrum for producing immunoglobulins on an industrial scale is interesting because it has high bioavailability and safety compared to blood products [23]. This makes bovine colostrum a source of bioactive compounds of interest for developing functional foods.

Colostrum shows a pH of 6.4 and acidity of 0.25 g lactic acid/100 mL (Table 1). The colostrum pH is higher than that reported by Saalfeld et al. [41] with a pH of 6.29, while the acidity value is similar to colostrum within 24 h from calving (0.25 g of lactic acid/100 mL; [41]). The acidity value is probably related to the presence of solids-not-fat, such as albumin, caseins, and phosphates; thus, when colostrum is characterized by high protein content, high acidity is expected [41]. Nardone et al. [42] also observe higher titratable acidity for colostrum when compared to milk and report a positive correlation with its protein content.

Regarding the composition of cheeses made by processing milk with colostrum (Table 2), moisture results ranged from 44.77 to 63.22 g 100 g^{-1}. Differences in moisture between the samples according to the colostrum level and the maturation time ($p < 0.05$) are observed. All the fresh cheese formulations with added colostrum (1F, 2F, and 3F) can be classified as very high moisture cheeses, as their moisture content is greater than 55.0 g 100 g^{-1}. The 0B and 1C cheeses are within the standards for medium-moisture cheeses, with percentages within 44.77 to 45.9 g 100 g^{-1} [23]. The remaining aged cheeses presented moisture ranging from 46.59 to 63.22 g 100 g^{-1}; therefore, they were classified as high-moisture (46–54.9 g 100 g^{-1}) to very high-moisture cheeses (above 55 g 100 g^{-1}) according to the Brazilian regulation [26]. As expected, the moisture content of the cheese samples is directly linked to the maturation time applied. Cheeses with a short maturation time tend to be moister than cheeses with a lengthier maturation time [43]. Evaluating the characteristics of rennet-type cheese produced using cow's milk, Roig et al. [44] and Santos et al. [45] observed that the high moisture of the cheeses was due to the greater presence of denatured whey proteins because of the use of pasteurized milk. When analyzing the variability of cheese moisture content in this study, the pasteurization effect seems to affect cheese obtained from milk added with colostrum. According to Souza and Saad [46] and Santos et al. [45], another explanation for the difference in cheese moisture would be related to milk and colostrum pH, as the concentration of hydrogen ions leads to a reduction in repulsive forces and a consequent increase in the aggregation of casein micelles, which may explain the lower moisture retention in cheese obtained from only milk.

Table 2. Composition of cheeses enriched with different levels of bovine colostrum and matured for 10, 20, and 40 days (mean ± standard deviation).

Treatment	Moisture	Ash	FDM	Protein	Acidity	pH	a_w
0—100:0 (milk:colostrum, *v:v*)							
F	47.52 ± 0.16 g	2.95 ± 0.00 f	41.09 ± 5.52 ab	22.24 ± 0.21 c	0.04 ± 0.00 b	5.39 ± 0.02 ef	0.95 ± 0.01 b
A	46.59 ± 0.29 h	3.53 ± 0.05 cde	31.00 ± 1.91 def	25.48 ± 0.57 abc	0.04 ± 0.01 b	5.65 ± 0.08 bc	0.95 ± 0.01 ab
B	44.77± 0.11 i	3.34 ± 0.06 ef	47.36 ± 0.98 a	23.06 ± 1.32 bc	0.05 ± 0.00 ab	5.50 ± 0.02 de	0.99 ± 0.01 a
C	49.11± 0.17 f	3.96 ± 0.04 bc	32.60 ± 0.19 cde	25.85 ± 0.49 ab	0.08 ± 0.01 a	5.22 ± 0.02 gh	0.97 ± 0.00 ab
1—85:15 (milk:colostrum, *v:v*)							
F	56.98 ± 0.22 b	3.44 ± 0.01 de	26.24 ± 1.33 efg	25.30 ± 0.16 abc	0.03 ± 0.01 b	5.61 ± 0.09 bc	0.96 ± 0.01 b
A	51.00 ± 0.13 e	3.32 ± 0.08 ef	32.58 ± 0.08 cde	23.45 ± 017 abc	0.02 ± 0.00 b	5.24 ± 0.03 gh	0.95 ± 0.02 ab
B	52.41 ± 0.41 d	3.63 ± 0.01 bcde	38.40 ± 0.96 bc	26.07 ± 2.40 ab	0.03 ± 0.01 b	5.68 ± 0.00 bc	0.96 ± 0.01 ab
C	45.99 ± 0.19 h	4.92 ± 0.02 a	26.87 ± 1.32 efg	24.00 ± 0.49 abc	0.08 ± 0.01 a	5.30 ± 0.02 fg	0.94 ± 0.01 b
2—80:20 (milk:colostrum, *v:v*)							
F	63.22 ± 0.16 a	3.21 ± 0.02 ef	31.56 ± 1.52 def	24.66 ± 0.25 abc	0.04 ± 0.00 b	5.73 ± 0.01 b	0.97 ± 0.01 ab
A	54.58 ± 0.11 c	4.01 ± 0.09 b	21.21 ± 0.65 g	26.65 ± 0.33 a	0.04 ± 0.02 ab	5.57 ± 0.03 cd	0.96 ± 0.00 ab
B	47.48 ± 0.37 g	3.85 ± 0.03 bcd	35.65 ± 0.34 bcd	25.58 ± 0.05 abc	0.02 ± 0.00 b	5.48 ± 0.02 de	0.97 ± 0.02 ab
C	49.42 ± 0.11 f	3.84 ± 0.09 bcd	35.13 ± 0.29 bcd	25.42 ± 0.03 abc	0.03 ± 0.01 b	5.03 ± 0.01 ij	0.95 ± 0.01 ab
3—75:25 (milk:colostrum, *v:v*)							
F	57.13 ± 0.69 b	3.23± 0.12 ef	40.29 ± 3.89 b	24.73 ± 1.51 abc	0.02 ± 0.00 b	6.23 ± 0.01 a	0.95 ± 0.03 b
A	53.06± 0.12 d	3.45 ± 0.02 de	32.68 ± 3.18 cde	23.83 ± 0.63 abc	0.04 ± 0.01 b	5.14 ± 0.01 hi	0.95 ± 0.01 ab
B	51.49 ± 0.14 e	4.90 ± 0.55 a	25.42 ± 3.16 fg	23.22 ± 0.25 bc	0.02 ± 0.00 b	5.46 ± 0.00 de	0.96 ± 0.00 ab
C	51.31 ± 0.27 e	3.53 ± 0.01 cde	29.65 ± 1.18 def	25.00 ± 0.44 abc	0.03 ± 0.01 b	5.02 ± 0.01 j	0.95 ± 0.01 b

FDM: Fat in Dry Matter; Moisture, ash, crude protein, and acidity are expressed in g 100 g^{-1}; 0: control samples (0 colostrum addition); 1: Cheese produced from milk with the addition of 15 mL 100 mL^{-1} colostrum; 2: Cheese produced from milk with the addition of 20 mL 100 mL^{-1} colostrum; 3: Cheese produced from milk with the addition of 25 mL 100 mL^{-1} colostrum; F: fresh cheese; A: Cheese matured for 10 days; B: Cheese matured for 20 days; C: Cheese matured for 40 days. Means with different letters in the same column differ from each other by the Tukey test ($p < 0.05$).

The treatments with the addition of 15 and 25 mL of colostrum per 100 mL of milk in cheeses matured for 20 and 40 days (3B and 1C, respectively) yielded the highest ash values. The simultaneous effect of colostrum addition and maturation promotes a higher concentration of this component in cheese. Indeed, significant differences ($p < 0.05$) are found between different concentrations of colostrum additions, as well as between maturation times ($p < 0.05$). Our cheeses added with colostrum reached a higher ash content than those reported by Assunção et al. (3.02 to 3.24 g 100 g^{-1}) for similar types of artisanal spicy cheeses without colostrum [47].

Among the treatments, type 0B cheese is classified as fatty cheese, as it has fat in the dry matter within the limits of 45.0 to 59.9 g 100 g^{-1} [26]. Cheeses 0B and 0F showed the highest percentages of FDM with significant differences ($p < 0.05$) compared to the other formulations. The 2A cheese sample has the lowest lipid content which is consequently reflected in the FDM (21.21 g 100 g^{-1}), constituting the lowest value among all treatments. Thus, according to Brazilian standards, cheese 2A is classified as reduced fat (fat content between 10.0 and 24.9 g 100 g^{-1}) [26]. All the other cheeses are classified as within the semi-fat criteria, including those with fat content in the dry matter between 25.0 and 44.9 g 100 g^{-1} [26]. The reduction in the lipid index observed in the cheeses with added colostrum can be explained by the differences between the high lipid content found in the fresh milk used for making the cheeses (5.4 g 100 g^{-1}) and the percentage of this component in the 24-h colostrum (4.05 g 100 g^{-1}). It may also be due to the formation of more fragile curds, which, according to Sousa et al. [48], affect the ability to retain fat, influencing the centesimal composition.

Cheeses produced from a different proportion of milk and colostrum are also significantly different regarding SNF ($p < 0.05$). This trait is important because it allows the assessment of fat expressed in relation to total dry matter (TDM), correcting for the variation that may occur because of moisture losses [49]. The control cheese (0B) shows the highest FDM, with this being justified by the higher fat content (5.40 g 100 g^{-1}) of milk compared to that in the colostrum (4.05 g 100 g^{-1}). According to Foley and Otterby [2], bovine colostrum fat decreases with the time interval from calving (fat equal to 6.7 g 100 g^{-1} postpartum, 5.4 g 100 g^{-1} at 24 h, and 3.9 g 100 g^{-1} at 72 h). Raimondo et al. [50] reported a high quantity of fat colostrum in Jersey cows, from 1.35 ± 1.17 g/dL on the first day from calving to 3.09 ± 2.19 g/dL on the third day of lactation.

The protein content of cheese is similar among treatments, although the percentage was slightly lower in samples without colostrum addition. This can be justified by the higher whey protein content in colostrum and the low proportion of milk's total protein content [51].

The titratable acidity of cheese (Table 2) is expressed as a percentage of lactic acid and varied between 0.02 to 0.08 among treatments, presenting a lower variability compared to that reported by Sousa et al. [48] for traditional rennet cheese (0.12 to 1.01). Those authors stated that the acidity resulting from lactic acid production has a direct influence on the expulsion of whey, especially during syneresis. In addition, acidity can influence the texture [52], microbial activity, and maturation index of cheese [48]. Thus, the low titratable acidity observed in this study reflects the pH values found in the cheese formulations.

Cheese pH mean values (Table 2) range from 5.02 (3C) to 6.23 (3F). The pH of the fresh cheeses among treatments gradually increased as the colostrum proportion in the formulation increased. A similar pH was found by Simon and coworkers for a cheese made with 100% colostrum (6, 15) [18]. The presence of immunoglobulins, considered the main antimicrobial factor of colostrum [50], possibly affected the fermentation process by the lactic acid bacteria added to the cheeses, even if the proteins that remain soluble (including immunoglobulins) are drained in the whey during curd formation. Further studies are needed to analyze the lactic acid bacteria action in cheese fermentation using milk added with colostrum.

Water activity values of cheese samples vary between 0.94 and 0.99 ($p < 0.05$; Table 2). It is known that cheeses with high water activity levels are more susceptible to high microbial development [53]. Thus, the activity is inversely proportional to the shelf life of cheeses. In general, the longer the ripening period, the lower the water activity; however, in our experiments, no significant differences were observed between ripe and fresh cheeses, meaning the maturation time was not long enough to produce changes in a_w.

Regarding the measurement of color, we report that the variation in the colostrum added to milk produces differences among cheese samples (Table 3). The cheeses of the present study generally show medium luminosity (L*), with a predominance of yellow (b*) and green (a*). Thus, the cheeses analyzed herein were characterized as yellowish-white, resembling the characteristics of rennet cheese. The L* brightness values ranged from 55.43 to 83.44. The negative a* value represents the intensity of the green color, while the positive value refers to the red one. The b* values ranged from 14.07 to 34.04, representing the intensity of the yellow color.

Table 3. Color characteristics of cheeses enriched with different levels of bovine colostrum and matured for 10, 20, and 40 days (mean ± standard deviation).

Treatments	L*	a*	b*	YI
0—100:0 (milk:colostrum, *v:v*)				
F	65.80 ± 7.87 def	−6.83 ± 8.78 abc	17.04 ± 5.68 d	37.24 ± 12.81 bc
A	62.38 ± 1.89 efg	−5.57 ± 5.13 abc	19.01 ± 4.50 bcd	43.62 ± 10.71 abc
B	64.65 ± 3.13 def	7.21 ± 8.62 a	17.83 ± 12.65 cd	39.35 ± 27.94 bc
C	75.66 ± 0.66 b	−7.76 ± 1.91 abc	30.73 ± 0.65 abc	58.01 ± 0.72 abc
1—85:15 (milk:colostrum, *v:v*)				
F	62.76 ± 0.86 defg	1.49 ± 5.22 ab	20.09 ± 1.88 bcd	45.69 ± 3.63 abc
A	59.77 ± 0.86 fg	−7.53 ± 1.36 abc	23.69 ± 0.27 abcd	56.62 ± 0.34 abc
B	63.26 ± 3.30 def	−10.28 ± 3.60 bc	15.30 ± 1.54 d	34.51 ± 1.88 c
C	64.48 ± 0.40 def	2.16 ± 1.49 ab	31.49 ± 1.80 ab	69.78 ± 4.40 a
2—80:20 (milk:colostrum, *v:v*)				
F	63.53 ± 1.84 def	−15.06 ± 6.09 c	16.70 ± 3.26 d	37.54 ± 7.03 bc
A	83.44 ± 1.03 a	−5.39 ± 4.22 abc	26.99 ± 3.92 abcd	46.16 ± 6.18 abc
B	66.51 ± 2.52 cdef	−8.60 ± 0.92 abc	25.25 ± 4.38 abcd	54.06 ± 7.47 abc
C	68.79 ± 1.07 bcde	7.04 ± 0.79 a	34.04 ± 2.71 a	70.64 ± 4.64 a
3—75:25 (milk:colostrum, *v:v*)				
F	55.43 ± 0.87 g	−4.10 ± 6.24 abc	14.07 ± 3.81 d	36.19 ± 9.30 c
A	74.16 ± 0.51 bc	0.02 ± 4.00 abc	33.66 ± 0.01 a	64.84 ± 0.44 ab
B	64.68 ± 1.88 def	3.69 ± 6.07 ab	24.82 ± 1.04 abcd	54.79 ± 0.92 abc
C	70.41 ± 0.51 bcd	1.89 ± 9.43 ab	26.06 ± 1.29 abcd	52.88 ± 2.65 abc

0: control samples (0 colostrum addition); 1: Cheese produced from milk with the addition of 15 mL 100 mL^{-1} colostrum; 2: Cheese produced from milk with the addition of 20 mL 100 mL^{-1} colostrum; 3: Cheese produced from milk with the addition of 25 mL 100 mL^{-1} colostrum; F: fresh cheese; A: Cheese matured for 10 days; B: Cheese matured for 20 days; C: Cheese matured for 40 days. Means with different letters in the same column differ from each other by the Tukey test ($p < 0.05$).

Generally, increased yellowing is directly related to the amount of fat due to the saturation of adipocytes with beta-carotene from animal metabolism [54,55]. During ripening, the concentration of non-aqueous components usually leads to color changes. Although slight differences in the yellowing index (YI) between fresh and matured cheeses can be observed in this study (40-day ripe cheeses achieved YI higher than the fresh formulations), there was no statistical relevance of maturation time on this parameter, except for the cheese produced from milk with the addition of 20 mL 100 mL^{-1} colostrum.

Table 4 presents the characteristics of the cheese texture profile among treatments. According to Fontan [56], cheese texture is affected by composition and maturation. The F0 and A0 samples showed greater firmness ($p < 0.05$). There is a tendency to lower firmness in cheeses with increasing colostrum levels in the formulation. These data are consistent with cheese moisture since they present a greater solids dispersion and, therefore, a lower firmness.

Table 4. Texture characteristics of cheeses enriched with different levels of bovine colostrum and matured for 10, 20, and 40 days (mean ± standard deviation).

Treatments	Firmness	Cohesivity	Chewability	Resilience
0—100:0 (milk:colostrum, *v:v*)				
F	58.99 ± 6.51 ab	0.60 ± 0.05 a	35.89 ± 6.96 ab	0.28 ± 0.03 b
A	65.20 ± 8.64 a	0.61 ± 0.05 a	40.09 ± 6.56 a	0.29 ± 0.04 b
B	50.03 ± 8.08 bc	0.55 ± 0.03 ab	27.41 ± 5.00 bcd	0.22 ± 0.03 b
C	46.31 ± 4.53 bcd	0.61 ± 0.05 a	28.14 ± 3.69 bc	0.20 ± 0.02 b
1—85:15 (milk:colostrum, *v:v*)				
F	33.44 ± 3.25 def	0.56 ± 0.03 ab	18.62 ± 1.15 def	0.31 ± 0.06 b
A	42.87 ± 4.82 cd	0.53 ± 0.05 ab	22.76 ± 3.82 cde	0.32 ± 0.08 b
B	26.37 ± 1.20 efg	0.50 ± 0.02 abc	13.22 ± 0.81 fg	0.44 ± 0.27 b
C	39.21 ± 6.91 cde	0.34 ± 0.06 d	13.39 ± 4.37 fg	0.39 ± 0.13 b
2—80:20 (milk:colostrum, *v:v*)				
F	11.75 ± 1.26 h	0.40 ± 0.08 cd	4.68 ± 0.80 g	0.39 ± 0.18 b
A	51.07 ± 4.5 bc	0.51 ± 0.04 abc	25.68 ± 3.72 cd	0.45 ± 0.18 b
B	45.34 ± 7.08 cd	0.57 ± 0.04 ab	25.64 ± 2.79 cd	0.22 ± 0.02 b
C	27.58 ± 2.26 efg	0.45 ± 0.03 bcd	12.56 ± 1.57 fg	0.40 ± 0.53 b
3—75:25 (milk:colostrum, *v:v*)				
F	16.46 ± 1.86 gh	0.55 ± 0.03 ab	8.99 ± 1.03 g	0.23 ± 0.04 b
A	23.51 ± 1.69 fgh	0.55± 0.06 ab	12.93 ± 2.22 fg	0.25 ± 0.11 b
B	24.24 ± 4.96 fgh	0.57 ± 0.04 ab	13.84 ± 3.53 efg	0.22 ± 0.03 b
C	23.34 ± 3.24 fgh	0.41 ± 0.05 cd	9.46 ± 0.69 fg	1.14 ± 0.23 a

0: control samples (0 colostrum addition); 1: Cheese produced from milk with the addition of 15 mL 100 mL^{-1} colostrum; 2: Cheese produced from milk with the addition of 20 mL 100 mL^{-1} colostrum; 3: Cheese produced from milk with the addition of 25 mL 100 mL^{-1} colostrum; F: fresh cheese; A: Cheese matured for 10 days; B: Cheese matured for 20 days; C: Cheese matured for 40 days. Means with different letters in the same column differ from each other by the Tukey test ($p < 0.05$).

Pereira et al. [57] studied the correlation between the instrumental and sensory texture of several similar commercial types of cheeses. They found that cheeses with low moisture content were generally the firmest. Such findings confirm the results of this work since the control cheese with the lowest moisture percentage (47.52 g 100 g^{-1}) was the firmest, while the cheeses with the colostrum addition had the highest moisture content and the lowest firmness.

Cohesiveness is instrumentally assessed as the amount of energy necessary to break the internal structure of the cheese, and in sensory analysis, it is defined as the degree to which cheese is compressed between the teeth before breaking [58]. In the present study, this parameter ranged from 0.34 to 0.57, with fresh control cheese having the lowest value ($p < 0.05$). Some of the values found in this study are close to the work of Andrade et al. [59] in evaluating industrial and artisanal cheeses, finding results ranging from 0.59 to 0.67 and 0.49 to 0.65, respectively.

Among all the treatments, 0F and 0A showed greater chewability ($p < 0.05$). According to the results obtained for this parameter, it is possible to observe that cheeses produced from formulations with higher amounts of colostrum were characterized by lower firmness. This phenomenon may possibly be associated with higher moisture content, as water in cheese acts together with fat as a lubricant between casein aggregates [58]. Thus, the reduction in moisture content results in an increase in cheese hardness and chewability.

Cheese yield is primarily due to the milk quantity and the recovery in the curd of the protein and fat from the milk to the cheese [53]. Figure 1 shows the cheese yield calculated in liters of milk per kilogram of cheese (L/kg), with the best performance for the formulation with the higher percentage of colostrum (6.08 L needed to produce one kg of cheese). There is also a significant difference ($p < 0.05$) between the formulations without colostrum and with the lowest percentage of colostrum (15 mL).

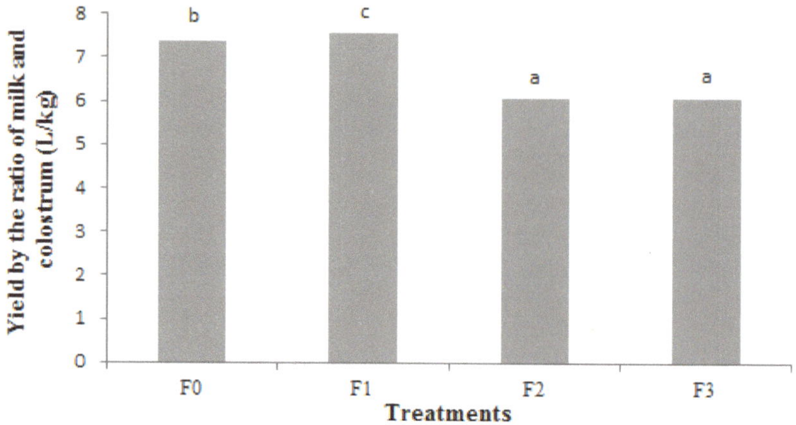

Figure 1. Average of fresh (F) cheese yield (L/kg) from cheese formulations added with different levels of bovine colostrum 0:100, 15:85, 20:80, and 25:75 mL 100 mL^{-1}, bovine colostrum:milk (*v:v*)—F0, F1, F2, and F3, respectively. Means identified with different letters differ from each other by the Tukey test ($p < 0.05$).

All the cheeses in this experiment were produced with similar production technology to rennet cheese processing and showed a ratio of the amount of milk and colostrum below 10 L of the total volume for each kilogram of cheese produced, which indicates a good yield when compared to cheeses with the same moisture content [60,61]. According to Dutra [62], the average yield for rennet cheese manufacturing by the traditional process is between 10.5 and 12.5 L of milk per kilogram of cheese.

The microbiological profile of milk, colostrum, and cheese samples met the Brazilian criteria determined by the national regulatory agency [26,33,34], which establishes the absence of *Salmonella* spp. and *Listeria monocytogenes* in a 25 g sample, and a tolerance limit of 1×10^3 MPN/g for coagulase-positive staphylococci, 1×10^4 MPN/g for total coliforms, 5×10^3 MPN/g for thermotolerant coliforms, and 5.0×10^3 UFC/g for molds and yeasts.

Table 5 presents the descriptive statistics of sensory attributes obtained for the fresh cheese among different treatments. The appearance and color parameters scores ranged from 7.67 to 8.10 and 7.54 to 7.81, respectively, with no statistical differences ($p < 0.05$) in these attributes between the samples.

Table 5. Scores attributed to appearance, color, aroma, texture, taste, overall evaluation (9-point hedonic scale), and purchase intention (5-point scale) for fresh cheeses with added bovine colostrum at the levels of 0:100, 15:85, 20:80, and 25:75, bovine colostrum:milk (*v:v*) (mean ± standard deviation).

Attributes	0F	1F	2F	3F
Appearance	8.10 ± 0.99^a	7.67 ± 1.34^a	7.86 ± 1.09^a	7.75 ± 1.22^a
Color	7.81 ± 1.31^a	7.54 ± 1.26^a	7.73 ± 1.17^a	7.64 ± 1.38^a
Aroma	7.52 ± 1.40^a	6.76 ± 1.63^b	7.07 ± 1.46^b	7.12 ± 1.62^b
Texture	7.83 ± 1.16^a	6.98 ± 1.97^b	6.98 ± 1.74^b	7.36 ± 1.61^b
Flavor	7.90 ± 1.32^a	6.63 ± 1.96^b	6.47 ± 1.93^b	7.07 ± 1.80^b
Overall evaluation	7.72 ± 1.02^a	7.08 ± 1.52^b	6.94 ± 1.56^b	7.41 ± 1.40^b
Purchase intent	4.38 ± 0.94^a	3.49 ± 1.16^b	3.50 ± 1.17^b	3.79 ± 1.28^b

0: control sample (0 colostrum addition); 1: Cheese produced from milk with the addition of 15 mL 100 mL^{-1} colostrum; 2: Cheese produced from milk with the addition of 20 mL 100 mL^{-1} colostrum; 3: Cheese produced from milk with the addition of 25 mL 100 mL^{-1} colostrum; F: fresh cheese. Means with different letters on the same line differ from each other by the Dunnet test ($p < 0.05$).

The control cheese had the highest average for aroma (7.52), texture (7.83), and flavor (7.90) compared to the other treatments ($p < 0.05$). The control cheese also obtained a significantly higher overall evaluation (7.72). It is worth noting that despite the statistical differences, cheese from milk added with 25 mL 100 mL^{-1} of colostrum is the treatment showing the closest scores to the control one in terms of overall evaluation and purchase intent.

Moreover, the control cheese obtained the highest AI for all evaluated attributes (Table 6). However, the 3F treatment scored the highest AI among cheeses produced from milk added with colostrum, except for the appearance and color attributes for which the 2F obtained the highest value.

Table 6. Acceptability Index (IA) of fresh cheeses obtained from milk with different colostrum additions.

Attributes	AI (%)			
	0F	1F	2F	3F
Appearance	90.08	85.30	87.33	86.14
Color	86.85	83.87	85.90	84.94
Aroma	83.63	75.14	78.49	79.21
Texture	87.09	75.12	77.65	81.83
Flavor	87.81	73.71	71.92	78.61
Overall evaluation	86.97	78.73	77.18	82.43

AI: Acceptability Index; 0F: fresh cheese with 0 colostrum; 1F, 2F and 3F: Fresh cheeses with the additions of 15, 20, and 25 mL 100 mL^{-1} colostrum, respectively.

It was further observed that all the evaluated cheese formulations obtained a satisfactory AI. Dutcosky [35] suggested that a food product with good acceptability presents average AI values above 70. Good acceptance of colostrum-added dairy products has previously been reported; Saalfeld et al. [41] observed sensory approval of dairy beverages enriched with colostrum silage. Mouton and Aryana [19] analyzed the influence of colostrum on the characteristics of ice cream and recommended such an industrial application based on a sensory analysis by trained panelists.

4. Conclusions

Part of the colostrum cows produce is a surplus dairy farm product. However, there are challenges to appropriately designate the daily surplus of colostrum production and a need for processing and product preservation technologies. Based on the overall assessment and the purchase intention of the fresh cheeses analyzed in this study, it is possible to speculate that all cheeses made using formulations with colostrum have the potential to be used for human consumption. Among the cheeses with colostrum, the fresh cheese with 75:25 milk:colostrum (*v:v*) received the highest scores for the aroma, texture, flavor, overall evaluation, and purchase intention parameters. Furthermore, this formulation (25% colostrum) achieved a better yield than those without or with 15% colostrum. Regarding texture measurements, after 40 days of maturation, the proportion of 25% colostrum reached cohesivity, resilience, and bright color significantly higher than its fresh correspondent.

This study contributes to knowledge on the use of colostrum in elaborating and enriching dairy derivatives intended for human consumption, as the use of colostrum is little known culturally in Brazil and many other countries. In addition, the results herein point to a product with innovative appeal, strengthening research in seeking technologies and innovations to use this raw material, which is still underutilized and wasted due to excess production.

Author Contributions: I.d.M.B.: Methodology, data curation, formal analysis, investigation, writing—original draft preparation. K.A.: Data curation, writing—review and editing, figures and illustrations, supervision. C.S.M., R.R.P.C., C.C.-G., E.G.d.S.O.S., N.G.A., B.M.E.d.C., J.P.F.d.O., C.A.B. and D.C.S.: writing—review and editing. E.d.O.M.A.: visualization, writing—original draft preparation, writing—review and editing. J.A.N.: Data curation, writing—review and editing. A.H.d.N.R.: Conceptualization, methodology, data curation, writing—review and editing, supervision. All authors have read and agreed to the published version of the manuscript.

Funding: This research received no external funding.

Data Availability Statement: The data used to support the findings of this study can be made available by the corresponding author upon request.

Acknowledgments: This work was supported by the Coordination for the Improvement of Higher Education Personnel (CAPES).

Conflicts of Interest: The authors declare no conflict of interest.

References

1. Mcgrath, B.A.; Fox, P.F.; Mcsweeney, P.L.H.; Kelly, A.L. Composition and properties of bovine colostrum: A review. *Dairy Sci. Technol.* **2016**, *96*, 133–158. [CrossRef]
2. Foley, J.A.; Otterby, D.E. Availability, storage, treatment, composition, and feeding value of surplus colostrum: A review. *J. Dairy Sci.* **1978**, *61*, 1033–1060. [CrossRef]
3. Silva, E.D.S.; Anaya, K.; Bezerra, M.D.F.; Borba, L.; Barbosa, I.D.M.; De Oliveira, J.P.F.; Urbano, S.A.; Macêdo, C.S.; de Lima Júnior, D.M.; Idiana, D.M.B.; et al. Physicochemical characterization and Brix in Jersey cow colostrum in tropical conditions. *Int. J. Agric. Biol.* **2021**, *26*, 139–144. [CrossRef]
4. Uruakpa, F.O.; Ismond, M.A.H.; Akobundu, E.N.T. Colostro e seus benefícios: Uma revisão. *Pesqui. Nutr.* **2002**, *22*, 755–767. [CrossRef]
5. Adar, T.; Ya'acov, A.B.; Lalazar, G.; Lichtenstein, Y.; Nahman, D.; Mizrahi, M.; Wong, V.; Muller, B.; Rawlin, G.; Ilan, Y. Oral administration of immunoglobulin G-enhanced colostrum alleviates insulin resistance and liver injury and is associated with alterations in natural killer T cells. *Clin. Exp. Dermatol.* **2012**, *167*, 252–260. [CrossRef]
6. Costa, A.; Sneddon, N.W.; Goi, A.; Visentin, G.; Mammi, L.M.E.; Savarino, E.V.; Zingone, F.; Formigoni, A.; Penasa, M.; De Marchi, M. Invited review: Bovine colostrum, a promising ingredient for humans and animals—Properties, processing technologies, and uses. *J. Dairy Sci.* **2023**, *106*, 5197–5217. [CrossRef]
7. Playford, R.J.; Cattell, M.; Marchbank, T. Marked variability in bioactivity between commercially available bovine colostrum for human use; implications for clinical trials. *PLoS ONE* **2020**, *15*, e0234719.
8. Galdino, A.B.d.S.; Rangel, A.H.D.N.; Buttar, H.S.; Nascimento, M.S.L.; Gavioli, E.C.; Oliveira, R.d.P.; Sales, D.C.; Urbano, S.A.; Anaya, K. Bovine colostrum: Benefits for the human respiratory system and potential contributions for clinical management of COVID-19. *Food Agric. Immunol.* **2021**, *32*, 143–162. [CrossRef]
9. Gomes, R.D.; Anaya, K.; Galdino, A.B.; Oliveira, J.P.; Gama, M.A.; Medeiros, C.A.; Gavioli, E.C.; Porto, A.L.F.; Rangel, A.H. Bovine colostrum: A source of bioactive compounds for prevention and treatment of gastrointestinal disorders. *NFS J.* **2021**, *25*, 1–11. [CrossRef]
10. Beltrão, F.A.S.; Moura, C.V.R.; Sousa, S.; Andrade, A.E.B.; Souza, W.F.C.; Santos, D.T.P. Caracterização físico química de queijo tipo "chevrotin" simbiótico. *Eng. Ambient.* **2017**, *14*, 128–136.
11. Cao, J.; Wang, X.; Zheng, H. Comparative studies on thermoresistance of protein G-binding region and antigen determinant region of immunoglobulin G in acidic colostral whey. *Food Agric. Immunol.* **2007**, *18*, 17–30. [CrossRef]
12. Sydney, A.C.N.; Ikeda, I.K.; Ribeiro, M.C.d.O.; Sydney, E.B.; Neto, D.P.d.C.; Karp, S.G.; Rodrigues, C.; Soccol, C.R. Colostrum new insights: Products and processes. In *Current Developments in Biotechnology and Bioengineering*; Elsevier: Amsterdam, The Netherlands, 2022; pp. 397–422.
13. Silva, E.G.d.S.O.; Anaya, K.; Bezerra, M.d.F.; Macêdo, C.S.; Urbano, S.A.; Borba, L.H.F.; Barbosa, I.d.M.; Ramalho, H.M.M.; Cipolat-Gotet, C.; Galdino, A.B.d.S.; et al. Physicochemical and sensory evaluation of greek style yoghurt with bovine colostrum. *J. Food Sci. Technol.* **2021**, *42*, e22121. [CrossRef]
14. Gomes, R.D.S.; Barbosa, I.M.; Ribeiro, C.V.D.M.; Anaya, K.; Silva, E.G.S.O.; Oliveira, C.A.A.; da Gama, M.A.S.; Oliveira, J.P.F.; Sale, D.C.; Araújo, E.O.M.; et al. Fatty acid profile of Greek yogurt with colostrum addition. *J. Food Sci. Technol.* **2023**, *43*, 26023. [CrossRef]
15. Galdino, A.B.S.; Anaya, K.; Barbosa, I.M.; Borba, L.H.F.; Silva, E.G.S.O.; Macêdo, C.S.; Ribeiro, C.V.D.M.; Oliveira, J.P.F.; Rangel, A.H.N. Nutritional and physicochemical quality of formulations based on colostrum and bovine whey. *PLoS ONE* **2022**, *17*, e0267409. [CrossRef]
16. Astuti, F.D.; Setyawardani, T.; Santosa, S.S. The physical characteristics of cheese made of milk, colostrum and both during the ripening. *J. Indones. Trop. Anim. Agric.* **2021**, *46*, 75–83. [CrossRef]
17. Herrera-Chávez, B.; Trujillo, A.J.; Calero, P.; Falconí, M.I.; Sánchez-Macías, D. Effects of colostrum in milk on the effectiveness of the pasteurization process and cheese milk quality. *J. Appl. Anim. Res.* **2022**, *50*, 246–253. [CrossRef]
18. Simon, R.; Gennari, A.; Kuhn, D.; Rama, G.R.; Souza, C.F.V.D. Making a fresh cheese using the colostrum surplus of dairy farms: An alternative aiming to minimize the waste of this raw material. *Braz. J. Food Technol.* **2022**, *25*, e2021125. [CrossRef]
19. Mouton, E.; Aryana, K. Influence of Colostrum on the Characteristics of Ice Cream. *Food Sci. Nutr.* **2015**, *6*, 480–484. [CrossRef]
20. Playford, R.J.; Weiser, M.J. Bovine colostrum: Its constituents and uses. *Nutrients* **2021**, *13*, 265. [CrossRef]
21. Feeney, E.L.; Lamichhane, P.; Sheehan, J.J. The cheese matrix: Understanding the impact of cheese structure on aspects of cardiovascular health: A food science and a human nutrition perspective. *Int. J. Dairy Technol.* **2021**, *74*, 656–670. [CrossRef]
22. Dzik, S.; Miciński, B.; Aitzhanova, I.; Miciński, J.; Pogorzelska, J.; Beisenov, A.; Kowalski, I.M. Properties of bovine colostrum and the possibilities of use. *Pol. Ann. Med.* **2017**, *24*, 295–299. [CrossRef]

23. Das, A.; Seth, R. Studies on quality attributes of skimmed colostrum powder. *Int. J. Chem. Stud.* **2017**, *5*, 17–20.
24. Normas Analíticas do Instituto Adolfo Lutz. *Métodos Químicos e Físicos para Análise de Alimentos*, 3rd ed.; IMESP: São Paulo, Brazil, 2008; Volume 1, pp. 233–234.
25. Gracioli, F.; Lehn, D.N.; Souza, C.F.V. Análise comparativa de custo e rendimento da fabricação de queijo tipo camembert e queijo colonial em pequena escala. *Rev. Destaques Acadêmicos* **2013**, *5*, 15–30.
26. Brasil. Ministério da Agricultura, Pecuária e Abastecimento. Secretaria de Defesa Agropecuária. *Portaria nº 146 de 7 de Março de 1996. Aprova os Regulamentos Técnicos de Identidade e Qualidade dos Produtos Lácteos*; Diário Oficial da União: Brasília, Brazil, 1996.
27. *Official Methods of Analysis of AOAC International*, 19th ed.; AOAC International: Gaithersburg, MD, USA, 2012.
28. Brasil. Ministério da Agricultura, Pecuária e Abastecimento. *Instrução Normativa 76 de 26 de Novembro de 2018. Dispõe Sobre Regulamentos Técnicos que Fixam a Identidade e as Características de Qualidade que Devem Apresentar o Leite Cru Refrigerado, o leite Pasteurizado e o Leite Pasteurizado Tipo A*; Seção 1; Diário Oficial da União: Brasília, Brazil, 2018.
29. Coimbra, P.T.; Bathazar, C.F.; Guimarães, J.T.; Coutinho, N.M.; Pimentel, T.C.; Neto, R.P.; Esmerino, E.A.; Freitas, M.Q.; Silva, M.C.; Tavares, M.I.; et al. Detection of formaldehyde in raw milk by time domain nuclear magnetic resonance and chemometrics. *Food Control* **2020**, *110*, 107006. [CrossRef]
30. Santurino, C.; Calvo, M.V.; Gómez-Candela, C.; Fontecha, J. Characterization of naturally goat cheese enriched in conjugated linoleic acid and omega-3 fatty acids for human clinical trial in overweight and obese subjects. *PharmaNutrition* **2017**, *5*, 8–17. [CrossRef]
31. Apha, A. *Standard Methods for the Examination of Water and Wastewater*, 21st ed.; American Public Health Association: Washington, DC, USA, 2005; p. 953.
32. Brasil. *Ministério da Agricultura, Pecuária e Abastecimento. Instrução Normativa n° 30, de 26 de Junho de 2001. Aprova os Regulamentos Técnicos de Identidade e Qualidade de Manteiga da terra ou Manteiga de Garrafa*; Queijo Coalho e Queijo de Manteiga; Diário Oficial da União: Brasília, Brazil, 2001.
33. Brasil. Ministério da Agricultura, Pecuária e Abastecimento. Secretaria de Defesa Agropecuária. *Instrução Normativa n° 60 de 23 de Dezembro de 2019. Estabelece as Listas de Padrões Microbiológicos para Alimentos Prontos para oferta ao Consumidor*; Seção 1; Diário Oficial da União: Brasília, Brazil, 2019; p. 133.
34. Brasil. Agência Nacional de Vigilância Sanitária. *Resolução RDC nº 331, de 23 de Dezembro de 2019. Aprova o Regulamento Técnico sobre Padrões Microbiológicos para Alimentos*; Seção 1; Diário Oficial da União: Brasília, Brazil, 2019; p. 96.
35. Dutcosky, S.D. *Análise Sensorial de Alimentos*, 4th ed.; Editora Universitária Champagnat: Curitiba, Brazil, 2013; 531p.
36. Conte, F.; Scarantino, S. A study on the quality of bovine colostrum: Physical, chemical and safety assessment. *Int. Food Res.* **2013**, *20*, 925–931.
37. Sobczuk-Szul, M.; Wielgoz-Groth, Z.; Wronski, M.; Rzemieniewski, A. Changes in the bioactive protein concentrations in the bovine colostrum of Jersey and Polish Holstein–Friesian cows. *Turk. J. Vet. Anim. Sci.* **2013**, *37*, 43–49. [CrossRef]
38. Morrill, N.K.M.; Conrad, E.; Lago, A.; Campbell, J.; Quigley, J.; Tyler, H. Nationwide evaluation of quality and composition of colostrum on dairy farms in the United States. *J. Dairy Sci.* **2012**, *95*, 3997–4005. [CrossRef]
39. Oliveira, E.N.A.; Santos, D.C.; Almeida, F.L.C.; Feitosa, B.F.; Feitosa, R.M. Caracterização de queijos artesanais comercializados em municípios do Ceará. *E-xacta* **2018**, *11*, 55–62. [CrossRef]
40. Patıroğlu, T.; Kondolot, M. The effect of bovine colostrum on viral upper respiratory tract infections in children with immunoglobulin a deficiency. *Clin. Respir. J.* **2013**, *7*, 21–26. [CrossRef]
41. Saalfeld, M.H.; Ira, K.R.K.; Diniz, G.L.; Kringel, D.H.; Alves, M.I.; Gularte, M.A.; Leite, F.P.L. Colostro: A redescoberta de um alimento saudável, nutritivo e com potencial probiótico. *Agroecologia e Desenvolvimento Rural Sustentável* **2012**, *5*, 18–24.
42. Nardone, A.; Lacetera, N.; Bernabucci, U.; Ronchi, B. Composition of colostrum from dairy heifers exposed to high air temperatures during late pregnancy and the early postpartum period. *J. Dairy Sci.* **1997**, *80*, 838–844. [CrossRef]
43. Louvatel, K.; Degenhardt, R. Caracterização bromatológica de queijos coloniais produzidos no distrito de Santa Lúcia, município de Ouro, SC. In *Anais da III Jornada Integrada em Biologia*; Universidade do Oeste de Santa Catarina Unoesc: Joaçaba, Brazil, 2016; pp. 37–46, ISSN 2358-0992.
44. Roig, S.; Narimatsu, A.; Dornelas, J.R.F.; Spadoti, L.M.; Pizaia, P.D.M. Avaliação da proteólise e do derretimento do queijo prato obtido por ultrafiltração. *Food Sci. Technol.* **2003**, *23*, 177–182.
45. Santos, B.M.; Oliveira, M.E.G.; Sousa, Y.R.F.; Madureira, A.R.M.F.M.; Pintado, M.M.E.; Gomes, A.M.P.; de Souza, E.L.; do Egypto, R.D.C.R. Caracterização físico-química e sensorial de queijo de coalho produzido com mistura de leite de cabra e de leite de vaca. *Revista do Instituto Adolfo Lutz* **2011**, *70*, 302–310.
46. Souza, C.H.B.; Saad, S.M.I. Viability of *lactobacillus acidophilus* La-5 added solely or in co-culture with a yoghurt starter culture and implications on physical-chemical and related properties of Minas fresh cheese during storage. *LWT-Food Sci. Technol.* **2009**, *42*, 633–640. [CrossRef]
47. Assunção, M.V.A.; Andrade, J.A.S.; Santos, T.T.; Lima, J.S.; Talma, S.V.; Machado, A.C.L.O.; Costa, L.P.; Barbosa, J.B. Elaboração e avaliação físico-química de queijo coalho condimentado artesanal no sertão sergipano. *Interfaces Científicas Saúde e Ambiente* **2018**, *7*, 79–86. [CrossRef]
48. Sousa, A.Z.B.; Abrantes, M.R.; Sakamoto, S.M.; Silva, J.B.A.; Lima, P.O.; Lima, R.N.; Rocha, M.O.C.; Sousa, D.B.P. Aspectos físico-químicos e microbiológicos do queijo tipo coalho comercializado em estados do nordeste do Brasil. *Arq. Inst. Biol.* **2014**, *81*, 30–35. [CrossRef]

49. Embrapa. *Validação e Transferência da Tecnologia do Queijo Coalho Caprino Maturado e Defumado*; Embrapa Agroindústria Tropical: Fortaleza, Brazil, 2018; Volume 152, p. 22. ISSN 1679-6543.
50. Raimondo, R.F.S.; Saut, P.E.; Souza, R.M.; Nunes, M.T.; Birgel Junior, E.H. Teores de proteína, gordura e sólidos totais no leite de vacas da raça Jersey criadas no Estado de São Paulo durante o primeiro mês de lactação. *Rev. Bras. Med. Vet.* **2009**, *46*, 355–362. [CrossRef]
51. Sgarbieri, V.C. Propriedades fisiológicas-funcionais das proteínas do soro de leite. *Rev. Nutr.* **2004**, *17*, 397–409. [CrossRef]
52. Mamede, M.E.D.O.; Viana, A.C.; Souza, A.L.C.; Farias, S.A.O.; Araújo, P.A. Estudo das características sensoriais e da composição química de queijo de Coalho industrializado. *Revista do Instituto Adolfo Lutz* **2010**, *69*, 364–370.
53. Borad, S.G.; Singh, A.K. Colostrum immunoglobulins: Processing, preservation and application aspects. *Int. Dairy J.* **2018**, *85*, 201–210. [CrossRef]
54. Pathare, P.B.; Opara, U.L.; Al-Said, F.A. Medição e análise de cor em alimentos in natura e processados: Uma revisão. *Tecnol. Bioprocessos Aliment.* **2012**, *6*, 36–60.
55. Melo, L.R.B.; Torres, F.R.; Guimarães, J.T.; Cortez, M.A.S. Development of processed low-sodium Maasdam cheese. *Arq. Bras. Med. Vet. Zootec.* **2022**, *74*, 1072–1082. [CrossRef]
56. Fontan, G.C.R. Queijo de Coalho Light: Produção, Caracterização Físico-Química, Sensorial e Reológica. Tese (Programa de Pós-graduação em Ciência e Tecnologia de Alimentos), Universidade Federal de Viçosa, 2013; 86f. Available online: https://www.locus.ufv.br/bitstream/123456789/491/1/texto%20completo.pdf (accessed on 6 October 2023).
57. Pereira, R.B.; Bennett, R.J.; Luckman, M.S. Correlation of sensory and instrumental texture evaluation in cheese analogues. *Aust. J. Dairy Technol.* **2002**, *57*, 154.
58. Bourne, M. *Food Texture and Viscosity: Concept and Measurement*, 2nd ed.; Elsevier Science & Technology Books: Amsterdam, The Netherlands, 2002; p. 416.
59. Andrade, A.S.A.; Rodrigues, M.D.C.P.; Tieko, R.; Nassu, M.A.D.S.N. Medidas instrumentais de cor e textura em queijo de coalho. In *Congresso Latino America de Analista de Alimentos*; LACEN, Anais: Fortaleza, Brazil, 2007.
60. Furtado, M.M. *Principais Problemas dos Queijos*, 3rd ed.; Setembro Editora: Paulo, Brazil, 2017; 225p.
61. Silva, M.C.D.; Ramos, A.C.S.; Moreno, I.; Moraes, J.d.O. Influência dos procedimentos de fabricação nas características físico-químicas, sensoriais e microbiológicas de queijo de coalho. *Revista do Instituto Adolfo Lutz* **2010**, *69*, 214–221.
62. Dutra, E.R.P. *Fundamentos Básicos da Produção de Queijos*; Templo: Juiz de Fora, Brazil, 2017; 160p.

foods
MDPI

Article

Effects of Cow's Milk Processing on MicroRNA Levels

Loubna Abou el qassim [1,*], Beatriz Martínez [2], Ana Rodríguez [2], Alberto Dávalos [3], María-Carmen López de las Hazas [3], Mario Menéndez Miranda [1] and Luis J. Royo [1,4,*]

1 Servicio Regional de Investigación y Desarrollo Agroalimentario (SERIDA), 33300 Villaviciosa, Spain; mmiranda@serida.org
2 Department of Technology and Biotechnology of Dairy Products, Instituto de Productos Lácteos de Asturias (IPLA-CSIC), 33300 Villaviciosa, Spain; bmf1@ipla.csic.es (B.M.); anarguez@ipla.csic.es (A.R.)
3 Laboratory of Epigenetics of Lipid Metabolism, Madrid Institute for Advanced Studies (IMDEA)-Food, CEI UAM+CSIC, 28049 Madrid, Spain; alberto.davalos@imdea.org (A.D.); mcarmen.lopez@imdea.org (M.-C.L.d.l.H.)
4 Department of Functional Biology, University of Oviedo, 33006 Oviedo, Spain
* Correspondence: loubna.ab.enam@gmail.com (L.A.e.q.); royoluis@uniovi.es (L.J.R.)

Abstract: MicroRNAs (miRNAs) regulate gene expression and might resist adverse physicochemical conditions, which makes them potential biomarkers. They are being investigated as biomarkers of dairy production systems, based on the variations in their levels in raw milk depending on animal diet and management. Whether miRNA levels can serve as biomarkers for dairy products remains unclear, since technological or culinary treatments, such as fermentation, may alter their levels. Here, 10 cow dairy farms were sampled in Asturias (north-west Spain) and milk samples were subjected to microwave heating or used to produce yogurt or cheese. Total RNA was isolated from raw milk and three derived products, and levels of seven miRNAs, selected based on previous studies as possible milk production system biomarkers, were assessed by RT-qPCR. The treatments decreased levels of all miRNAs to some extent. These results also imply that cheesemaking increases the concentration of miRNAs in this product; raw milk and cheese supposedly may provide similar concentrations of miRNAs, higher than those of yogurt and microwaved milk. They also indicate that the content of certain miRNAs in raw milk cannot necessarily be extrapolated to other dairy products.

Keywords: cow's milk; dairy products; microRNA; biomarkers

Citation: Abou el qassim, L.; Martínez, B.; Rodríguez, A.; Dávalos, A.; López de las Hazas, M.-C.; Menéndez Miranda, M.; Royo, L.J. Effects of Cow's Milk Processing on MicroRNA Levels. *Foods* **2023**, *12*, 2950. https://doi.org/10.3390/foods12152950

Academic Editors: Michele Faccia and Giuseppe Natrella

Received: 5 July 2023
Revised: 30 July 2023
Accepted: 2 August 2023
Published: 4 August 2023

1. Introduction

MicroRNAs (miRNAs) are non-coding RNAs, only 21–25 nucleotides long, endogenously synthesized, and specific to eukaryotic cells. They are involved in a vast coordination of gene expression regulatory networks. So far, it is known that they mediate post-transcriptional regulation by degrading mRNA or repressing its transcription which results in an attenuation of protein translation [1,2]. miRNAs regulate genes not only in the cells that produce them, but they may regulate genes in other cells too [3]. They have been detected in body fluids such as blood, saliva, and milk [4], which is particularly rich in miRNAs [5]. A comparison between serum and milk miRNAs in humans has concluded that most milk miRNAs are not provided by the blood circulation [6] but originate from their biogenesis in mammary alveolar epithelial cells [7].

In milk, miRNAs are found packaged in vesicles such as milk exosomes and fat globules [8,9]. After milk consumption by human adults, bovine milk exosomes can enhance the miRNA resistance under gastrointestinal digestion and transport to the human colon, and at least some of them transferred to the bloodstream [10–12], where they may affect gene expression in humans [13,14]. In other words, milk miRNAs might have bioactive effects in humans, although there are many obstacles and challenges to reach the target tissues [15].

The production of miRNAs depends on numerous factors, both within the organism [16] and external in the environment [17–19]. For example, miRNA expression differs with milk fraction (fat, whey, and epithelial cells), reflecting differences in several metabolic pathways [20,21]. Characteristics of dairy production systems also influence miRNA profiles and therefore the functional properties of bovine milk [22,23]. The sensitivity of miRNA levels to numerous aspects of animal physiology and farm conditions, coupled with their strong resistance to adverse conditions, including temperature variation, RNase, low pH [24], and even pasteurization [25], and the fact that they can be sampled in a non or minimally invasive manner, make them ideal biomarkers [26].

Milk miRNAs vary their levels based on diet and production system. Many studies on miRNAs used as biomarkers have focused on raw milk. However, milk is usually submitted to technological processes before human consumption such as pasteurization, fermentation, or many others. Whether miRNA levels can serve as biomarkers for dairy products remains unclear. Therefore, in the present study, we compared the levels of seven miRNAs in raw milk with the levels in milk subjected to microwave heating, fermentation, or ripening. The concentration of the studied miRNAs in these products was also estimated, for further assay to assess their potential bio-functionality.

2. Materials and Methods

2.1. Milk Sample Collection and Treatments

Raw tank milk was sampled on 10 dairy farms in Asturias during June and July 2021. The selected farms are included in different production systems (Table S1).

The 10 samples were transported to the laboratory at 4 °C and then processed the following day. To generate control samples (n = 10), 2 mL of QIAzol lysis reagent was added to 1 g of raw milk, and then samples were mixed and stored at −80 °C. To obtain microwaved samples (n = 10), 50 mL of raw milk was heated in a 700-W microwave oven for 1 min, and then allowed to cool to room temperature. An aliquot of this milk (1 g) was transferred to a new RNase-free tube, 2 mL of QIAzol lysis reagent was added, and samples were mixed and stored at −80 °C.

To obtain yogurt samples (n = 10), 200 mL of raw milk was pasteurized in a thermostatic bath at 85 °C for 30 min, allowed to cool to 42 °C, and then inoculated with a commercial yogurt starter, which contains *Streptococcus thermophilus* and *Lactobacillus delbrueckii subsp. bulgaricus*, at the dose recommended by the manufacturer (50 units/250 L of milk). Once inoculated, aliquots (100 mL) were transferred into two containers, which were incubated in a water bath at 42 °C until the pH reached 4.5, which occurred after approximately 4 h. Then, an aliquot of yogurt (1 g) was transferred to a RNase-free tube, 2 mL of QIAzol lysis reagent was added, and samples were mixed and stored at −80 °C. The pH and product weight were monitored at all stages.

Cheese samples (n = 10) were prepared as described by Hynes et al. [27]. A total of 500 mL of raw milk was inoculated with 10 mL of the starter, prepared according to the dose recommended by the manufacturer (10 units/100 L of milk). The starter culture consisted of mesophilic strains *of Lactococcus lactis subsp. lactis* and *cremoris*. Calcium chloride (1 mL, 20% w/v, final concentration = 0.02%) was added, the mixture was homogenized by stirring, and then aliquots (200 mL) were transferred into two centrifuge flasks (250 mL volume) and incubated at 26–30 °C for 45 min in a water bath. Then, 65 μL of rennet (Nievi, Bizkaia, Spain; 1 × 10,000) was added, and the milk mixture was allowed to coagulate in a water bath at 30 °C until reaching the appropriate consistency after approximately 90 min. The curd was cut into 5 mm cubes using a sterile stainless steel knife, stirred for 20 min, and centrifuged at 220× g at room temperature for 10 min. The entire aqueous phase (whey) was removed, and then 35 mL of saturated brine (NaCl 330 g/L, pH 5.4) was added to curd and kept for 5 min. The mini cheeses were ripened at 10–12 °C in a ripening chamber for one week. After this time, 1 g of cheese was transferred to a Falcon tube, 2 mL of QIAzol lysis reagent was added, and samples were mixed and stored at −80 °C. As during yogurt manufacturing, pH and product weight were monitored at all stages.

2.2. Total RNA Extraction and Spike in

Prior to RNA extraction, raw milk and dairy product samples were spiked with defined concentrations of synthetic standard miRNAs. To measure the losses of miRNAs due to different milk processing, 6 fmol of cel-miR-238 (Norgen, Thorold, Canada) was added to the mixture of sample + QIAzol in the case of raw milk, microwaved milk, and yogurt. In the case of cheese, 54 fmol of cel-miR-238 was added to 1 g of cheese in 2 mL of QIAzol, based on our observation (from 10 one-week-old cheeses) that 9 g of milk was necessary to obtain 1 g of cheese. In addition, to compare the amounts of miRNAs between the different products, (1 g raw milk vs. 1 g microwaved milk vs. 1 g yogurt vs. 1 g cheese), 6 fmol of cel-miR-39 (Norgen, Thorold, Canada) was added to the mix (sample + QIAzol) of raw milk and cheese. The use of external synthetic reference miRNAs has been reported since the early work on circulating miRNAs [28].

Total RNA was extracted from aliquots (2 mL) of each mix (sample + QIAzol), which amounted to 40 samples, using the mirVana miRNA isolation kit according to the manufacturer's instructions. RNA was eluted with 100 μL of RNase-free water. RNA concentration and purity (ratio of absorbance at 260 to 280 nm) were assessed using a Nano-Drop spectrophotometer.

2.3. Real Time–Quantitative PCR (RT-qPCR)

Total RNA was used for complementary DNA (cDNA) synthesis using the TaqMan Advanced miRNA cDNA Synthesis Kit, and the resulting cDNA was stored at −20 °C until use. Seven miRNAs were chosen due to their expression level in milk estimated from previous sequencing results [23]. Three miRNAs with high expression levels (more than 190,000 rpm) were chosen: bta-mir-148a, bta-mir-30a5p, and bta-mir-21a5p. Three miRNA with low expression levels (between 150 and 500 rpm) were chosen: bta-mir-451, bta-mir-29b and bta-mir-215. Finally, one miRNA with a limited expression level was chosen: bta-mir-7863. The mature sequence of the miRNA used is shown in Table 1. The levels of these seven miRNAs were determined by RT-qPCR in a StepOne thermocycler The final reaction solution contained 10 μL of 2× TaqMan Fast Advanced Master Mix, 1 μL of 20× TaqMan Advanced miRNA Assay, 4 μL of RNase free water, and 5 μL of 1:10 diluted cDNA. The thermocycler program was set at 95 °C for 20 s, followed by 40 cycles at 95 °C for 1 s, and 60 °C for 20 s. All RT-qPCR reactions were performed in duplicate, and the results were averaged only when the duplicates differed within the 0.5 threshold cycle. To assess miRNA losses due to different milk manipulations, miRNA levels were normalized to those of cel-miR-238, while cel-miR-39 was used to compare the concentration of raw milk and cheese. Then, miRNA levels were estimated using qbase+ 3.1 software and expressed using the $\triangle\triangle$Ct method in base log^2 [29].

Table 1. Mature sequence of miRNA used.

MiRNA	Mature Sequence
bta-mir-148a	UCAGUGCACUACAGAACUUUGU
bta-mir-21-5p	UAGCUUAUCAGACUGAUGUUGA
bta-mir-215	AUGACCUAUGAAUUGACAGACA
bta-mir-29b	UAGCACCAUUUGAAAUCAGUGUU
bta-mir-30a-5p	UGUAAACAUCCUCGACUGGAAGCU
bta-mir-451	AAACCGUUACCAUUACUGAGUU
bta-mir-7863	AUGGACUGUCACCUGAGGAGC
cel-mir-238	UUUGUACUCCGAUGCCAUUCAGA
cel-mir-39	UCACCGGGUGUAAAUCAGCUUG

2.4. Prediction of miRNA Structure

Predictions on the secondary structure of selected bovine miRNAs were obtained by using an online application (http://rna.urmc.rochester.edu/RNAstructure.html, accessed on 28 September 2022), using default input conditions. The structures with the lowest free energy of formation (ΔG) were selected for each specific miRNA.

2.5. Statistical Analyses

Data were expressed using mean ± standard deviation (SD). Because sample sizes were small and some data showed a skewed distribution based on the Shapiro test, non-parametric statistical analysis was carried out. Pairwise comparisons of miRNA levels among raw milk, heated milk, yogurt, and cheese were performed using the Wilcoxon test for paired data. Significance was defined as $p \leq 0.05$. All analyses were performed using IBM SPSS Statistics for Windows version 22.0.

3. Results

3.1. Validation of Milk Treatments

As expected, in yogurt fermentation, a reduction in pH, from 6.63 ± 0.19 to 4.53 ± 0.14, was observed after 4 h of the starter culture addition (Figure S1).

During cheese manufacturing, pH was measured at five timepoints: raw milk, immediately after starter addition, after coagulation, as well as before and after ripening. The pH decreased significantly from the moment the ferment was inoculated. The pH decreased strongly after seven days of ripening. Overall, pH fell from 6.23 to 4.61 (Figure S2), confirming lactic fermentation.

When 206 g of raw milk was used to prepare cheese, the average weight of the fresh cheese, after the removal of whey, was 60.13 ± 7.63 g, and the ripened cheese (after 7 days ripening) weighed 22.26 ± 4.07 g. This means that coagulation and whey removal reduced the weight by approximately 69%, and moisture decrease during ripening reduced weight by an additional 18%. Altogether, cheese yield averaged 11%.

3.2. miRNA Losses Due to Milk Processing

Relative levels of bta-miR-148a, bta-miR-21-5p, bta-miR-215, bta-miR-29b, bta-miR-30a-5p, bta-miR-451, and bta-miR-7863, normalized to the levels of spiked cel-miR-238, were compared across raw milk, microwaved milk, yogurt, and ripened cheese (Table 2) to assess the losses of miRNAs after the different treatments. All treatments decreased the levels of all seven miRNAs: approximately a 31% decrease by microwave treatment and yogurt fermentation and approximately a 43% decrease during cheese production. However, not all miRNAs were affected in the same way (Table 3, Figure 1). The reductions after microwave heating varied from 17.20% for bta-miR-30a-5p to 39.42% for bta-miR-451; after yogurt fermentation, from 21.45% for bta-miR-21-5p to 41.62% for bta-miR-451; and after cheese production, from 32.73% for bta-miR-30a-5p to 56.32% for bta-miR-215. The reductions were significant for all seven miRNAs in the case of yogurt and cheese, and only for four of seven miRNAs in the case of microwaving (bta-miR-148a, bta-miR-21-5p, bta-miR-215, and bta-miR-451).

Table 2. Relative levels of the seven miRNAs of interest in raw milk, microwave-treated milk, yogurt, and ripen cheese *.

MiRNA	Raw Milk		Microwave		Yogurt		Cheese	
	Mean	SD	Mean	SD	Mean	SD	Mean	SD
bta-miR-148a	4.62	0.72	3.44	0.87	3.04	0.16	2.17	0.63
bta-miR-21-5p	4.04	0.62	2.86	1.08	3.18	0.23	2.97	0.33
bta-miR-215	3.90	0.85	2.60	1.24	2.86	0.28	1.70	0.38
bta-miR-29b	4.20	0.34	2.98	1.64	2.63	0.73	2.68	0.83
bta-miR-30a-5p	3.61	0.91	2.99	0.94	2.49	0.15	2.43	0.35
bta-miR-451	5.24	0.42	3.17	1.99	3.06	0.28	2.46	0.40
bta-miR-7863	4.28	0.39	2.33	1.75	3.21	0.18	2.39	0.36

SD, standard deviation, * Levels were normalized to those of spiked cel-miR-238 (see Section 2.3).

Table 3. Decrease in miRNA levels after treatment of raw milk *.

miRNA	Microwaved Milk		Yogurt		Cheese	
	Decrease (%)	*p*-Value	Decrease (%)	*p*-Value	Decrease (%)	*p*-Value
bta-miR-148a	25.53	0.043	34.22	0.005	53.15	0.005
bta-miR-21-5p	29.18	0.012	21.45	0.013	26.51	0.005
bta-miR-215	33.20	0.012	26.58	0.013	56.32	0.005
bta-miR-29b	29.07	0.080	37.46	0.028	36.15	0.028
bta-miR-30a-5p	17.20	0.063	31.04	0.013	32.73	0.013
bta-miR-451	39.42	0.018	41.62	0.005	53.09	0.005
bta-miR-7863	45.45	0.123	24.84	0.005	44.01	0.005
Mean	31.29		31.03		43.14	

Decreases are expressed as the percentage in raw milk. Levels of miRNA were normalized to those of spiked cel-miR-238, * *p*-value vs. raw milk (Wilcoxon test).

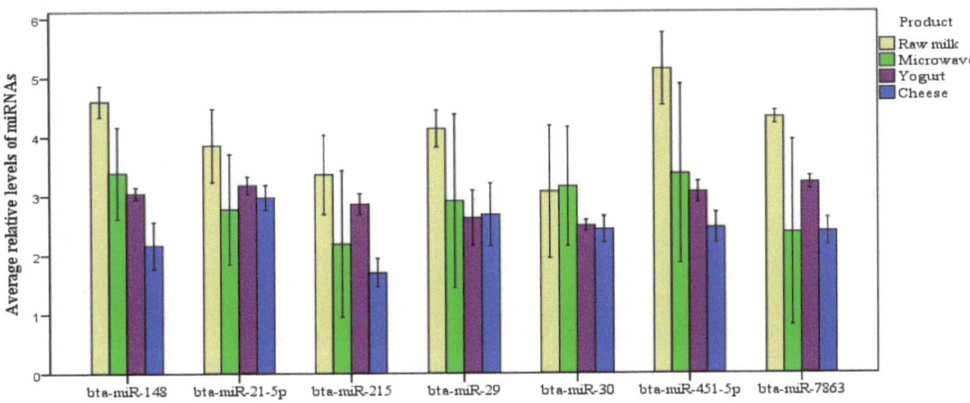

Figure 1. Loss of miRNA content in raw milk after microwave heating, yogurt fermentation, and cheese manufacture. Average relative levels of bta-miR-148a, bta-miR-21-5p, bta-miR-215, bta-miR-29b, bta-miR-30a-5p, bta-miR-451, and bta-miR-7863 in raw milk (n = 10), microwave-heated milk (n = 10), yogurt (n = 10), and cheese (n = 10). Levels were normalized to those of spiked cel-miR-238. The bar chart shows the average miRNA level for each product and the standard error bars.

3.3. miRNA Concentrations in Milk and Cheese

Relative levels of bta-miR-148a, bta-miR-21-5p, bta-miR-215, bta-miR-29b, bta-miR-30a-5p, bta-miR-451, and bta-miR-7863, normalized to the levels of spiked cel-miR-39, were compared across raw milk and ripened cheese to assess the concentration of miRNAs in raw milk and ripened cheese. The differences in levels between the same amount of raw milk and cheese were not significant for the studied miRNAs, which might indicate that their concentrations were similar between raw milk and cheese (Table 4).

Table 4. Relative abundance levels of seven miRNAs in raw milk and cheese *.

MiRNA	Raw Milk		Cheese		*p*-Value (Wilcoxon Test)
	Mean	SD	Mean	SD	
bta-miR-148a	2.32	1.35	2.10	0.30	0.76
bta-miR-21-5p	2.30	2.01	2.97	1.64	0.33
bta-miR-215	1.68	2.06	1.60	0.17	>0.99
bta-miR-29b	1.72	2.26	3.16	1.63	0.11
bta-miR-30a-5p	1.77	1.95	2.89	0.33	0.13
bta-miR-451	2.24	2.58	1.50	0.11	0.31
bta-miR-7863	1.99	1.46	2.43	1.21	0.40

* Levels were normalized to those of spiked cel-miR-39.

3.4. Determination of Secondary Structure of miRNA

The secondary structure of each miRNA was predicted using a bioinformatic tool. We have selected the structures with the lowest free energy of formation (Figure 2). The negative values of ΔG were predicted for bta-miR-29b and bta-miR-30a-5p, indicating that the formation of the secondary structure is most suitable. By contrast, the largest section of the stem-loop structure was in bta-miR-21-5p followed by bta-miR-451.

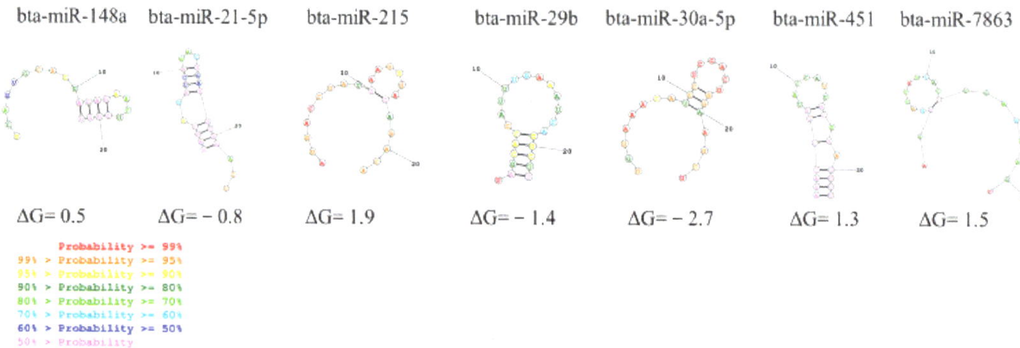

Figure 2. Predicted secondary structures of bovine miRNAs. The motifs with the lowest predicted free energy of formation (ΔG) were selected as the most likely structures.

4. Discussion

Here, we evaluated the effects of different cow's milk treatments on miRNA content in order to guide efforts to define biomarkers for assessing the quality or provenance of dairy products destined for human consumption and also to evaluate the contribution of different dairy products in miRNAs, considering them bioactive compounds [30]. Our results confirmed the presence of miRNAs in the studied dairy products and suggest that thermal treatment of raw milk by the microwave as well as yogurt or cheese production can substantially reduce miRNA levels, indicating that the levels of potential biomarker miRNAs in raw milk cannot be necessarily extrapolated to dairy products derived from that milk.

Microwave heating has been shown to affect certain physical and chemical characteristics of milk [31] and to damage DNA [32], which led us to hypothesize that this treatment could affect miRNAs. Indeed, we found that microwave treatment of raw milk significantly decreased the amounts of bta-miR-148a, bta-miR-215, bta-miR-21-5p, and bta-miR-451. A previous study also described a significant decrease in bta-miR-21-5p but not bta-miR-29b [33]. We did not observe a significant decrease in bta-miR-29b, yet other work reported a 40% loss in miR-29b [34]. This discrepancy may be due to the storage of raw milk prior to treatment. Howard et al. [34] studied the stability of miRNAs in milk after 15 days of cold storage, being after that heated by microwave (n = 3). No significant difference was seen in miR-29b after 15 days of cold storage, but the significant difference appeared after microwave heating [34]. This family of milk miRNA has also been reported as sensitive to high pressure processing (HPP) [35].

Uneven stability of individual miRNAs to milk processing methods has been previously pointed out [35]. The fact that milk treatment significantly affects some miRNAs and not others may depend on the different fractions of milk where the miRNAs are found [21]. Intracellular miRNAs are less stable than extracellular miRNAs within exosomes, microvesicles, apoptotic bodies, high-density lipoproteins, or protein complexes [36]. In the case of milk fat, the fat globule membrane appears to be more resistant to gastrointestinal enzymes [37] and microwave heating [38] than other milk components.

In contrast to microwave heating, milk pasteurization has been reported not to significantly affect the miRNA content of fat or milk cells [25] or extracellular vesicles in milk [39].

Although the homogenization process can cause a significant loss of miRNAs [34], the secondary structure might influence each miRNA stability.

As cheese and yogurt can be presented in different forms on the market, a simple model of these treatments was elaborated, representing the general characteristics of each product. In contrast to the effect of microwaving, fermentation of previously pasteurized milk to make yogurt significantly reduced levels of many miRNAs [34,40]. The reduction in miRNAs during milk processing is not surprising given the range of changes that occur during milk fermentation to make yogurt and cheese.

In yogurt production, the starter culture can lower the pH below 4.6, leading to aggregation of caseins [41]. Lactose is converted into lactic acid and several amino acids and fatty acids (especially stearic and oleic) are released into yogurt. During bacterial fermentation, vitamin B content increases and minerals are converted into an ionic form [42]. This study revealed that fermentation of previously pasteurized milk to make yogurt significantly reduced the levels of many miRNAs, with a loss of up to 41.62% in some cases.

We suspect that much of the miRNA loss in our study can be attributed to degradation of exosomes, as others have proposed [34]. In one study, fermentation was found to reduce by 90% the protein content of milk exosomes; assuming that, under the effect of bacterial proteases, exosomes can be altered, consequently the miRNAs contained in these exosomes can be easily degraded [40]. Assuming that pasteurization does not significantly reduce the miRNA content [25], most of the reduction in the miRNA content could be attributed to exosome degradation, as described before.

In our study, we prepared cheese following the modified protocol of Hynes et al. [27], obtaining similar yields to those reported in the original work. We observed a more acidic pH on day 7 (4.61) than in that work (5.21), perhaps because we used raw milk for cheese making, so the natural bacteria in the milk could also metabolize lactose. Pasteurization, in contrast, destroys most bacteria, limiting the acidification [43]. The pH can also vary depending on the starter culture used [44].

As it was expected, the greatest losses of miRNAs occurred during cheese manufacturing, given the losses due to fermentation (as in yogurt) but also because of the removal of whey, which is known to contain a wide variety of miRNAs [20,21].

However, we found that concentrations in the final product of the seven miRNAs (1 g of milk vs. 1 g of cheese, spiked by cel-miR-39) did not differ significantly between raw milk and cheese (Table 4). Similarly, two studies reported even higher miRNA concentrations in two types of cheese (camembert and Fresco queso dip) than in raw milk [34,45]. We suspect that fermentation and whey drainage may reduce absolute miRNA levels, but that the subsequent water loss during ripening increases their concentrations.

Apart from the milk fraction where the miRNAs are found, the differences in miRNA losses in the studied processes could also be attributed to the secondary structure of the individual miRNAs. However, although the stem-loop structure predicted for both *bta*-miR-21-5p and bta-miR-451 could indicate higher stability than the loop structure presented in other miRNA as, for example, bta-miR-148a, bta-miR-215, bta-miR-30a-5p, and 7863 (Figure 2), no association was found between the predicted secondary structure and the decrease in miRNA levels. Although the secondary RNA structure could be relevant in miRNA resistance to degradation, there are other structure aspects that need to be considered, including the content in GC and the specific sequence [46,47]. However, it has been proposed that beyond GC content, changes in sequence, structure, and putative RNase A substrate motifs can impact the stability of dietary small RNAs [47], suggesting, overall, that the stability of dietary miRNAs should be experimentally validated one by one. No differences were also found due to the putative level of expression in raw milk (Tables 2 and 3).

Most of the dairy products with quality labels are processed agro-food products. Previous works have pointed to miRNAs as possible traceability biomarkers in raw cow's milk because of their sensitivity to farm conditions [22,23] or cow breed [48,49]. Unfortunately, technological processes reduce the amount of miRNAs unevenly, complicating the process of identifying traceability biomarkers for the different dairy products, since not all technological processes reduce miRNAs equally. Further analysis, with larger sampling, should be designed to compare dairy products made from milk produced under different farm managements or breed, for example.

Two milk miRNA we used in this study, bta-mir-21-5p and bta-mir-30a-5p, have been demonstrated to increase their plasma concentration in humans after bovine milk consumption [50], although it remains unknown whether this miRNA concentration might be sufficient to produce gene modulation in the consumer [12]. Since technological processes reduce the amount of miRNAs (Table 3), we can assert that more research is needed on the functionality and bioavailability of miRNAs in each of the different dairy products.

5. Conclusions

We confirmed that bovine milk contains several miRNAs even after microwaving, pasteurization followed by fermentation to make yogurt, and cheese manufacturing. We showed that these treatments decreased all miRNA levels.

Our results clearly argue for caution in efforts to identify miRNAs that may be useful biomarkers; they may need to be assessed in the final products for human consumption, and not merely extrapolated from assays of the raw milk from which they are produced.

Finally, considering miRNAs as bioactive components in milk and dairy products, raw milk and cheese supposedly may provide similar concentrations of miRNAs, higher than those of yogurt and microwaved milk, although the miRNA profile may differ between these two products. Additional studies are needed to explore the complete profiles and availability of miRNAs in dairy products and, subsequently, their putative functionality in human cells.

Supplementary Materials: The following supporting information can be downloaded at: https://www.mdpi.com/article/10.3390/foods12152950/s1, Table S1: Characteristics of the sampled farms; Figure S1: Mean pH and standard deviation for milk and yogurt; Figure S2: Evolution of pH in milk, during cheese manufacture and ripening. * $p < 0.05$, ** $p < 0.01$, *** $p < 0.001$.

Author Contributions: Conceptualization, L.J.R. and L.A.e.q.; methodology, L.J.R., L.A.e.q., B.M., A.R., A.D., M.-C.L.d.l.H. and M.M.M.; software, L.J.R., L.A.e.q., A.D. and M.-C.L.d.l.H.; validation, L.J.R. and L.A.e.q.; formal analysis, L.J.R., L.A.e.q., A.D. and M.-C.L.d.l.H.; investigation, L.J.R. and L.A.e.q.; resources, L.J.R., L.A.e.q., B.M., A.R. and A.D.; writing—original draft preparation, L.J.R. and L.A.e.q.; writing—review and editing, L.J.R., L.A.e.q., B.M., A.R., A.D., M.-C.L.d.l.H. and M.M.M.; visualization, L.J.R. and L.A.e.q.; supervision, L.J.R.; project administration, L.J.R.; funding acquisition, L.J.R. All authors have read and agreed to the published version of the manuscript.

Funding: This research was funded by Principado de Asturias Regional Government (project IDI/2021/000102), co-financed by the European Union through the ERDF (European Regional Development Fund) and Ministry of Science and Innovation (AEI) project PID2021-126010OR-I00. Loubna Abou el qassim was funded by FICYT Severo Ochoa Grant (BP17-49).

Data Availability Statement: The data that support the findings of this study are available from the corresponding author upon reasonable request.

Acknowledgments: We thank all the farmers in the Asturias region who contributed to this work by providing milk samples.

Conflicts of Interest: The authors declare no conflict of interest.

References

1. He, L.; Hannon, G.J. MicroRNAs: Small rnas with a big role in gene regulation. *Nat. Rev. Genet.* **2004**, *5*, 522–531.
2. Gebert, L.F.R.; MacRae, I.J. Regulation of microrna function in animals. *Nat. Rev. Mol. Cell Biol.* **2019**, *20*, 21–37. [CrossRef] [PubMed]
3. Chen, X.; Liang, H.; Zhang, J.; Zen, K.; Zhang, C.Y. Horizontal transfer of micrornas: Molecular mechanisms and clinical applications. *Protein Cell* **2012**, *3*, 28–37. [CrossRef] [PubMed]
4. Weber, J.A.; Baxter, D.H.; Zhang, S.; Huang, D.Y.; Huang, K.H.; Lee, M.J.; Galas, D.J.; Wang, K. The MicroRNA spectrum in 12 body fluids. *Clin. Chem.* **2010**, *56*, 1733–1741. [CrossRef] [PubMed]

5. Carrillo-Lozano, E.; Sebastián-Valles, F.; Knott-Torcal, C. Circulating microRNAs in breast milk and their potential impact on the infant. *Nutrients* **2020**, *12*, 3066. [CrossRef]
6. Alsaweed, M.; Lai, C.T.; Hartmann, P.E.; Geddes, D.T.; Kakulas, F. Human Milk MiRNAs Primarily originate from the mammary gland resulting in unique mirna profiles of fractionated milk. *Sci. Rep.* **2016**, *6*, 20680. [CrossRef]
7. Hata, T.; Murakami, K.; Nakatani, H.; Yamamoto, Y.; Matsuda, T.; Aoki, N. Isolation of bovine milk-derived microvesicles carrying mRNAs and microRNAs. *Biochem. Biophys. Res. Commun.* **2010**, *396*, 528–533. [CrossRef]
8. Munch, E.M.; Harris, R.A.; Mohammad, M.; Benham, A.L.; Pejerrey, S.M.; Showalter, L.; Hu, M.; Shope, C.D.; Maningat, P.D.; Gunaratne, P.H.; et al. Transcriptome profiling of microrna by next-gen deep sequencing reveals known and novel miRNA species in the lipid fraction of human breast milk. *PLoS ONE* **2013**, *8*, e50564. [CrossRef]
9. Izumi, H.; Kosaka, N.; Shimizu, T.; Sekine, K.; Ochiya, T.; Takase, M. Time-dependent expression profiles of microRNAs and mRNAs in rat milk Whey. *PLoS ONE* **2014**, *9*, e88843. [CrossRef]
10. Wolf, T.; Baier, S.R.; Zempleni, J. The Intestinal Transport of Bovine Milk Exosomes is mediated by endocytosis in human colon carcinoma caco-2 cells and rat small intestinal iec-6 cells1-3. *J. Nutr.* **2015**, *145*, 2201–2206. [CrossRef]
11. Baier, S.R.; Nguyen, C.; Xie, F.; Wood, J.R.; Zempleni, J. MicroRNAs are absorbed in biologically meaningful amounts from nutritionally relevant doses of cow milk and affect gene expression in peripheral blood mononuclear cells, hek-293 kidney cell cultures, and mouse livers. *J. Nutr.* **2014**, *144*, 1495–1500. [CrossRef] [PubMed]
12. López de Las Hazas, M.C.; Del Pozo-Acebo, L.; Hansen, M.S.; Gil-Zamorano, J.; Mantilla-Escalante, D.C.; Gómez-Coronado, D.; Marín, F.; Garcia-Ruiz, A.; Rasmussen, J.T.; Dávalos, A. Dietary Bovine Milk MiRNAs transported in extracellular vesicles are partially stable during gi digestion, are bioavailable and reach target tissues but need a minimum dose to impact on gene expression. *Eur. J. Nutr.* **2022**, *61*, 1043–1056. [CrossRef]
13. Zempleni, J.; Baier, S.R.; Howard, K.M.; Cui, J. Gene regulation by dietary microRNAs. *Can. J. Physiol. Pharmacol.* **2015**, *93*, 1097–1102. [CrossRef] [PubMed]
14. Del Pozo-Acebo, L.; de las Hazas, M.C.L.; Tomé-Carneiro, J.; Gil-Cabrerizo, P.; San-Cristobal, R.; Busto, R.; García-Ruiz, A.; Dávalos, A. Bovine milk-derived exosomes as a drug delivery vehicle for mirna-based therapy. *Int. J. Mol. Sci.* **2021**, *22*, 1105. [CrossRef] [PubMed]
15. Tomé-Carneiro, J.; Fernández-Alonso, N.; Tomás-Zapico, C.; Visioli, F.; Iglesias-Gutierrez, E.; Dávalos, A. Breast milk microRNASs harsh journey towards potential effects in infant development and maturation. Lipid encapsulation can help. *Pharmacol. Res.* **2018**, *132*, 21–32. [CrossRef]
16. Cammaerts, S.; Strazisar, M.; De Rijk, P.; Del Favero, J. Genetic variants in microRNA genes: Impact on microRNA expression, function, and disease. *Front. Genet.* **2015**, *6*, 186. [CrossRef]
17. Colitti, M.; Sgorlon, S.; Licastro, D.; Stefanon, B. Differential expression of miRNAs in milk exosomes of cows subjected to group relocation. *Res. Vet. Sci.* **2019**, *122*, 148–155. [CrossRef]
18. Li, R.; Beaudoin, F.; Ammah, A.A.; Bissonnette, N.; Benchaar, C.; Zhao, X.; Lei, C.; Ibeagha-Awemu, E.M. Deep sequencing shows microRNA involvement in bovine mammary gland adaptation to diets supplemented with linseed oil or safflower oil. *BMC Genom.* **2015**, *16*, 884. [CrossRef]
19. Muroya, S.; Shibata, M.; Hayashi, M.; Oe, M.; Ojima, K. Differences in circulating microRNAs between grazing and grain-fed wagyu cattle are associated with altered expression of intramuscular microRNA, the potential target PTEN, and lipogenic genes. *PLoS ONE* **2016**, *11*, e0162496. [CrossRef]
20. Alsaweed, M.; Hepworth, A.R.; Lefèvre, C.; Hartmann, P.E.; Geddes, D.T.; Hassiotou, F. Human Milk microRNA and total RNA differ depending on milk fractionation. *J. Cell. Biochem.* **2015**, *116*, 2397–2407. [CrossRef]
21. Li, R.; Dudemaine, P.L.; Zhao, X.; Lei, C.; Ibeagha-Awemu, E.M. Comparative analysis of the miRNome of bovine milk fat, whey and cells. *PLoS ONE* **2016**, *11*, e0154129. [CrossRef]
22. Abou El Qassim, L.; Le Guillou, S.; Royo, L.J. Variation of miRNA content in cow raw milk depending on the dairy production system. *Int. J. Mol. Sci.* **2022**, *23*, 11681. [CrossRef] [PubMed]
23. Abou El Qassim, L.; Alonso, J.; Zhao, K.; Guillou, S.L.; Diez, J.; Vicente, F. Differences in the microRNAs levels of raw milk from dairy cattle raised under extensive or intensive production systems. *Vet. Sci.* **2022**, *9*, 661. [CrossRef]
24. Izumi, H.; Kosaka, N.; Shimizu, T.; Sekine, K.; Ochiya, T.; Takase, M. Bovine milk contains microRNA and messenger RNA that are stable under degradative conditions. *J. Dairy Sci.* **2012**, *95*, 4831–4841. [CrossRef]
25. Abou El Qassim, L.; Royo, L.J. The effect of pasteurization in the expression of bovine milk microRNA. In *Book of Abstracts of the 72nd Annual Meeting of the European Federation of Animal Science, Davos, Switzerland, 30 August–3 September 2021*; European Federation of Animal Science: Rome, Italy, 2021; Volume 27, p. 637.
26. Buschmann, D.; Haberberger, A.; Kirchner, B.; Spornraft, M.; Riedmaier, I.; Schelling, G.; Pfaffl, M.W. Toward reliable biomarker signatures in the age of liquid biopsies—How to standardize the small RNA-seq workflow. *Nucleic Acids Res.* **2016**, *44*, 5995–6018. [CrossRef] [PubMed]
27. Hynes, E.; Ogier, J.C.; Delacroix-Buchet, A. Protocol for the Manufacture of miniature washed-curd cheeses under controlled microbiological conditions. *Int. Dairy J.* **2000**, *10*, 733–737. [CrossRef]
28. Mitchell, P.S.; Parkin, R.K.; Kroh, E.M.; Fritz, B.R.; Wyman, S.K.; Pogosova-Agadjanyan, E.L.; Peterson, A.; Noteboom, J.; O'Briant, K.C.; Allen, A.; et al. Circulating microRNAs as stable blood-based markers for cancer detection. *Proc. Natl. Acad. Sci. USA* **2008**, *105*, 10513–10518. [CrossRef]

29. Hellemans, J.; Mortier, G.; De Paepe, A.; Speleman, F.; Vandesompele, J. qBase relative quantification framework and software for management and automated analysis of real-time quantitative PCR data. *Genome Biol.* **2008**, *8*, R19. [CrossRef]
30. Rani, P.; Yenuganti, V.R.; Shandilya, S.; Onteru, S.K.; Singh, D. MiRNAs: The hidden bioactive component of milk. *Trends Food Sci. Technol.* **2017**, *65*, 94–102. [CrossRef]
31. Musto, M.; Faraone, D.; Cellini, F.; Musto, E. Changes of DNA quality and meat physicochemical properties in bovine supraspinatus muscle during microwave heating. *J. Sci. Food Agric.* **2014**, *94*, 785–791. [CrossRef]
32. Dehghan, A.; Jamalian, J.; Farahnaky, A.; Mesbahi, G.; Moosavi-Nasab, M. The effect of microwave pasteurization on some physical and chemical characteristics of milk. *Int. J. Food Eng.* **2012**, *8*, 1–12. [CrossRef]
33. Zhao, Z.; Yu, S.; Xu, M.; Li, P. Effects of microwave on extracellular vesicles and microRNA in milk. *J. Dairy Sci.* **2018**, *101*, 2932–2940. [CrossRef] [PubMed]
34. Howard, K.M.; Jati Kusuma, R.; Baier, S.R.; Friemel, T.; Markham, L.; Vanamala, J.; Zempleni, J. Loss of miRNAs during processing and storage of cow's (Bos taurus) Milk. *J. Agric. Food Chem.* **2015**, *63*, 588–592. [CrossRef] [PubMed]
35. Smyczynska, U.; Bartlomiejczyk, M.A.; Stanczak, M.M.; Sztromwasser, P.; Wesolowska, A.; Barbarska, O.; Pawlikowska, E.; Fendler, W. Impact of processing method on donated human breast milk microRNA content. *PLoS ONE* **2020**, *15*, e0236126. [CrossRef]
36. Sohel, M.H. Extracellular/Circulating microRNAs: Release mechanisms, functions and challenges. Achiev. *Life Sci.* **2016**, *10*, 175–186. [CrossRef]
37. Le, T.T.; Van de Wiele, T.; Do, T.N.H.; Debyser, G.; Struijs, K.; Devreese, B.; Dewettinck, K.; Van Camp, J. Stability of milk fat globule membrane proteins toward human enzymatic gastrointestinal digestion. *J. Dairy Sci.* **2012**, *95*, 2307–2318. [CrossRef]
38. Rodríguez-Alcalá, L.M.; Alonso, L.; Fontecha, J. Stability of fatty acid composition after thermal, high pressure, and microwave processing of cow milk as affected by polyunsaturated fatty acid concentration. *J. Dairy Sci.* **2014**, *97*, 7307–7315. [CrossRef]
39. Hansen, M.S.; Gregersen, S.B.; Rasmussen, J.T. Bovine milk processing impacts characteristics of extracellular vesicle isolates obtained by size-exclusion chromatography. *Int. Dairy J.* **2022**, *127*, 105212. [CrossRef]
40. Yu, S.; Zhao, Z.; Sun, L.; Li, P. Fermentation results in quantitative changes in milk-derived exosomes and different effects on cell growth and survival. *J. Agric. Food Chem.* **2017**, *65*, 1220–1228. [CrossRef]
41. Settachaimongkon, S.; Nout, M.J.R.; Antunes Fernandes, E.C.; Hettinga, K.A.; Vervoort, J.M.; van Hooijdonk, T.C.M.; Zwietering, M.H.; Smid, E.J.; Van Valenberg, H.J.F. Influence of different proteolytic strains of Streptococcus thermophilus in co-culture with Lactobacillus delbrueckii subsp. bulgaricus on the metabolite profile of set-yoghurt. *Int. J. Food Microbiol.* **2014**, *177*, 29–36. [CrossRef]
42. Sfakianakis, P.; Tzia, C. Conventional and innovative processing of milk for yogurt manufacture; development of texture and flavor: A review. *Foods* **2014**, *3*, 176–193. [CrossRef] [PubMed]
43. Dos Santos, S.; Ressutte, J.; Bánkuti, S.; Bánkuti, F.; Pozza, M.; Madrona, G.S. Características tecnológicas, de qualidade e potencialidades da cadeia produtiva de queijo colonial na região sul do brasil: Uma Revisão. *FTT J. Eng. Bus.* **2017**, *1*, 50–64.
44. Ressutte, J.B.; Rodrigues, T.S.; dos Santos Pozza, M.S.; Madrona, G.S. Application of Lactococcus Lactis Subsp. Lactis and Cremoris as starter culture in the colonial cheese production. *J. Agric. Stud.* **2020**, *8*, 561. [CrossRef]
45. Oh, S.; Park, M.R.; Ryu, S.; Maburutse, B.E.; Kim, J.U.; Kim, Y. Quantitative analysis of milk-derived microRNAs and microbiota during the manufacturing and ripening of soft cheese. *J. Microbiol. Biotechnol.* **2017**, *27*, 1566–1575. [CrossRef] [PubMed]
46. Zhou, Z.; Li, X.; Liu, J.; Dong, L.; Chen, Q.; Liu, J.; Kong, H.; Zhang, Q.; Qi, X.; Hou, D.; et al. Honeysuckle-encoded atypical microRNA2911 directly targets influenza A viruses. *Cell Res.* **2015**, *25*, 39–49. [CrossRef]
47. Yang, J.; Elbaz-Younes, I.; Primo, C.; Murungi, D.; Hirschi, K.D. Intestinal permeability, digestive stability and oral bioavailability of dietary small RNAs. *Sci. Rep.* **2018**, *8*, 10253. [CrossRef]
48. Billa, P.A.; Faulconnier, Y.; Ye, T.; Chervet, M.; LeProvost, F.; Pires, J.A.A.; Leroux, C. Deep RNA-Seq reveals miRNome differences in mammary tissue of lactating Holstein and Montbéliarde cows. *BMC Genom.* **2019**, *20*, 621. [CrossRef]
49. LeGuillou, S.; Leduc, A.; Laubier, J.; Barbey, S.; Rossignol, M.N.; Lefebvre, R.; Marthey, S.; Laloë, D.; LeProvost, F. Characterization of Holstein and Normande whole milk miRNomes highlights breed specificities. *Sci. Rep.* **2019**, *9*, 20345. [CrossRef]
50. Wang, L.; Sadri, M.; Giraud, D.; Zempleni, J. RNase H2-Dependent Polymerase Chain Reaction and elimination of confounders in sample collection, storage, and analysis strengthen evidence that microRNAs in bovine milk are bioavailable in humans. *J. Nutr.* **2018**, *148*, 153–159. [CrossRef]

foods

MDPI

Article

Physicochemical, Microbiological and Sensory Characteristics of White Brined Cheese Ripened and Preserved in Large-Capacity Stainless Steel Tanks

Theofilos Massouras [1,*], Evangelia Zoidou [1], Zinovia Baradaki [2] and Marianna Karela [1]

[1] Laboratory of Dairy Science, Department of Food Science and Human Nutrition, Agricultural University of Athens, Iera Odos 75, Votanikos, 11855 Athens, Greece
[2] Galaktokomiki Kritis Dairy S.A., Selia, 74053 Rethymno, Greece
* Correspondence: theomas@aua.gr; Tel.: +30-210-529-4675

Abstract: The objective of the present study was to investigate the effect of ripening and preservation containers on the physico-chemical, microbiological, and textural characteristics, and volatile profile of white cheese. White cheeses were manufactured on an industrial scale using large-capacity stainless steel tanks (SST) of 500 kg, and the respective control samples in tin containers (TC) of 17 kg. No significant differences ($p > 0.05$) in fat in dry matter and total protein content were observed at 60 days of ripening between the TC and SST cheeses. After 60 days, of ripening, the moisture of the cheeses in SST and TC did not show significant statistical differences ($p > 0.05$). No significant differences ($p > 0.05$) were observed between the TC and SST cheeses in the mineral concentration (Ca, Mg, K, and Na) and textural characteristics. Similar results of pH and bacterial counts, as well as absence of yeasts and molds, were observed during ripening and preservation time in both groups of cheeses. Furthermore, proteolysis was not affected statistically significantly ($p > 0.05$). A moderately increased rate of ripening for the cheeses in TC was observed up to 90 days but, at 180 days, proteolysis was similar in both groups of cheeses. Regarding the SFA, MUFA, and PUFA content, no significant differences ($p > 0.05$) were observed between the TC and SST cheeses. A total of 94 volatile compounds were identified in the volatile fraction of both the SST and TC cheeses. Organic acids and alcohols were the most abundant classes of volatile compounds that were identified. The flavor and texture scores in the TC and SST cheeses were similar ($p > 0.05$). Overall, the TC and SST cheeses did not show any significant statistical difference in any of the analyzed parameters.

Keywords: white cheese in brine; large-capacity stainless steel tanks; tin containers; physico-chemical characteristics

Citation: Massouras, T.; Zoidou, E.; Baradaki, Z.; Karela, M. Physicochemical, Microbiological and Sensory Characteristics of White Brined Cheese Ripened and Preserved in Large-Capacity Stainless Steel Tanks. *Foods* **2023**, *12*, 2332. https://doi.org/10.3390/foods12122332

Academic Editors: Michele Faccia and Giuseppe Natrella

Received: 23 April 2023
Revised: 7 June 2023
Accepted: 8 June 2023
Published: 9 June 2023

1. Introduction

Different types of cheese have been produced in wide-ranging textures, flavors, and forms, in different regions with unique cultures and environments. Different cheesemaking techniques have been developed over time in response to new technologies and consumer demands. Cheeses can be grouped or classified according to criteria such as the production method, animal milk used, ripening process, country of origin, texture, fat and moisture content, etc. However, key differences in cheese characteristics can generally be attributed to the origin of the milk, moisture content, variation in the container for ripening and preservation, as well as lengths of aging [1].

White brine cheeses (WBC) undergo a ripening process during which they develop microbiological and technological characteristics [1]. WBC are particularly popular in Balkan, Middle Eastern, and Mediterranean regions, and in North Africa and Eastern Europe. They include a large number of varieties, such as Feta (Greece), Telemea or Telemes (Romania, Greece), Akawi (Lebanon, Syria), Halloumi (Cyprus) etc., and are produced by different processing methods, so they have differences in their physicochemical, textural, and sensory properties [2]. Their manufacture dates back thousands of years (approximately 8000 years ago) [3].

WBC, in general, have a texture that varies from soft to semi-hard. These cheeses have no rind, a slightly acidic taste, due to the action of lactic acid bacteria during ripening, and a salty taste, which arises from storage in brine. Therefore, salt and acid are the critical parameters for the conservation of these types of cheese [4,5]. The main differences among the cheeses are observed in the manufacturing process (for example, milk type, coagulation time and temperature, pressure during draining, shape and size of the curd, and salting of the curd before brining). Sheep or goat milk, or their mixture, is commonly used in their production, however, due to the high purchasing demand, other types of milk, such as cow's or buffalo's, can be used [4]. Raw (according to legislation) or pasteurized (72 °C for 15 s or 63 °C for 30 min) milk cheese is used in WBC manufacture [6]. Thermophilic or mesophilic lactic acid bacteria, or combination of them, are commonly used as starter cultures. Non-starter lactic acid bacteria (NSLAB) are predominant when using unpasteurized milk in production.

WBC are white in color, except for those made from cow's milk that gives a yellowish color due to the presence of carotenoids. These cheeses have no gas holes but they sometimes develop small mechanical openings [3]. The texture of WBC is smooth, soft, and crumbly but still sliceable, and some of them may become brittle when old [6]. Their flavor is slightly sour to very salty and, for some varieties, mildly acidic and piquant. The shape varies but is usually produced in rectangular or cubic blocks that weigh 250–1000 g. The cheeses are packed in containers of various capacities. The most common are rectangular tin cans, lacquered metal or plastic containers with up to 15–16 kg capacity or, traditionally, wooden barrels of 40–50 kg capacity [3,6]. Most WBC are consumed fresh after ripening for 2 months or more in brine (8–10% NaCl concentration) [3,6]. Post-ripening, WBC are repacked in plastic bags under vacuum (without brine) or in plastic containers with brine [6].

Tin cans are usually used for the ripening and preservation of WBC, before their sub-packaging in smaller containers. However, in recent years, stainless steel tanks of large capacity have been used more and more, mainly in large cheese factories. More specifically, the use of tin containers (TC) was one of the most suitable for the ripening, packaging, and handling of brine cheeses due to their low cost, light weight per unit area, easy handling, and O_2 product protection. Over the years, TC have been partially replaced by other materials, such as plastic and stainless steel. The latter is resistant to corrosion and low temperatures, accepts electro-polishing, and offers the product protection from O_2, light, odors, and microorganisms. An advantage of the stainless steel tanks (SST) is their reusability and ease of cleaning, even with a CPI system. Thus, the use of SST containers is constantly increasing and mainly by cheese factories, where a large part of their production is intended for repackaging.

Although there are many research studies that refer to the characteristics of WBC matured and preserved in tin or wooden containers, no research to date provide data on the characteristics of cheeses that are ripened and preserved in large capacity stainless steel tanks (SST) (approximately 500 kg).

Considering the growing interest in the use of large capacity containers for ripening and preservation, this study aims to investigate the use of stainless steel tanks (SST), and their effect on the quality of WBC, comparing this to those ripened in tin containers (TC). The physico-chemical, microbiological, textural, and sensory characteristics of WBC, made from sheep's milk, were studied during their preservation over 180 days in SST and TC.

2. Materials and Methods

2.1. Cheese Manufacture

The production of the WBC was carried out on an industrial scale, using 5000 L of sheep's milk, according to the procedure described in Figure 1. Sheep's milk, with an average composition of 5.78 ± 1.06% fat, 4.87 ± 0.62% proteins, and 4.02 ± 0.55% lactose, was pasteurized in a plate heat exchanger at 72 °C for 10 sec and cooled down to 35 °C. Commercially available freeze-dried direct vat set (DVS) cultures, containing *Lactococcus lac-*

tis (25%), *Lactococcus lactis* subsp. *cremoris* (15%), and *Streptococcus thermophilus/Lactobacillus delbrueckii* subsp. *bulgaricus* (60%) (Chr. Hansen, Copenhagen, Denmark), were used as starter cultures. $CaCl_2$ solution (10% w/v), at a rate of 100 mL/100 kg milk, and powdered calf rennet (HA-LA, Hansen's Laboratorium, Copenhagen, Denmark) were then added to achieve coagulation in approximately 40 min. The cheese curd was cut into small cubes (2.0 cm^3), allowed to rest for 10 min and was transferred into rectangular plastic multi-molds (capacity of 2–2.5 kg drained curd). After draining (a day after) the curd was placed into two different ripening–preservation containers; TC (17 kg capacity) and SST (500 kg capacity), respectively, for 180 days. Preservation in TC was used as a control. Three individual replicates were carried out. It is important to mention that the brine that was used had a salt concentration of 7%, with a pH of 7, and a $CaCl_2$ concentration of 0.02%.

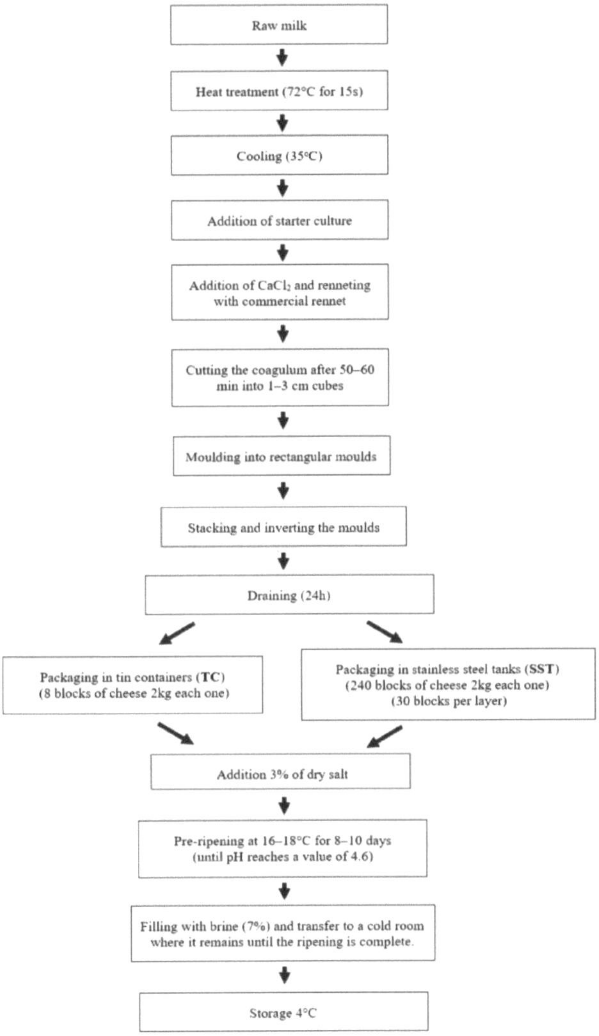

Figure 1. Flow chart for white brine cheeses (WBC) ripened and preserved in tin containers (TC) and stainless-steel tanks (SST).

2.2. Cheese and Brine Samples

WBC samples were taken from TC and SST containers and analyzed at 1, 10, 60, 90, and 180 days for physico-chemical composition, microbiological, textural, and sensory characteristics. To establish the interaction between the cheeses and the respective brines, the determination of major minerals composition (calcium, magnesium, potassium, and sodium), as well as the total protein and nitrogen fraction content, of the brines was carried out. All samples were analyzed in triplicate.

2.3. Analytical Methods

2.3.1. Physico-Chemical Analysis

The pH was measured using an electronic pH meter (Orion 720 A, Orion Pacific Pty Ltd., Frankston, Vic, Australia). Total solids content was determined by oven drying in a laboratory oven at 105 °C for 24 h, according to the method of I.D.F. [7]. Fat content was measured according to the volumetric method of Gerber [8]. Chloride content determination was performed by the potentiometric titration method according to the I.S.O. [9], and the ash content was determined by dry ashing the samples in a muffle furnace at 550 °C for 24 h, according to the method of I.D.F. [10]. The major minerals (Ca, Mg, K, and Na) concentrations of the samples (both cheeses and respective brines) was determined by Atomic Absorption Spectrometry on a Shimadzu AA-6800 atomic absorption spectrophotometer (Shimadzu AA-6800, Kyoto, Japan) equipped with the autosampler Shimadzu ASC-6100 and the software Wizard v.2.30., according to the method of I.D.F [11].

The extent of proteolysis of the cheeses during ripening was monitored by measuring the levels of total nitrogen (TN), soluble nitrogen (SN) fractions (i.e., water-soluble nitrogen) (WSN), and nitrogen soluble in 12% trichloroacetic acid (TCA-N) or in 5% phosphotungstic acid (PTA-N)), using the Kjeldahl nitrogen determination method [12]. Total protein was calculated by % total N × factor of 6.38. The RP-HPLC peptide profiles of SN, TCA–N, and PTA-N were conducted based on the methodology of Nega and Moatsou [13].

2.3.2. Fatty Acids Composition

For fatty acids (FAs) composition analysis, lipid extraction was performed with solvents after suitable preparation of the samples (WBC) according to I.D.F. [14]. The fat residue extracted was stored in amber vials, exposed to a stream of N_2 and frozen at −20 °C until analysis. Fatty acids were methylated according to Massouras et al. [15] with some modifications. Briefly, cheese lipid extract of 100 mg was methylated in a screw-cap Pyrex culture tube with the addition of 2 mL of 0.5 M sodium methylate at 50 °C for 30 min, followed by 2 mL of 140 g L^{-1} boron trifluoride in methanol (BF3) at 50 °C for 30 min. Fatty acid methyl esters (FAMEs) were recovered in hexane (2 mL). Each sample (1 μL) was injected by Shimadzu GC-2014 GC AOC-20i autosampler into a Shimadzu gas chromatograph (model GC-17A, Columbia, MD, USA), equipped with a flame ionization detector (FID), and analyzed in duplicate. Separation of fatty acid methyl esters was achieved on a SP-2560 fused silica capillary column (75 m × 0.18 mm I.D., 0.14 μm; Supelco Inc., Bellefonte, PA, USA). Helium (purity N5) was used as a carrier gas with a flow rate of 1 mL·min^{-1} at a split ratio of 1:50 with constant flow control. The injection and detector temperatures used were 250 °C and 270 °C, respectively. The oven temperature program was as follows: the initial temperature was held at 75 °C for 5 min after injection, then programmed to increase at 5 °C/min to 150 °C, to hold for 5 min, and then to increase to 220 °C at 7 °C/min and hold for 20 min. Fatty acid peaks were recorded and integrated using a Shimadzu GC solution software (Shimadzu Corporation, Kyoto, Japan). Individual fatty acids were identified by their retention times and their comparison with known fatty acid methyl ester standards (Supelco 37 Component FAME Mix, purchased from Sigma–Aldrich, Taufkirchen, Germany). Amounts of fatty acids were expressed as a weight percentage of total methyl esters of fatty acid (g·100 g^{-1} of total FAMES).

2.3.3. Analysis of Volatile Compounds by Solid-Phase Microextraction (SPME) and Gas Chromatography–Mass Spectrometry (GC–MS)

Volatile compounds of WBC at the 60th, 90th, and 180th day of ripening were determined using solid-phase microextraction (SPME) combined with GC/MS. Cheese samples (4 g) were homogenized with 2 mL of saturated Na_2SO_4 aqueous solution and 100 μL of an internal standard (IS) aqueous solution containing 0.77 g L^{-1} cyclohexanone (Sigma–Aldrich Quvmica, Alcobendas, Spain). Aliquots (3 g) of the homogenates were placed into 22 mL vials sealed with PTFE/silicone septa (Supelco, Bellefonte, PA, USA) through which the SPME syringe needle (bearing a 50/30 μm DVB/CAR/PDMS fiber Supelco, Bellefonte, PA, USA) was introduced. The samples were stirred continuously on a stir plate revolving at 750 rpm. Fiber was exposed to the headspace above the sample for 30 min at 65 °C. The absorbed volatiles were immediately desorbed at 250 °C for 3 min, in splitless mode, into the injection port of a GC–MS system (Shimadzu GC-17 A, MS QP5050). Volatile compounds were separated by a capillary column HP-INNOWax 60 m, 0.25 mm i.d., 0.25 μm film thickness (J&W Scientific, Agilent Technologies Palo Alto, CA, USA). The temperatures for the ion source, quadrupole, and interface were set at 230, 150, and 280 °C, respectively. The oven temperature was held at 45 °C for 5 min, increased to 150 °C at a rate of 5 °C min^{-1}, then raised at 7 °C min^{-1} to 220 °C, and held at 250 °C for 20 min. Helium was used as carrier gas at a flow rate of 1.0 mL min^{-1}. Electron impact ionization of MS was used at a voltage of 70 eV with a scan range from 40 to 500 m/z. The volatile compounds were identified by comparing their spectra with those from the NIST (National Institute of Standards and Technology, Gaithersburg, MD, USA) MS library. The volatile compounds were quantified by dividing the peak areas of the compounds of interest by the peak area of the IS, multiplying this ratio by the initial concentration of the IS (expressed as ppm). The peak areas were measured from the full scan chromatograph using total ion current (TIC).

2.3.4. Texture Profile Analysis

Textural profile analysis of the cheeses was assessed with a Shimadzu testing instrument, model AGS-500 NG (Shimadzu Corporation, Kyoto, Japan) equipped with a 5 kg load cell. A plunger with a diameter of 6 mm was attached to the moving crosshead. The speed of the crosshead was set at 2.5 cm mid in both upward and downward directions. The cheese sample was placed on a flat holding plate at 20 °C and the plunger was inserted 20 mm below the cheese surface. Two consecutive bites were taken. The analysis was conducted as described by Kaminarides and Stachtiaris [16]. The following six textural parameters were calculated: Hardness (N), defined as the peak force (H) during the first compression cycle (first bite), is the force necessary to attain a given deformation. Cohesiveness (N mm), defined as the ratio of the positive area under the curve during the second compression to that during the first compression. Adhesiveness (N mm), defined as the negative force area for the first bite, is the work necessary to overcome the attractive forces between the surfaces of the cheese and the plunger with which the cheese comes into contact. Elasticity (mm), defined as the ratio of the base line of the positive curve during the second compression to that during the first compression, is the height that the cheese recovers during the time that elapses between the end of the first and the start of the second bite. Gumminess (N), which is the product of hardness X cohesiveness, is the energy required to disintegrate the cheese to a state ready for swallowing. Chewiness (N), which is the product of gumminess X elasticity is the energy required to masticate a cheese to a state ready for swallowing [16].

2.3.5. Microbiological Analysis

Samples of curd (1 day after the draining), cheeses, and brines at different ripening and storage times (10, 60, 90, and 180 days) were examined for total viable count (TVC), yeasts, and molds, following the I.D.F. [17] and I.D.F. [18] methods, respectively. All the counts were expressed as colony-forming units per gram of cheese (CFU g^{-1}).

2.4. Sensory Evaluation

Sensory evaluation of the WBC samples was carried out at 60, 90, and 180 days of storage by a trained taste panel of the Dairy Laboratory of the Agricultural University of Athens. Panel members evaluated the cheeses for appearance, flavor, and body-texture using a 10-point scale. More importance was given to flavor and to body/texture than to appearance of the cheese, as advised by I.D.F. (1997) [19], by multiplying their scores by five and four, respectively. Total score was obtained by the addition of scores of the three attributes. Excellent cheese received a total score of 100.

2.5. Statistical Analysis

Physico-chemical, microbiological, and sensory parameters of two groups of cheese were subjected to analysis of variance (ANOVA) using Statgraphics Centurion XVII software (Statpoint Technologies, Inc., Warrenton, VA, USA). The difference of the means of the results of the analyses for each component was checked separately by the method of the least significant difference at a significance level of 95% (LSD, $p < 0.05$). The model used was:

$$Yi = \mu + \text{Treatment (Ti)} + ei$$

where:

μ = the mean
Ti = the fixed effect of treatment with i = 1: SST, 2:TC
ei = the random error, assumed to be normally and independently distributed with zero expectation and common variance.

3. Results and Discussion

3.1. Physico-Chemical Composition

The results of the physico-chemical analysis of the experimental cheeses and brines during the ripening–preservation period are shown in Tables 1 and 2. The type of containers (TC and SST) did not have a statistically significant ($p > 0.05$) effect on the content of total solids, moisture, fat, protein, ash, salt, and pH between both the WBC and the brines of the two groups (TC and SST).

Table 1. Physico-chemical composition (%) and pH of cheeses during ripening–preservation in tin containers (TC) and stainless steel tanks (SST) (Means ± S.D.).

	Ripening Time (Days)								
Composition	1 *	10		60		90		180	
		TC	SST	TC	SST	TC	SST	TC	SST
Moisture	56.29 ± 1.30	55.13 ± 1.53 [a]	52.96 ± 1.00 [a]	53.04 ± 1.39 [a]	54.32 ± 0.57 [a]	56.34 ± 2.00 [a]	54.13 ± 1.68 [a]	54.89 ± 1.84 [a]	54.63 ± 1.24 [a]
Fat	22.79 ± 1.86	23.23 ± 0.62 [a]	23.47 ± 0.92 [a]	24.25 ± 1.21 [a]	23.84 ± 0.87 [a]	23.36 ± 1.83 [a]	24.58 ± 0.40 [a]	23.31 ± 1.67 [a]	24.38 ± 1.90 [a]
Fat (in dry matter)	52.15 ± 4.25	51.77 ± 1.37 [a]	49.90 ± 1.96 [a]	51.64 ± 2.59 [a]	52.18 ± 1.91 [a]	53.50 ± 4.20 [a]	53.59 ± 0.87 [a]	51.67 ± 3.70 [a]	53.44 ± 4.20 [a]
Protein	16.23 ± 1.46	15.76 ± 0.63 [a]	16.15 ± 0.20 [a]	16.09 ± 0.35 [a]	15.90 ± 0.19 [a]	15.65 ± 1.09 [a]	16.62 ± 0.59 [a]	15.93 ± 1.17 [a]	15.86 ± 0.58 [a]
Protein (in dry matter)	37.14 ± 3.34	35.12 ± 1.40 [a]	34.34 ± 0.42 [a]	34.26 ± 0.75 [a]	34.80 ± 0.41 [a]	35.84 ± 2.49 [a]	36.24 ± 1.29 [a]	35.32 ± 2.59 [a]	34.95 ± 1.28 [a]
Ash	1.45 ± 0.06	3.30 ± 0.34 [a]	3.38 ± 0.36 [a]	3.29 ± 0.23 [a]	3.44 ± 0.32 [a]	3.37 ± 0.15 [a]	3.56 ± 0.31 [a]	3.74 ± 0.17 [a]	3.81 ± 0.19 [a]
Salt	0.23 ± 0.03	2.56 ± 0.43 [a]	2.74 ± 0.62 [a]	2.77 ± 0.29 [a]	2.94 ± 0.30 [a]	2.79 ± 0.02 [a]	2.90 ± 0.24 [a]	3.19 ± 0.21 [a]	3.24 ± 0.39 [a]
pH	4.72 ± 0.04	4.43 ± 0.12 [a]	4.45 ± 0.08 [a]	4.40 ± 0.10 [a]	4.46 ± 0.15 [a]	4.35 ± 0.12 [a]	4.43 ± 0.16 [a]	4.44 ± 0.06 [a]	4.42 ± 0.05 [a]

[a] Means in each parameter, with the same superscripts for TC and SST values, do not differ statistically significantly ($p > 0.05$) in each time interval. * Before packaging.

Table 2. pH, proteins, and salt concentration of brine (%) during ripening–preservation in tin containers (TC) and stainless steel tanks (SST). (Means ± S.D.).

Composition	Ripening Time (Days)							
	10		60		90		180	
	TC	SST	TC	SST	TC	SST	TC	SST
pH	4.53 ± 0.14 [a]	4.49 ± 0.15 [a]	4.57 ± 0.07 [a]	4.59 ± 0.09 [a]	4.40 ± 0.16 [a]	4.48 ± 0.12 [a]	4.31 ± 0.08 [a]	4.35 ± 0.05 [a]
Protein	1.36 ± 0.17 [a]	0.95 ± 0.30 [a]	2.21 ± 0.58 [a]	1.91 ± 0.30 [a]	2.72 ± 0.53 [a]	2.43 ± 0.52 [a]	2.40 ± 0.71 [a]	2.23 ± 0.65 [a]
Salt	4.85 ± 1.08 [a]	5.67 ± 0.55 [a]	4.97 ± 0.98 [a]	4.63 ± 0.52 [a]	4.36 ± 0.19 [a]	4.56 ± 0.59 [a]	5.10 ± 0.52 [a]	5.09 ± 0.65 [a]

[a] Means in each parameter, with the same superscripts for TC and SST values, do not differ significantly ($p > 0.05$) in each time interval.

After 10 days of ripening, the pH of the cheeses kept in TC and SST decreased, at 4.43 and 4.45, respectively, which is desirable. At 60 days onwards, both white brine cheeses (kept in TC and SST) had the desired pH (up to 4.5). The same pH (<4.6) was observed in the respective brines too. This pH value is particularly important in order to prevent microbial spoilage.

At 60 days onwards, when the cheese is considered ready for consumption, the average moisture content of the cheeses being ripened in TC and SST ranged from 53.04 to 56.34% and 54.13 to 54.63%, respectively. Both cheeses (TC and SST ripened), are classified as quality A, as set out in Greek legislation [20].

During the ripening–preservation process, the fat content (expressed on dry matter) of both cheeses was similar. At 60 days, the fat content of TC- and SST-ripened cheeses ranged from 51.64 to 53.50% and 52.18 to 53.74%, and that remained constant up to 90 days. Based on their fat content, the cheeses from both groups are classified as quality A based on Greek legislation [20].

The protein content (expressed on dry matter) of both TC- and SST-ripened cheeses was similar ($p > 0.05$); during ripening–preservation, and after the 60th day, this ranged from 34.26 to 35.84% and 34.80 to 36.24%, respectively. Both groups presented their maximum value at day 90. In addition, the protein content of the respective brine, was seen to increase continuously up to day 90, when a maximum value for both brines was recorded. This probably indicates the transfer of proteins from the cheese to the brine.

The salt concentration in both cheese groups increased over time and, on the 60th day onwards, ranged from 2.77 to 3.19% and 2.90 to 3.24% in TC and SST, respectively. No significant statistical differences ($p > 0.05$) were observed in the salt concentration of SST-ripened cheeses compared to the TC cheeses. This can be confirmed by the salt content of the brine which, in both cases, decreased up to the 90th day.

The ash content of the cheeses (in TC and SST) was similar ($p > 0.05$) and increased over time and varied from 3.29 to 3.74% and 3.44 to 3.81% for TC and SST, respectively. The physico-chemical characteristics and the pH of TC- and SST-ripened cheeses, throughout their ripening–preservation process, agree with those described in recent research works [21–24].

3.2. Major Mineral Composition

Tables 3 and 4 show the composition of the major minerals (Ca, Mg, K, and Na) in TC-/SST-ripened cheeses, and the respective brines, during their ripening–preservation. The different ripening–preservation containers did not statistically significantly ($p > 0.05$) affect the concentration of mineral in both cheese groups and the respective cheese brines throughout the ripening–preservation process.

Table 3. Major mineral concentration (mg/100 g) of cheeses during ripening and preservation in tin containers (TC) and stainless steel tanks (SST). (Means ± S.D.).

Mineral	Ripening Time (Days)								
	1 *	10		60		90		180	
		TC	SST	TC	SST	TC	SST	TC	SST
Calcium	358.04 ± 78.03	270.51 ± 37.39 [a]	242.27 ± 80.76 [a]	267.70 ± 42.22 [a]	243.94 ± 25.43 [a]	313.65 ± 44.41 [a]	280.59 ± 63.60 [a]	223.96 ± 56.13 [a]	235.05 ± 62.78 [a]
Magnesium	23.26 ± 4.11	19.43 ± 1.93 [a]	17.60 ± 3.60 [a]	17.40 ± 1.53 [a]	16.58 ± 1.01 [a]	19.39 ± 1.75 [a]	17.23 ± 3.16 [a]	14.62 ± 2.79 [a]	14.30 ± 1.73 [a]
Potassium	77.36 ± 12.73	63.11 ± 1.49 [a]	53.11 ± 10.19 [a]	62.29 ± 4.66 [a]	59.42 ± 5.23 [a]	86.95 ± 8.57 [a]	75.72 ± 8.65 [a]	48.65 ± 18.44 [a]	59.79 ± 16.92 [a]
Sodium	148.13 ± 88.97	1241.20 ± 424.97 [a]	1188.02 ± 283.23 [a]	1053.40 ± 129.96 [a]	1201.44 ± 167.23 [a]	1084.63 ± 54.49 [a]	1167.88 ± 137.49 [a]	1215.50 ± 77.37 [a]	1266.06 ± 125.15 [a]

[a] Means in each parameter with the same superscripts for TC and SST values do not differ statistically significantly ($p > 0.05$) in each time interval. * Before packaging.

Table 4. Major mineral concentration (mg/100g) of brines during ripening and preservation in tin containers (TC) and stainless steel tanks (SST). (Means ± S.D.).

Mineral	Ripening Time (Days)							
	10		60		90		180	
	TC	SST	TC	SST	TC	SST	TC	SST
Calcium	432.58 ± 19.34 [a]	290.67 ± 81.27 [b]	415.74 ± 93.71 [a]	363.99 ± 94.00 [a]	450.60 ± 29.00 [a]	413.09 ± 74.07 [a]	372.33 ± 52.00 [a]	359.97 ± 28.60 [a]
Magnesium	26.24 ± 1.42 [a]	21.47 ± 4.01 [a]	28.18 ± 3.91 [a]	24.60 ± 3.28 [a]	29.91 ± 1.47 [a]	28.42 ± 4.04 [a]	27.85 ± 3.57 [a]	25.92 ± 1.99 [a]
Potassium	98.09 ± 4.65 [a]	82.68 ± 10.67 [a]	104.80 ± 22.94 [a]	92.73 ± 10.86 [a]	108.54 ± 0.19 [a]	108.46 ± 11.64 [a]	120.09 ± 22.39 [a]	130.12 ± 4.05 [a]
Sodium	2421.42 ± 539.25 [a]	2797.89 ± 164.04 [a]	2471.68 ± 407.09 [a]	2371.95 ± 169.13 [a]	2046.38 ± 57.98 [a]	2224.34 ± 218.37 [a]	2255.00 ± 171.84 [a]	2234.93 ± 369.30 [a]

[a,b] Means in each parameter with different superscripts for TC and SST values differ statistically significantly ($p > 0.05$) in each time interval.

From the 60th day onwards, the concentration of Ca and K was similar, for TC (223.96–313.67 mg 100 g^{-1} and 48.65–86.95 mg 100 g^{-1}) and SST (235.05–280.59 mg 100 g^{-1} and 59.42–75.73 mg 100 g^{-1}). The concentration of Mg did not have significant differences between the TC and SST cheeses. This was observed throughout the ripening and preservation of the cheeses, as shown in the Table 3. The concentration of Na was similar for both groups of cheeses (1053.40–1215.50 mg 100 g^{-1} for TC cheeses and 1167.88–1266.06 mg 100 g^{-1} for SST cheeses) with the highest being noted at day 180, as was also observed with the salt concentration (Table 1).

The composition of the inorganic elements in both TS- and SST-ripened cheeses, throughout the ripening–preservation process, agrees with previous findings. The amount of Ca and Mg found in TC and SST cheeses were slightly lower than those reported by Abou Jaoude et al. and Barać et al. [25,26], while the concentration of K and Na agreed with those reported by Barać et al. [25].

Regarding the brines, the concentrations of Ca and Mg were similar in both groups (with the exception of the 10th day, when Ca was statistically significantly higher in SST, p-value = 0.042). In the TC brine, Ca and Mg ranged between 372.33 and 450.60 mg 100 g^{-1} and 27.85 and 29.91 mg 100 g^{-1}, respectively, while, in SST brine, Ca and Mg ranged between 354.97 and 413.09 mg 100 g^{-1} and 24.60 and 28.42 mg 100 g^{-1}. The concentration of K ranged from 104.80 to 120.09 and 92.73 mg 100 g^{-1} to 130.12 for TC and SST brines, respectively, with the highest values being noted on the 180th day. Finally, the concentration of Na was similar in the two groups of brines throughout the ripening–preservation process, with values ranging from 2046.38 to 2255.00 mg 100 g^{-1} and 2234.93 to 2371.95 mg 100 g^{-1} for TC and SST brines, respectively.

3.3. Proteolysis

Proteolysis is the most important biochemical event during the ripening of most rennet-coagulated cheese varieties, with a major impact on flavor and texture. During the ripening of the cheeses, caseins are hydrolyzed, resulting in water-soluble nitrogenous fractions (WSN). These fractions are indicators of the proteolysis that the cheeses undergo, which describes the speed and manner of ripening of the cheeses. Proteolysis, in terms of both SN and low-molecular-weight nitrogen fractions (i.e., TCA–SN and PTA–SN) expressed on cheese weight, is lower in brined cheeses compared to semi-hard and hard cheeses [10]. The main reason is the higher moisture content (53–56%) compared to the other cheeses groups and the very high salt-in-moisture content. Furthermore, the migration of whey proteins and soluble proteolysis products into the brine limits the concentration of these products in the cheese mass [10]. Table 5 shows in detail the course of the WSN/TN, TCA-N/TN, and TCA-N/WSN indices of the cheeses ripened and preserved in TC and SST containers. The water-soluble nitrogen ratio (WSN/TN) increased from 8.19% on day 1 to 10.29% and 10.41% for (TC) and (SST), respectively, on day 10, and remained at these levels until, on the 180th day of preservation of the cheeses, they showed small, but not statistically significant, differences ($p > 0.05$). No significant difference ($p > 0.05$) was found for ripening index between the two groups of cheeses during ripening–preservation. The values of the WSN/TN index found for both the TS and SST groups of cheese during the ripening–preservation process agreed with those reported by Zoidou, et al. and by Abd El Salam and Alichanidis [22,27]. The TCA-N/TN index showed a similar trend and no significant differences ($p > 0.05$) between the two groups of cheeses. The TCA-N/TN values in the TS and SST cheeses are in accordance with those of other research studies [22,27]. Finally, the TCA-N/WSN index showed no statistically significant differences ($p > 0.05$) between the two groups during ripening–preservation. Therefore, considering the above, we can draw the conclusion that there was a slightly increased ripening speed in the TC cheeses, up to 90 days; however, at 180 days, the level of proteolysis of both groups of cheeses was similar.

Table 5. Nitrogenous fractions of WSN/TN, TCA-N/TN, and TCA-N/WSN of cheeses during ripening and preservation in tin containers (TC) and stainless steel tanks (SST). (Means ± S.D.).

Nitrogenous Fractions	Ripening Time (Days)								
	1 *	10		60		90		180	
		TC	SST	TC	SST	TC	SST	TC	SST
%WSN/TN	8.19 ± 088 [b]	10.29 ± 2.51 [a]	10.41 ± 2.46 [a]	9.99 ±2.45 [a]	9.48 ± 1.80 [a]	11.69 ± 2.30 [a]	10.67 ± 2.84 [a]	10.21 ± 4.89 [a]	9.99 ± 3.81 [a]
%TCA-N/TN	3.84 ± 0.15 [b]	7.31 ± 1.53 [a]	6.33 ± 1.23 [a]	6.82 ± 1.69 [a]	6.00 ± 0.78 [a]	10.03 ± 1.07 [a]	8.05 ± 2.40 [a]	9.30 ± 3.31 [a]	8.97 ± 2.72 [a]
%TCA-N/WSN	47.31 ± 6.34 [b]	71.50 ± 6.19 [a]	61.35 ± 5.28 [a]	68.20 ± 2.71 [a]	64.16 ± 8.34 [a]	87.07 ± 9.92 [a]	75.61 ± 9.05 [a]	94.40 ± 11.43 [a]	91.76 ± 11.56 [a]

[a,b] Means in each parameter with different superscripts for TC and SST values differ statistically significantly ($p > 0.05$) in each time interval. * Before packaging.

Figure 2 and Tables 6–8 show the evolution of the characteristic regions of the RP-HPLC profiles. Chromatograms show some free amino acids and non-nitrogenous soluble components (eluted in the 0–10 min interval), followed by small peptides and the majority of free amino acids (proteolysis products) (in the 10–40 min interval), the water-soluble components (in the 40–70 min interval), and, finally, the hydrophobic components, large peptides, and whey proteins (in the 70–100 min interval). Comparing the chromatographic analysis of the nitrogen fractions of the cheeses and brines in the two ripening-preservation media (TC and SST), no differences were found between them. Specifically, the percentages of the chromatographic surfaces in the time intervals 0–10, 10–40, 40–70, and 70–100 min, of the chromatograms and the ratios (55–100 min)/(10–55 min), and (70–100 min)/(0–70 min) of the nitrogen fractions of cheeses (WSN/TN and TCA-N/TN Tables 7 and 8, respectively) and brines (Table 6), were similar during ripening–preservation in TC and SST, and did not present any statistically significant difference ($p > 0.05$).

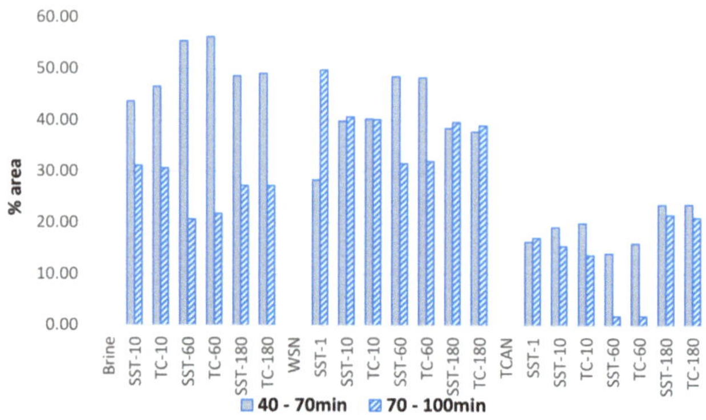

Figure 2. Evolution of characteristic regions of RP-HPLC profiles of brine and WSN/TN and TCA-N/TN of cheeses.

Table 6. Peptide areas in RP-HPLC profiles of cheese brine during ripening and preservation in tin containers (TC) and stainless steel tanks (SST). (Means ± S.D.).

RP-HPLC Profiles	Ripening Time (Days)					
	10		60		180	
	TC	SST	TC	SST	TC	SST
0–10 min	9.71 ± 1.73 [a]	10.65 ± 1.02 [a]	9.63 ± 1.68 [a]	10.63 ± 2.31 [a]	7.06 ± 0.77 [a]	6.88 ± 0.94 [a]
10–40 min	9.89 ± 2.26 [a]	10.85 ± 1.90 [a]	11.96 ± 2.81 [a]	12.75 ± 2.18 [a]	13.31 ± 1.29 [a]	13.45 ± 1.57 [a]
40–70 min	46.60 ± 1.87 [a]	43.68 ± 3.74 [a]	56.25 ± 2.46 [a]	55.41 ± 2.70 [a]	49.14 ± 3.92 [a]	48.65 ± 3.59 [a]
70–100 min	30.62 ± 2.25 [a]	31.17 ± 2.90 [a]	21.81 ± 1.97 [a]	20.69 ± 1.41 [a]	27.27 ± 1.33 [a]	27.25 ± 0.39 [a]
HB/HL [1]	1.45 ± 0.21 [a]	1.51 ± 0.24 [a]	0.96 ± 0.21 [a]	0.90 ± 0.14 [a]	1.07 ± 0.06 [a]	1.08 ± 0.04 [a]
HB/HL [2]	0.46 ± 0.06 [a]	0.48 ± 0.07 [a]	0.28 ± 0.03 [a]	0.26 ± 0.02 [a]	0.39 ± 0.03 [a]	0.40 ± 0.01 [a]

[a] Means in each parameter with the same superscripts for TC and SST values do not differ statistically significantly ($p > 0.05$) in each time interval. 1: Ratio of the area of peaks eluted from 55 to 100 min (hydrophobic peptides (HB)), to those eluted from 10 to 55 min (hydrophilic peptides (HL)). 2: Ratio of the area of peaks eluted from 70 to 100 min (hydrophobic peptides (HB)), to those eluted from 0 to 70 min (hydrophilic peptides (HL)).

Table 7. Peptide areas in RP-HPLC profiles of the WSN/TN fraction of the cheeses during ripening and preservation in tin containers (TC) and stainless steel tanks (SST). (Means ± S.D.).

RP-HPLC Profiles	Ripening Time (Days)					
	10		60		180	
	TC	SST	TC	SST	TC	SST
0–10 min	5.18 ± 1.55 [a]	4.91 ± 1.63 [a]	8.05 ± 1.14 [a]	8.43 ± 0.67 [a]	3.85 ± 0.48 [a]	3.97 ± 0.11 [a]
10–40 min	7.79 ± 1.76 [a]	7.88 ± 1.77 [a]	10.60 ± 5.08 [a]	10.50 ± 4.27 [a]	10.85 ± 0.43 [a]	10.84 ± 0.91 [a]
40–70 min	40.31 ± 5.11 [a]	39.84 ± 4.59 [a]	48.33 ± 0.80 [a]	48.56 ± 1.45 [a]	37.81 ± 3.59 [a]	38.49 ± 2.05 [a]
70–100 min	40.20 ± 3.70 [a]	40.73 ± 2.71 [a]	32.01 ± 4.10 [a]	31.56 ± 2.79 [a]	39.05 ± 1.91 [a]	39.63 ± 1.00 [a]
HB/HL [1]	2.06 ± 0.28 [a]	2.23 ± 0.15 [a]	1.24 ± 0.35 [a]	1.23 ± 0.17 [a]	1.75 ± 0.18 [a]	1.79 ± 0.08 [a]
HB/HL [2]	0.76 ± 0.13 [a]	0.78 ± 0.11 [a]	0.48 ± 0.09 [a]	0.47 ± 0.06 [a]	0.75 ± 0.09 [a]	0.74 ± 0.03 [a]

[a] Means in each parameter with the same superscripts for TC and SST values do not differ statistically significantly ($p > 0.05$) in each time interval. 1: Ratio of the area of peaks eluted from 55 to 100 min (hydrophobic peptides (HB)), to those eluted from 10 to 55 min (hydrophilic peptides (HL)). 2: Ratio of the area of peaks eluted from 70 to 100 min (hydrophobic peptides (HB)), to those eluted from 0 to 70 min (hydrophilic peptides (HL)).

Table 8. Peptide areas in RP-HPLC profiles of the TCA-N/TN fraction of the cheeses during ripening and preservation in tin containers (TC) and stainless steel tanks (SST). (Means ± S.D.).

RP-HPLC Profiles	Ripening Time (Days)					
	10		60		180	
	TC	SST	TC	SST	TC	SST
0–10 min	52.16 ± 1.00 [a]	51.27 ± 0.73 [a]	72.31 ± 7.14 [a]	74.51 ± 8.40 [a]	40.40 ± 1.41 [a]	39.96 ± 0.60 [a]
10–40 min	7.40 ± 0.55 [a]	7.18 ± 0.18 [a]	9.21 ± 2.83 [a]	9.07 ± 3.39 [a]	6.95 ± 0.26 [a]	6.85 ± 0.16 [a]
40–70 min	20.00 ± 1.05 [a]	19.16 ± 1.02 [a]	16.07 ± 4.01 [a]	14.12 ± 3.73 [a]	23.61 ± 1.94 [a]	23.59 ± 2.32 [a]
70–100 min	13.79 ± 1.07 [a]	15.55 ± 1.24 [a]	1.87 ± 1.36 [a]	1.86 ± 1.63 [a]	21.07 ± 1.09 [a]	21.60 ± 1.84 [a]
HB/HL [1]	1.46 ± 0.08 [a]	1.66 ± 0.17 [a]	0.45 ± 0.06 [a]	0.45 ± 0.05 [a]	1.71 ± 0.17 [a]	1.83 ± 0.23 [a]
HB/HL [2]	0.17 ± 0.01 [a]	0.20 ± 0.02 [a]	0.02 ± 0.01 [a]	0.02 ± 0.02 [a]	0.30 ± 0.02 [a]	0.31 ± 0.04 [a]

[a] Means in each parameter with the same superscripts for TC and SST values do not differ statistically significantly ($p > 0.05$) in each time interval. 1: Ratio of the area of peaks eluted from 55 to 100 min (hydrophobic peptides (HB)), to those eluted from 10 to 55 min (hydrophilic peptides (HL)). 2: Ratio of the area of peaks eluted from 70 to 100 min (hydrophobic peptides (HB)), to those eluted from 0 to 70 min (hydrophilic peptides (HL)).

At the 70–100 min range, hydrophobic peptides and whey proteins were eluted. Much of the nitrogenous components of the brine consist of such components (20–30%) throughout the ripening of both cheese types. This shows how whey proteins diffuse into the brine. However, more than 50% of the brine peptides appear to be composed of the soluble components of the 40–70 range, following the course of their increase in the cheese. Therefore, the level of proteolysis and the rate of ripening were similar in TC and SST cheeses.

3.4. Fatty Acids Profile

Individual fatty acids and the proportion of fatty acid groups (saturated, SFA; monounsaturated, MUFA; and polyunsaturated, PUFA) found in both the TC and SST WBC, at the 60th, 90th, and 180th day of ripening–preservation, are shown in Table 9. The most abundant FAs throughout the ripening–preservation period were palmitic acid (C16:0), oleic (C18:1n-9), and myristic (C14:0) acids. The values for palmitic acid and oleic acid for TC cheeses ranged from 30.74 to 31.14% and 19.14 to 20.72%, respectively, while, for SST, they ranged from 30.24 to 31.08% and 19.14 to 20.72%, respectively. Myristic acid, the third most abundant FA, showed a statistically significant difference between the TC and SST cheeses on day 60 (p-value = 0.027) and day 90 (p-value = 0.017). For the other identified FAs, similar values with no statistically significant differences ($p > 0.05$) were found between the two groups of cheeses, with the exception of the caprylic acid which presented a statistically significant difference (p-value = 0.038) on day 90, with SST having a higher value than TC. Overall, the values of the cheeses, regarding their SFA, MUFA, and PUFA content, were similar between the TC and SST cheeses, and no statistical difference was observed between them ($p > 0.05$). Therefore, we can claim that the different ripening–preservation containers did not affect their fatty acid content. The fatty acid composition reported here was similar to that reported by other studies concerning white brine cheeses [21,23,28].

Table 9. Means ± S.D of fatty acid composition (g/100g) of cheeses during ripening and preservation in tin containers (TC) and stainless steel tanks (SST).

Fatty Acids	Ripening Time (Days)					
	60		90		180	
	TC	SST	TC	SST	TC	SST
C4	3.36 ± 0.90 [a]	2.59 ± 0.30 [a]	2.56 ± 0.30 [a]	4.26 ± 2.52 [a]	2.78 ± 0.32 [a]	2.93 ± 0.30 [a]
C6	1.98 ± 0.56 [a]	1.83 ± 0.16 [a]	1.87 ± 0.43 [a]	2.12 ± 0.50 [a]	2.01 ± 0.32 [a]	2.10 ± 0.33 [a]

Table 9. *Cont.*

Fatty Acids	Ripening Time (Days)					
	60		90		180	
	TC	SST	TC	SST	TC	SST
C8	2.97 ± 0.99 [a]	2.27 ± 0.13 [a]	2.41 ± 0.19 [a]	2.76 ± 0.06 [b]	2.77 ± 0.15 [a]	2.96 ± 0.08 [a]
C10	10.28 ± 3.56 [a]	8.35 ± 0.20 [a]	8.59 ± 0.45 [a]	9.05 ± 0.52 [a]	9.34 ± 0.49 [a]	9.77 ± 0.12 [a]
C12	5.05 ± 0.91 [a]	4.78 ± 0.31 [a]	4.72 ± 0.24 [a]	5.03 ± 0.35 [a]	5.24 ± 0.25 [a]	5.42 ± 0.22 [a]
C14	13.84 ± 0.14 [a]	13.34 ± 0.22 [b]	13.42 ± 0.15 [a]	13.78 ± 0.05 [b]	14.21 ± 0.38 [a]	14.01 ± 0.26 [a]
C15	0.82 ± 0.14 [a]	1.77 ± 1.35 [a]	0.83 ± 0.13 [a]	0.93 ± 0.12 [a]	1.05 ± 0.03 [a]	1.01 ± 0.06 [a]
C16	30.87 ± 2.08 [a]	30.72 ± 0.65 [a]	31.14 ± 0.83 [a]	31.08 ± 1.87 [a]	30.67 ± 0.64 [a]	30.24 ± 0.37 [a]
C17	0.40 ± 0.02 [a]	0.39 ± 0.12 [a]	0.41 ± 0.06 [a]	0.50 ± 0.25 [a]	0.48 ± 0.03 [a]	0.46 ± 0.03 [a]
C18	7.69 ± 1.04 [a]	8.24 ± 0.33 [a]	8.55 ± 0.62 [a]	8.20 ± 0.67 [a]	7.67 ± 0.37 [a]	7.51 ± 0.24 [a]
C14:1	0.39 ± 0.09 [a]	0.33 ± 0.13 [a]	0.35 ± 0.06 [a]	0.38 ± 0.05 [a]	0.44 ± 0.07 [a]	0.46 ± 0.03 [a]
C16:1	1.71 ± 0.26 [a]	1.54 ± 0.41 [a]	1.36 ± 0.23 [a]	1.79 ± 0.43 [a]	1.52 ± 0.06 [a]	1.49 ± 0.05 [a]
C18:1 n9	19.14 ± 2.48 [a]	23.84 ± 5.08 [a]	20.72 ± 1.10 [a]	19.66 ± 1.07 [a]	19.42 ± 0.63 [a]	19.13 ± 0.54 [a]
C18:2 n6 t	1.62 ± 0.39 [a]	1.88 ± 0.21 [a]	1.81 ± 0.13 [a]	1.77 ± 0.15 [a]	1.91 ± 0.10 [a]	1.87 ± 0.12 [a]
C18:2 n6 c	0.27 ± 0.10 [a]	0.31 ± 0.08 [a]	0.34 ± 0.04 [a]	0.31 ± 0.04 [a]	0.36 ± 0.02 [a]	0.35 ± 0.02 [a]
C18:3 n3	0.04 ± 0.08 [a]	0.10 ± 0.03 [a]	0.09 ± 0.08 [a]	0.13 ± 0.17 [a]	0.15 ± 0.02 [a]	0.08 ± 0.07 [a]
CLA	0.45 ± 0.32 [a]	1.76 ± 1.21 [a]	1.29 ± 1.11 [a]	2.02 ± 1.36 [a]	0.50 ± 0.03 [a]	0.63 ± 0.44 [a]
SFA	77.26 ± 3.94 [a]	74.28 ± 0.77 [a]	74.49 ± 0.44 [a]	77.71 ± 5.20 [a]	76.21 ± 0.64 [a]	76.41 ± 0.54 [a]
MUFA	21.25 ± 2.26 [a]	25.70 ± 5.27 [a]	22.43 ± 0.82 [a]	21.82 ± 1.06 [a]	21.39 ± 0.61 [a]	21.08 ± 0.51 [a]
PUFA	2.38 ± 0.82 [a]	4.05 ± 0.94 [a]	3.52 ± 0.99 [a]	4.24 ± 1.26 [a]	2.91 ± 0.15 [a]	2.92 ± 0.43 [a]

[a,b] Means in each parameter with different superscripts for TC and SST values differ statistically significantly ($p > 0.05$) in each time interval.

3.5. Volatile Compounds

The results of the volatile compounds analysis of WBCs are shown in Table 10. A total of 94 volatile compounds were identified and grouped into the following chemical classes: organic acids (15), alcohols (16), aldehydes (10), esters (13), ketones (11), lactones (5), terpens (6), alcanes (8), and amines (3). Organic acids, alcohols, and esters, both in number and amount, are the dominant chemical groups. Regarding the individual identified volatile compounds, no significant differences were observed between the TC and SST cheeses. Organic acids were the most abundant chemical class in both cheeses (TC and SST) with acetic acid, decanoic acid, hexanoic acid, and octanoic acid representing about 78% of the total amount of organic acids. Each of them gives a characteristic flavor note [21]. In general, fatty acids, having between 4 and 20 carbon atoms, are formed through lipolysis by microbial lipases. The shorter fatty acids may come (originate) from the degradation of lactose and amino acids, as well as from the oxidation of ketones, esters, and aldehydes [21,29]. As expected, the samples (TC and SST) had the highest content of total organic acids on the 180th day (3828.29 mg/kg for TC and 3846.02 mg/kg for SST). From the chemical class of alcohols, the major alcohol was ethanol (both in TC and SST) throughout their ripening–preservation period, with the maximum value in both cases being noted on the 180th day (530.41 and 549.44 for TC and SST, respectively), which was expected, considering that its formation is due to lactose fermentation, proteolysis, and reduction in acetaldehyde [21,29]. Esters can be produced enzymatically or chemically through the reaction of short to medium chain fatty acids with primary and secondary alcohols that both derive from lactose fermentation and amino acid catabolism [21,30].

Table 10. Means ± S.D of volatile compounds (mg/kg) identified in cheeses during ripening and preservation in tin containers (TC) and stainless steel tanks (SST).

Volatiles Compounds	Ripening Time (Days)					
	60	90	180	60	90	180
		TC			SST	
Organic Acids (15)						
9-Decenoic acid	0.36 ± 0.62	18.65 ± 11.17	38.16 ± 10.33	31.85 ± 7.89	26.62 ± 8.62	48.49 ± 6.41
Acetic acid	281.13 ± 35.88	354.16 ± 70.56	653.86 ± 100.41	250.82 ± 180.86	251.14 ± 89.48	726.87 ± 141.69
Benzoic acid	16.58 ± 13.47	58.28 ± 23.05	62.76 ± 23.37	57.54 ± 28.03	55.34 ± 12.06	112.70 ± 42.80
Butanoic acid	132.32 ± 64.25	424.67 ± 47.91	337.25 ± 93.49	217.51 ± 121.15	521.14 ± 129.51	352.88 ± 142.39
Decanoic acid	224.62 ± 69.49	621.97 ± 32.47	782.20 ± 124.09	199.63 ± 171.66	676.52 ± 154.63	708.49 ± 165.78
Dodecanoic acid	23.02 ± 16.72	83.74 ± 19.74	130.33 ± 89.16	214.73 ± 45.25	144.07 ± 77.19	250.67 ± 56.09
Heptanoic acid	2.68 ± 2.92	52.98 ± 22.32	12.29 ± 4.11	8.89 ± 9.22	17.74 ± 6.84	25.74 ± 11.79
Hexanoic acid	276.36 ± 28.18	660.92 ± 136.16	706.56 ± 63.05	269.36 ± 120.58	737.93 ± 148.76	808.06 ± 142.36
Isovaleric acid	0.00 ± 0.00	12.12 ± 10.25	1.54 ± 1.07	1.03 ± 1.47	3.05 ± 1.60	1.13 ± 0.88
Nonanoic acid	4.12 ± 2.68	14.96 ± 6.94	17.38 ± 9.23	27.38 ± 13.04	22.95 ± 5.67	37.33 ± 10.19
Octanoic acid	251.58 ± 41.99	925.58 ± 69.82	1013.75 ± 181.24	365.49 ± 119.99	744.87 ± 137.96	655.13 ± 160.15
Pentanoic acid	1.22 ± 1.04	7.71 ± 11.36	6.46 ± 3.37	10.01 ± 4.38	8.91 ± 4.81	12.32 ± 4.90
Propanoic acid	1.20 ± 2.08	4.79 ± 6.96	12.16 ± 5.16	35.49 ± 3.74	2.67 ± 8.21	26.75 ± 11.62
Tetradecanoic acid	1.38 ± 0.39	20.63 ± 13.19	42.10 ± 9.73	63.51 ± 7.38	2.98 ± 36.20	71.59 ± 7.56
Undecanoic acid	0.00 ± 0.00	4.35 ± 3.52	11.49 ± 9.43	11.56 ± 9.00	6.77 ± 108.10	7.87 ± 2.21
Total acids	**1216.57**	**3265.51**	**3828.29**	**1764.80**	**3222.70**	**3846.02**
Alcohols (16)						
Ethanol	257.21 ± 28.63	371.16 ± 37.90	420.53 ± 77.87	260.03 ± 48.85	377.96 ± 54.26	406.17 ± 133.74
1-Propanol	6.06 ± 2.85	18.00 ± 10.08	47.12 ± 22.04	15.18 ± 34.20	22.99 ± 1.92	43.63 ± 29.57
2-Methyl-1-propanol	10.29 ± 3.59	10.05 ± 3.32	8.77 ± 2.47	8.96 ± 2.48	8.95 ± 2.65	8.40 ± 2.40
2-Propen-1-ol	0.51 ± 0.17	4.27 ± 2.34	5.02 ± 3.46	5.72 ± 2.79	2.87 ± 2.34	6.81 ± 3.60
1-Pentanol	10.43 ± 2.14	13.19 ± 4.87	3.29 ± 11.04	6.70 ± 13.79	4.56 ± 11.60	4.49 ± 11.13
1-Hexanol	4.22 ± 4.04	10.71 ± 14.36	9.46 ± 6.37	13.01 ± 7.38	11.91 ± 7.81	15.32 ± 7.90
1-Heptanol	0.43 ± 0.14	3.19 ± 2.87	1.29 ± 1.04	4.70 ± 3.79	2.56 ± 1.60	2.49 ± 1.13
1-Octen-3-ol	0.28 ± 0.17	0.74 ± 0.63	0.89 ± 0.25	1.67 ± 0.79	1.48 ± 0.74	2.90 ± 1.54
2-Nonanol	0.00 ± 0.00	1.80 ± 1.15	0.75 ± 0.46	3.50 ± 2.81	2.11 ± 1.22	1.91 ± 1.67
Decanol	2.29 ± 1.59	2.05 ± 1.32	0.77 ± 0.47	0.96 ± 0.48	0.95 ± 0.65	0.40 ± 0.40
Dodecanol	0.59 ± 0.41	0.67 ± 0.65	2.49 ± 1.51	5.52 ± 5.80	1.49 ± 1.28	9.85 ± 5.65
Tridecanol	0.18 ± 0.11	0.81 ± 0.53	2.59 ± 1.06	3.95 ± 4.18	2.14 ± 2.69	3.94 ± 2.42
Tetradecanol	0.00 ± 0.00	1.58 ± 1.30	1.61 ± 1.31	2.10 ± 1.72	0.87 ± 0.66	4.37 ± 3.70
Hexadecanol	0.31 ± 0.18	1.38 ± 1.62	1.77 ± 1.25	0.64 ± 0.26	0.94 ± 0.24	2.68 ± 1.97
Benzene ethanol	12.76 ± 4.40	95.47 ± 46.65	21.58 ± 4.31	19.99 ± 11.25	12.73 ± 5.88	32.29 ± 17.39
2-Phenyl ethanol	0.43 ± 0.17	1.75 ± 1.01	2.48 ± 1.76	1.73 ± 1.41	2.35 ± 3.42	3.79 ± 3.94
Total Alcohol	**305.99**	**536.82**	**530.41**	**354.36**	**456.86**	**549.44**
Aldehydes (10)						
Acetaldehyde	1.82 ± 0.52	1.28 ± 0.33	1.13 ± 0.65	1.92 ± 1.02	0.44 ± 0.66	0.96 ± 0.54
Furfural	0.02 ± 0.10	0.25 ± 0.10	0.82 ± 0.22	0.52 ± 0.14	0.22 ± 0.25	0.61 ± 0.25
Hexanal	3.83 ± 2.20	3.59 ± 2.50	2.31 ± 1.45	2.50 ± 1.02	2.49 ± 1.35	1.94 ± 2.02
Heptanal	0.00 ± 0.00	0.60 ± 0.54	1.44 ± 1.08	0.65 ± 0.81	0.70 ± 0.26	1.49 ± 1.99
Nonanal	1.69 ± 0.59	2.68 ± 1.30	3.87 ± 3.82	1.80 ± 1.66	3.36 ± 3.58	3.91 ± 3.54
Decanal	1.20 ± 1.05	1.82 ± 1.43	1.57 ± 1.13	2.39 ± 1.02	2.52 ± 2.07	3.55 ± 2.44
2,4-Dimethylpentanal	0.89 ± 0.25	1.96 ± 1.35	2.35 ± 1.20	1.22 ± 1.20	1.52 ± 1.32	3.26 ± 1.88
3-Hydroxybutanal	1.97 ± 1.25	1.70 ± 1.84	0.85 ± 0.55	0.86 ± 0.22	1.03 ± 1.20	0.78 ± 0.33
2-methylpentanal	0.79 ± 0.55	1.03 ± 0.88	1.89 ± 1.33	1.36 ± 1.25	1.66 ± 1.35	6.03 ± 3.25
Benzaldehyde	0.62 ± 0.23	1.63 ± 1.46	3.55 ± 2.28	2.71 ± 2.35	1.54 ± 3.37	5.27 ± 2.54
Total Aldehydes	**12.83**	**16.54**	**19.78**	**15.93**	**15.48**	**27.80**

Table 10. *Cont.*

Volatiles Compounds	Ripening Time (Days)					
	60	90	180	60	90	180
		TC			SST	
Esters (13)						
2-Hydroxy-propanoic acid. ethyl ester	1.52 ± 0.64	2.56 ± 1.56	1.32 ± 0.99	0.03 ± 0.03	3.80 ± 2.29	8.64 ± 4.40
Aceticacid. 2-phenylethyl ester	1.86 ± 1.03	86.44 ± 24.37	60.60 ± 10.04	26.77 ± 15.73	15.91 ± 6.88	35.22 ± 15.46
Citronellylformate	0.92 ± 0.37	2.94 ± 1.01	0.76 ± 0.66	4.15 ± 2.77	1.81 ± 0.97	2.49 ± 2.13
Decanoic acid. Ethylester	9.29 ± 1.95	37.54 ± 14.44	37.84 ± 11.69	28.22 ± 19.96	39.01 ± 24.71	115.30 ± 60.57
Dihydrocitronellol acetate	0.42 ± 0.16	0.35 ± 0.16	1.62 ± 0.54	0.94 ± 0.26	1.19 ± 0.77	0.51 ± 0.13
Ethyl acetate	1.00 ± 0.73	11.22 ± 11.41	19.59 ± 16.54	7.76 ± 10.04	1.52 ± 1.21	18.48 ± 15.57
Hexadecanoic acid. ethylester	0.55 ± 0.48	1.09 ± 0.96	2.45 ± 1.30	1.71 ± 3.57	1.70 ± 1.88	2.86 ± 3.87
Hexanoic acid. ethylester	1.77 ± 2.81	3.09 ± 1.69	3.99 ± 2.80	3.94 ± 2.56	3.77 ± 2.81	8.73 ± 3.81
Isopentyl formate	3.66 ± 1.34	0.50 ± 0.20	0.86 ± 0.25	0.55 ± 1.05	1.98 ± 2.52	0.95 ± 0.53
Isopentyl-isovalerate	0.30 ± 0.13	0.79 ± 0.23	2.07 ± 1.83	2.81 ± 1.31	2.17 ± 2.13	5.48 ± 4.04
Octanoic acid. ethylester	4.85 ± 2.93	6.45 ± 3.64	7.54 ± 4.11	9.27 ± 5.31	8.36 ± 6.71	34.37 ± 18.52
Tetradecanoi acid. ethylester	2.15 ± 0.97	3.54 ± 2.49	3.77 ± 3.09	6.28 ± 8.87	1.16 ± 1.34	9.36 ± 5.65
2-Ethenyloxy ethanol	0.75 ± 0.29	4.27 ± 2.09	77.59 ± 35.09	6.98 ± 4.52	20.30 ± 8.44	33.09 ± 67.68
Total Esters	**29.04**	**160.78**	**220.00**	**99.41**	**102.68**	**275.48**
Ketones (11)						
2-Butanone	4.90 ± 2.85	5.79 ± 2.50	7.04 ± 3.59	7.72 ± 4.25	7.51 ± 2.55	9.09 ± 3.55
2-Heptanone	2.68 ± 1.23	3.02 ± 1.85	3.76 ± 2.05	4.49 ± 3.20	3.26 ± 1.25	3.61 ± 1.64
2-Octanone	3.43 ± 2.22	4.24 ± 2.89	5.05 ± 3.66	6.34 ± 3.65	5.59 ± 2.38	6.63 ± 2.22
2-Nonanone	5.72 ± 3.25	7.77 ± 2.01	1.60 ± 0.22	1.34 ± 1.02	1.74 ± 0.80	1.04 ± 0.55
2-Decanone	2.30 ± 1.55	3.19 ± 2.99	4.44 ± 1.85	5.12 ± 3.81	4.91 ± 2.46	6.49 ± 5.02
2-Undecanone	0.08 ± 0.04	0.42 ± 0.37	1.16 ± 1.13	1.89 ± 1.79	0.66 ± 1.06	1.01 ± 1.67
2-Dodecanone	0.83 ± 0.31	1.64 ± 1.10	2.45 ± 2.83	3.74 ± 2.79	2.99 ± 1.19	4.03 ± 2.60
5-Methyl-2-Hexanone	5.12 ± 3.54	5.17 ± 2.95	2.75 ± 0.84	5.93 ± 2.23	2.19 ± 1.14	7.45 ± 4.25
5-Methyl-3-Heptanone	2.28 ± 1.95	6.78 ± 2.44	8.38 ± 5.12	12.76 ± 6.89	8.71 ± 4.88	19.46 ± 10.17
2-Methyl-4-heptanone	3.41 ± 1.99	7.71 ± 2.81	8.77 ± 10.99	14.84 ± 10.08	8.31 ± 6.71	19.46 ± 13.17
2-Piperidinone	0.14 ± 0.14	1.57 ± 1.27	1.35 ± 1.41	0.96 ± 0.22	1.06 ± 1.63	2.32 ± 2.00
Total ketones	**35.41**	**54.22**	**50.25**	**67.75**	**47.73**	**95.96**
Lactones (5)						
Gamma Nonalactone	1.02 ± 0.55	0.25 ± 0.22	1.28 ± 0.69	2.55 ± 0.65	2.05 ± 1.55	0.88 ± 0.33
Gamma-decalactone	0.37 ± 0.33	0.95 ± 0.28	1.83 ± 1.27	1.58 ± 1.00	0.91 ± 0.65	1.02 ± 1.07
Gamma-dodecalactone	4.52 ± 3.82	4.67 ± 2.04	7.36 ± 4.52	10.16 ± 6.28	4.77 ± 4.82	20.77 ± 13.87
Delta-nonalactone	2.95 ± 1.52	3.03 ± 1.48	4.00 ± 2.46	12.93 ± 9.83	5.51 ± 4.56	19.46 ± 13.17
Delta-decalactone	1.22 ± 0.80	2.05 ± 1.02	2.02 ± 1.08	0.58 ± 0.33	0.95 ± 0.88	1.64 ± 0.45
Total Lactones	**10.08**	**10.95**	**16.49**	**27.80**	**14.19**	**43.77**
Terpens (6)						
Dehydro-apofarnesol	0.30 ± 0.22	1.48 ± 0.83	0.75 ± 0.63	2.44 ± 1.13	2.72 ± 2.58	3.82 ± 3.17
Dihydrocitronellol	0.51 ± 0.47	4.27 ± 2.34	5.02 ± 3.46	2.55 ± 1.25	2.87 ± 2.34	3.88 ± 1.05
Farnesol	0.43 ± 0.24	3.19 ± 2.87	1.29 ± 1.04	4.70 ± 3.79	2.56 ± 1.60	2.49 ± 2.13
Sesquilavandulol	0.00 ± 0.00	1.80 ± 0.95	0.75 ± 0.16	3.50 ± 2.81	2.11 ± 1.22	1.91 ± 1.67
Tetrahydro-lavandulol	0.43 ± 0.17	1.75 ± 0.81	2.48 ± 0.76	1.73 ± 1.41	2.35 ± 1.42	3.79 ± 2.94
Tetrahydro-citronellene	0.46 ± 0.19	1.06 ± 0.27	1.47 ± 1.46	0.97 ± 0.81	0.77 ± 0.17	3.56 ± 2.84
Total Terpens	**2.13**	**13.55**	**11.76**	**15.89**	**13.38**	**19.45**
Alcanes (8)						
2.2-Dimethylbutane	0.30 ± 0.26	0.85 ± 0.59	1.01 ± 0.45	0.76 ± 1.18	0.41 ± 0.19	0.62 ± 0.10
3-Methyl-hexane	0.56 ± 0.49	2.11 ± 1.43	0.83 ± 0.88	1.89 ± 0.96	0.41 ± 0.14	2.94 ± 3.39
Decane	1.46 ± 1.17	1.56 ± 0.69	11.75 ± 3.61	1.39 ± 4.79	2.95 ± 1.83	3.20 ± 1.20

Table 10. *Cont.*

Volatiles Compounds	Ripening Time (Days)					
	60	90	180	60	90	180
		TC			SST	
Dodecane	0.92 ± 0.43	1.04 ± 1.29	3.02 ± 0.46	2.72 ± 4.15	1.61 ± 1.09	2.54 ± 1.41
Nonane	0.91 ± 0.88	1.33 ± 1.76	1.77 ± 1.01	2.92 ± 1.46	2.20 ± 1.08	2.90 ± 2.70
Tetradecane	0.54 ± 0.73	0.61 ± 0.15	0.81 ± 0.22	1.22 ± 0.59	0.80 ± 0.44	3.40 ± 2.10
Tridecane	1.00 ± 0.52	0.84 ± 1.53	1.42 ± 1.12	1.79 ± 1.89	0.85 ± 0.30	1.40 ± 2.47
Undecane	1.89 ± 1.14	4.11 ± 2.43	1.80 ± 0.76	2.60 ± 0.62	1.63 ± 1.10	1.39 ± 1.98
Total Alcanes	**7.58**	**12.45**	**22.41**	**15.29**	**10.86**	**18.39**
Amines (3)						
Amide	0.12 ± 0.15	0.05 ± 0.06	1.01 ± 1.02	1.06 ± 0.02	1.11 ± 1.02	0.32 ± 0.16
Piperidini	0.00 ± 0.00	0.16 ± 0.14	1.32 ± 1.24	0.02 ± 0.03	0.01 ± 0.02	1.24 ± 1.13
Dimethyl-amine	0.06 ± 0.03	0.11 ± 0.16	0.23 ± 0.12	0.03 ± 0.12	0.41 ± 0.55	0.24 ± 0.11
Total Amines	**0.182**	**0.326**	**2.564**	**1.093**	**1.536**	**1.806**
Other aroma compounds (6)						
2 H-Pyran-2-one tetrahydro-6-pentyl	2.28 ± 1.95	1.78 ± 0.44	0.38 ± 0.12	1.76 ± 1.89	1.71 ± 1.88	1.46 ± 1.17
2 H-Pyran-2-one tetrahydro-6-propyl	0.41 ± 1.99	0.71 ± 0.81	0.77 ± 1.99	1.84 ± 1.08	2.31 ± 0.71	1.46 ± 1.17
Phenol	0.28 ± 0.17	0.74 ± 0.63	0.89 ± 0.25	1.67 ± 0.79	1.48 ± 0.74	2.90 ± 1.54
Camphor	0.83 ± 0.58	1.83 ± 0.67	0.68 ± 4.34	1.97 ± 5.98	1.01 ± 4.13	1.63 ± 0.19
2 H-Pyran-2-one tetrahydro-6-pentyl	2.28 ± 1.95	0.78 ± 0.44	0.38 ± 5.12	1.76 ± 6.89	2.71 ± 2.88	1.46 ± 10.17
Styrene	2.62 ± 1.42	0.95 ± 0.54	1.79 ± 1.90	0.90 ± 0.63	0.82 ± 0.24	1.75 ± 8.51
Total other aroma compounds	**8.70**	**6.79**	**4.89**	**9.90**	**10.04**	**10.66**

Regarding the chemical class of esters, it was observed that their total amount increased with time, with the maximum value being noted on the 180th day in both TC- (220 mg/kg) and SST- (275.48 mg/kg) ripened cheeses. The esters detected in large quantities in all cheese samples were acetic acid 2-phenylethylester, decanoic acid ethylester, and 2-ethenyloxyethanol. Among the less abundant groups, alkanes and terpenes, lactones, and ketones were more represented in both cheeses at 180 days. The volatile profile of white brine cheese in this study is close to those reported by other studies concerning this type of cheese [21,31].

3.6. Texture Profile Analysis

Table 11 shows the textural characteristics of the TC- and SST-ripened cheeses during the ripening–preservation period. The ripening–preservation containers did not significantly ($p > 0.05$) affect the textural characteristics of the cheeses, as their values had been similar over time.

Table 11. Means ± S.D. of textural characteristics (hardness (N), adhesiveness (J), elasticity (mm), cohesiveness, gumminess (N), and chewiness (J)) of cheeses during ripening and preservation in tin containers (TC) and stainless steel tanks (SST).

Textural Parameters	Ripening Time (Days)							
	10		60		90		180	
	TC	SST	TC	SST	TC	SST	TC	SST
Hardness	8.28 ±2.05 ^{a}	7.58 ± 0.91 ^{a}	6.04 ± 1.90 ^{a}	7.14 ± 2.47 ^{a}	7.48 ± 2.58 ^{a}	8.41 ± 1.40 ^{a}	9.66 ± 1.10 ^{a}	8.45 ± 2.29 ^{a}
Adhesiveness	−31.38 ± 15.29 ^{a}	−23.72 ± 6.41 ^{a}	−22.50 ± 9.59 ^{a}	−18.84 ± 12.90 ^{a}	−24.95 ± 9.72 ^{a}	−28.32 ± 8.12 ^{a}	−33.39 ± 7.17 ^{a}	−40.66 ± 13.86 ^{a}
Elasticity	1.02 ± 0.02 ^{a}	1.22 ± 0.36 ^{a}	1.09 ± 0.09 ^{a}	1.21 ± 0.20 ^{a}	1.24 ± 0.34 ^{a}	1.18 ± 0.30 ^{a}	1.07 ± 0.06 ^{a}	2.11 ± 1.14 ^{a}
Cohesiveness	0.37 ± 0.03 ^{a}	0.37 ± 0.07 ^{a}	0.35 ± 0.03 ^{a}	0.32 ± 0.06 ^{a}	0.32 ± 0.04 ^{a}	0.34 ± 0.04 ^{a}	0.36 ± 0.02 ^{a}	0.35 ± 0.01 ^{a}

Table 11. *Cont.*

Textural Parameters	Ripening Time (Days)							
	10		60		90		180	
	TC	SST	TC	SST	TC	SST	TC	SST
Gumminess	3.09 ± 0.92 [a]	2.85 ± 0.88 [a]	2.14 ± 0.83 [a]	2.36 ± 1.16 [a]	2.45 ± 1.07 [a]	2.85 ± 0.70 [a]	3.44 ± 0.37 [a]	2.99 ± 0.86 [a]
Chewiness	3.15 ± 0.93 [a]	3.31 ± 0.47 [a]	2.37 ± 1.07 [a]	2.70 ± 0.86 [a]	2.81 ± 0.67 [a]	3.21 ± 0.13 [a]	3.67 ± 0.26 [a]	5.89 ± 2.48 [a]

[a] Means in each parameter with the same superscripts for TC and SST values do not differ statistically significantly ($p > 0.05$) in each time interval.

The hardness of both groups of cheeses increased with time and specifically in the period of 60–180 days. This increase was enhanced by the progressive increase in the salt and ash content of the respective cheeses during the 60th–180th days [16], and the decrease in the fat content during the 90th–180th days [32]. Both cheese groups showed the lowest (6.04 ± 1.90 N for TC and 7.14 ± 2.47 N for SST) and the highest (9.66 ± 1.10 N for TC and 8.45 ± 2.29 N for SST) values on the 60th and 180th days, respectively. Due to the increase in hardness, the gumminess and chewiness of both cheeses also increased during their ripening–preservation. The cheeses (TC and SST) showed the highest value in gumminess and chewiness on day 180 (3.44 ± 0.37 N and 3.67 ± 0.26 J for TC, and 2.99 ± 0.86 N and 5.89 ± 2.48 J for SST, respectively). The value of the adhesiveness (in both groups of cheeses) increased (in absolute value). Finally, the cohesiveness of the cheeses increased in both groups, showing their maximum value on the 180th day (0.36 ± 0.02 for TC and 0.35 ± 0.01 for SST). As an exception, at day 90, the cohesiveness of the TC cheese decreased. Overall, the values of the textural characteristics of both TC- and SST-ripened cheeses reported here are similar to those reported by Kaminarides et al. [33], with the exception of the adhesiveness and gumminess parameters.

3.7. *Microbiological Evolution during Cheese Ripening–Preservation*

The microbiological results, total viable count (TVC), and molds/yeasts concerning the cheeses of the two groups and the respective brines during their ripening–preservation are presented in Tables 12 and 13.

Table 12. Microbial counts (log CFU/g) of cheeses for total viable count and molds/yeasts during ripening and preservation in tin containers (TC) and stainless steel tanks (SST). (Means ± S.D.).

Microbial Groups	Ripening Time (Days)								
	1 **	10		60		90		180	
		TC	SST	TC	SST	TC	SST	TC	SST
Total viable count	9.40 ± 0.25	8.75 ± 0.21 [a]	8.45 ± 0.33 [a]	7.56 ± 0.29 [a]	6.62 ± 1.40 [a]	7.39 ± 1.11 [a]	5.92 ± 0.74 [a]	6.50 ± 1.08 [a]	5.47 ± 0.79 [a]
Molds/Yeasts	1.63 ± 1.41 [a]	3.44 ± 1.22 [a]	3.16 ± 1.00 [a]	3.24 ± 0.78 [a]	3.45 ± 0.51 [a]	2.34 ± 0.96 [a]	3.18 ± 0.83 [a]	1.87 ± 1.68 [a]	2.67 ± 0.37 [a]

[a] Means in each parameter with the same superscripts for TC and SST values do not differ statistically significantly ($p > 0.05$) in each time interval. ** Before packaging.

Table 13. Microbial counts (log cfu/g) of brines for total viable count and molds/yeasts during ripening and preservation in tin containers (TC) and stainless steel tanks (SST). (Means ± S.D.).

Microbial Groups	Ripening Time (Days)							
	10		60		90		180	
	TC	SST	TC	SST	TC	SST	TC	SST
Total viable count	7.13 ± 0.57 [a]	7.13 ± 1.37 [a]	7.37 ± 0.15 [a]	6.90 ± 1.33 [a]	7.51 ± 0.28 [a]	6.19 ± 0.72 [b]	6.81 ± 0.58 [a]	5.29 ± 0.75 [a]
Molds/Yeasts	4.35 ± 0.32 [a]	3.51 ± 1.00 [a]	4.18 ± 0.10 [a]	4.77 ± 0.44 [a]	3.44 ± 1.06 [a]	4.15 ± 1.01 [a]	3.17 ± 1.02 [a]	3.83 ± 0.13 [a]

[a,b] Means in each parameter with different superscripts for TC and SST values differ statistically significantly ($p > 0.05$) in each time interval.

The TVCs in both cheeses and brines decreased over the ripening–preservation period. The mesophilic flora had quite similar values in both the cheeses and brines of the two groups. No statistically significant differences ($p > 0.05$) were found between the cheeses of the two groups and the respective brines, except for the 90th day, when the brine at TC showed a significantly ($p = 0.041$) higher TVC than that of the SST brine.

Molds/yeasts counts were low in both TC- and SST-ripened cheeses and brines. Different ripening–preservation containers did not statistically significantly ($p > 0.05$) affect the population of molds/yeasts, neither in the cheeses nor in the brines. The results regarding the TVC and the molds/yeasts, observed in both TC and SST cheeses during their ripening–preservation, agree with those reported in other studies [22,34].

3.8. Sensory Evaluation

The results of the sensory evaluation of the cheeses are presented in the Table 14. Specifically, the TC- and SST-ripened cheeses did not show significant differences ($p > 0.05$) between them regarding their appearance/color, structure/texture, and flavor/odor, with their scores being similar throughout their ripening–preservation. On days 60 and 90, the cheeses (of both groups) had the characteristics of mature cheeses, with a semi-hard texture and a pleasant acidic taste. Both cheese groups obtained the highest total score on day 90 (91.45% ± 1.97 for TC and 90.95% ± 4.60 for SST, with coefficients of 4 and 5, respectively). The high score of the structure parameter, on day 90, was expected due to the fact that the gradual accumulation of total solids and large water-soluble peptides showed their maximum value for both groups of cheeses on day 90 as well. However, the cheeses of both groups received lower scores on day 180, which was expected since proteolysis and lipolysis are more intense during that time. The cheeses were characterized by a pleasant, acidic taste.

Table 14. Sensory evaluation of cheeses during their ripening and preservation in tin containers (TC) and stainless steel tanks (SST). (Means ± S.D.).

Sensory Parameters	**Ripening Time (Days)**					
	60		**90**		**180**	
	TC	**SST**	**TC**	**SST**	**TC**	**SST**
Appearance/Colour (0–10)	9.14 ± 0.14 [a]	9.33 ± 0.08 [a]	9.21 ± 0.37 [a]	9.17 ± 0.22 [a]	9.42 ± 0.30 [a]	9.28 ± 0.25 [a]
Body/Texture (0–40)	35.14 ± 0.29 [a]	35.14 ± 0.29 [a]	36.76 ± 0.66 [a]	36.67 ± 1.84 [a]	34.89 ± 1.54 [a]	33.56 ± 2.04 [a]
Flavor/Odor (0–50)	42.98 ± 2.03 [a]	44.52 ± 0.55 [a]	45.48 ± 1.15 [a]	45.12 ± 2.58 [a]	40.00 ± 2.89 [a]	39.86 ± 3.34 [a]
Total (0–100)	87.26 ± 1.76 [a]	89.00 ± 0.75 [a]	91.45 ± 1.97 [a]	90.95 ± 4.60 [a]	84.31 ± 4.65 [a]	82.69 ± 5.51 [a]

[a] Means in each parameter with the same superscripts for TC and SST values do not differ statistically significantly ($p > 0.05$) in each time interval.

4. Conclusions

The results presented in this study showed that the material, and the capacity, of the ripening–preservation containers did not statistically significantly affect the physicochemical, textural, microbiological, and sensory characteristics of the white brine cheeses. It should be noted that those white brine cheeses that are ripened and preserved in tin containers, and those that are ripened and preserved in stainless steel tanks, did not differ from each other in the two main parameters—their fat and moisture content—that determine the quality of cheese, according to the Greek legislation. Although no significant differences were observed between the white cheeses in brine that ripened and preserved in tin containers and stainless steel tanks, we consider that the latter can be used by cheese factories where a considerable part of their production is planned for repackaging as a SST container has many advantages, such as its reusability, resistance to corrosion and low temperatures, contribution to the product's hygiene and protection, and minimizing of losses during repackaging.

Author Contributions: Conceptualization, T.M., E.Z., Z.B. and M.K.; methodology, T.M., E.Z. and Z.B.; statistical analysis, M.K. and T.M.; data curation, T.M., E.Z. and Z.B.; writing—original draft preparation, T.M., E.Z., Z.B. and M.K.; validation, T.M. and M.K.; formal analysis, T.M. and M.K.; writing—review editing and supervision, T.M. All authors have read and agreed to the published version of the manuscript.

Funding: This research has been co-financed by the European Regional Development Fund of the European Union and Greek national funds through the Operational Program Competitiveness, Entrepreneurship and Innovation, under the call RESEARCH—CREATE—INNOVATE (project code: T1EDK04595).

Data Availability Statement: All related data and methods are presented in this paper. Additional inquiries should be addressed to the corresponding author.

Acknowledgments: The authors gratefully acknowledge, for their advice, for their enthusiasm and support of this work, and for their experimental assistance, G. Moatsou, E. Moschopoulou and L. Sakkas; the coordinator of the project INOX DESIGN KATERIS S.A., as well as the technical assistance and collaboration of J. Kyrikos, Z. Baradaki, J. Karavas, and all the technical staff working at the dairies of Galaktokomiki S.A. and Tzafetas S.A., at whose facilities all experiments were performed. The authors also thank K. Angelis and M. Adamou for their administrative and technical support.

Conflicts of Interest: The authors declare no conflict of interest. "Galaktokomiki kritis Dairy" is the industrial partner of the funded program quoted in this article. Zinovia Baradaki is a member of the research team affiliated to "Galaktokomiki kritis Dairy", thus no conflict of interest is to be declared.

References

1. International Dairy Federation. *Cheese and Varieties Part II: Cheese Styles*; International Dairy Federation: Brussels, Belgium, 2021.
2. McSweeney, P.L.H. Chesse problems solved. In *Cheeses Ripened in Brine*; Alichanidis, E., Ed.; Woodhead Publishing Limited: Cambridge, UK, 2007; pp. 330–335.
3. Geronikou, A.; Srimahaeak, T.; Rantsiou, K.; Triantafillidis, G.; Larsen, N.; Jespersen, L. Occurrence of Yeasts in White-Brined Cheeses: Methodologies for Identification, Spoilage Potential and Good Manufacturing Practices. *Front. Microbiol.* **2020**, *11*, 582–778. [CrossRef] [PubMed]
4. Soltani, M.; Saremnezhad, S.; Faraji, A.R.; Hayaloglu, A.A. Perspectives and recent innovations on white cheese produced by conventional methods or ultrafiltration technique. *Int. Dairy J.* **2022**, *125*, 105–232. [CrossRef]
5. Gursoy, O.; Kesenkas, H.; Yilmaz, Y. White cheese. In *Handbook of Cheese in Health: Production, Nutrition and Medical Sciences*; Preedy, V.R., Watson, R.R., Patel, V.B., Eds.; Wageningen Academic Publishers: Wageningen, The Netherlands, 2013; pp. 184–188.
6. Alichanidis, E.; Polychroniadou, A. Characteristics of major traditional regional cheese varieties of East-Mediterranean countries: A review. *Dairy Sci. Technol.* **2008**, *88*, 495–510. [CrossRef]
7. *IDF 4A:1982*; Cheese and Processed Cheese: Determination of the Total Solids Content (Reference Method). International Dairy Federation: Brussels, Belgium, 1986.
8. *ISO 19662 | IDF 238:2018*; Milk—Determination of Fat Content—Acido-Butyrometric (Gerber Method). International Dairy Federation: Brussels, Belgium, 2018.
9. *ISO 5943 | IDF 88:2006*; Cheese and Processed Cheese Products—Determination of Chloride Content—Potentiometric Titration Method. International Dairy Federation: Brussels, Belgium, 2006.
10. *IDF 27:1964*; Determination of the Ash Content of Processed Cheese Products (Reference Method). International Dairy Federation: Brussels, Belgium, 1964.
11. *IDF 154:1992*; Milk and Dried Milk–Determination of Calcium Content–Flame Atomic Absorption Spectrometric Method. International Dairy Federation: Brussels, Belgium, 1992.
12. *ISO 17837 | IDF 25:2008*; Processed Cheese Products—Determination of Nitrogen Content and Crude Protein Calculation—Kjeldahl Method. International Dairy Federation: Brussels, Belgium, 2008.
13. Nega, A.; Moatsou, G. Proteolysis and related enzymatic activities in ten Greek cheese varieties. *Dairy Sci. Technol.* **2012**, *92*, 57–73. [CrossRef]
14. *IDF 172:1995*; Milk and Milk Products-Extraction Methods for Lipids and Liposoluble Compounds. International Dairy Federation: Brussels, Belgium, 1995.
15. Massouras, T.; Pappa, E.C.; Mallatou, H. Headspace analysis of volatile flavour compounds of teleme cheese made from sheep and goat milk. *Int. J. Dairy Technol.* **2006**, *59*, 250–256. [CrossRef]
16. Kaminarides, S.; Stachtiaris, S. Production of processed cheese using kasseri cheese and processed cheese analogues incorporating whey protein concentrate and soybean oil. *Int. J. Dairy Technol.* **2000**, *53*, 69–74. [CrossRef]
17. *IDF 100B:1991*; Milk and Milk Products–Enumeration of Microorganisms–Colony Count Technique at 30 °C. International Dairy Federation: Brussels, Belgium, 1991.

18. *IDF 94B:1991*; Milk and Milk Products–Enumeration of Yeasts and Moulds–Colony Count Technique at 25 °C. International Dairy Federation: Brussels, Belgium, 1991.
19. *IDF 99C:1997*; Sensory Evaluation of Dairy Products by Scoring—Reference Method. International Dairy Federation: Brussels, Belgium, 1997.
20. *Greek Food Code*; General Chemical State Laboratory, Ministry of Economy and Finance, Hellenic Republic, Greek Ministry of Rural Development and Food: Athens, Greece, 2009; Article 83.
21. Gatzias, I.S.; Karabagias, I.K.; Kontominas, M.G.; Badeka, A.V. Geographical differentiation of feta cheese from northern Greece based on physicochemical parameters, volatile compounds and fatty acids. *LWT* **2021**, *131*, 109615. [CrossRef]
22. Zoidou, E.; Kandarakis, I.; Anyfantakis, E.; Massouras, T.; Fragoulaki, M.; Kritakis, G. Evaluation of the use of a new type of rennet and starter culture on the quality of Feta cheese. *Greek J. Dairy Sci. Technol.* **2007**, *1*, 42–58.
23. Georgala, A.; Moschopoulou, E.; Aktypis, A.; Massouras, T.; Zoidou, E.; Kandarakis, I.; Anifantakis, E. Evolution of lipolysis during the ripening of traditional Feta cheese. *Food Chem.* **2005**, *93*, 73–80. [CrossRef]
24. Robinson, R.K.; Tamime, A.Y. Feta and related cheeses. In *Traditional Feta Cheese*; Anifandakis, E., Ed.; Ellis Horwood Limited: Chichester, UK, 1991; pp. 49–68.
25. Abou Jaoude, D.; Olabi, A.; Ouyoun Najm, N.E.; Malek, A.; Saadeh, C.; Baydoun, E.; Toufeili, I. Chemical composition, mineral content and cholesterol levels of some regular and reduced-fat white brined cheeses and strained yogurt (Labneh). *Dairy Sci. Technol.* **2010**, *90*, 699–706. [CrossRef]
26. Barać, M.; Kresojević, M.; Špirović-Trifunović, B.; Pešić, M.; Vučić, T.; Kostić, A.; Despotović, S. Fatty acid profiles and mineral content of Serbian traditional white brined cheeses. *Mljekarstvo* **2018**, *68*, 37–45. [CrossRef]
27. Abd El-Salam, M.H.; Alichanidis, E. *Cheese Varieties Ripened in Brine*; Elsevier: Amsterdam, The Netherlands, 2004; Volume 2, pp. 227–249.
28. Temiz, H.; Tarakci, Z.; Aykut, U.; Turhan, S. The fatty acid levels and physicochemical properties of herby brined cheese, a traditional Turkish cheese. *Int. J. Dairy Technol.* **2009**, *62*, 56–62. [CrossRef]
29. Molimard, P.; Spinnler, H.E. Review: Compounds involved in the flavor of surface mold-ripened cheeses: Origins and properties. *J. Dairy Sci.* **1996**, *79*, 169–184. [CrossRef]
30. Engels, W.J.M.; Dekker, R.; de Jong, C.; Neeter, R.; Visser, S. A comparative study of volatile compounds in the water soluble fraction of various types of ripened cheese. *Int. Dairy J.* **1997**, *7*, 255–263. [CrossRef]
31. Kondyli, E.; Pappa, E.C.; Vlachou, A.M. Effect of package type on the composition and volatile compounds of Feta cheese. *Small Rumin. Res.* **2012**, *108*, 95–101. [CrossRef]
32. Kaminarides, S.; Zagari, H.; Zoidou, E. Effect of whey fat content on the properties and yields of whey cheese and serum. *J. Hell. Vet. Med. Soc.* **2020**, *71*, 2149–2156. [CrossRef]
33. Kaminarides, S.; Moschopoulou, E.; Karali, F. Influence of salting method on the chemical and texture characteristics of ovine Halloumi cheese. *Foods* **2019**, *8*, 232. [CrossRef] [PubMed]
34. Ammar, E.-T.; Khalel, A.E.; Mostafa, M.S. Effect of type of milk on the properties of traditional feta cheese. *J. Food Dairy Sci.* **2014**, *5*, 315–327. [CrossRef]

foods

MDPI

Article

Application of Commercial Biopreservation Starter in Combination with MAP for Shelf-Life Extension of Burrata Cheese

Giuseppe Natrella *, Giuseppe Gambacorta and Michele Faccia

Department of Soil, Plant and Food Sciences, University of Bari, Via Amendola 165/A, 70126 Bari, Italy
* Correspondence: giuseppe.natrella@uniba.it

Abstract: Burrata is a fresh pasta filata cheese manufactured in Italy. Its demand on the worldwide market is constantly growing, and prolonging its shelf-life is an important challenge for the Italian dairy industry. In the present study, combining a commercial bio-protective starter and modified atmosphere packaging (MAP) was evaluated as a strategy to delay the spoilage of product quality. Three experimental samples of burrata were produced by experimental trials at the industrial level and stored for 28 days under refrigerated conditions. Two samples contained the protective starter but were packaged differently (under MAP and immersed in water), and one did not contain the starter and was packaged under MAP. A sample of burrata without a starter and immersed in water was also prepared and used as a control. The combination of MAP and bio-protective starter delayed the degradation of lactose and citric acid, used as indices of microbial activity. In fact, lower counts of *Enterobacteriaceae* and *Pseudomonas* were observed in this sample. In contrast, control burrata had the highest level of total Volatile Organic Compounds (VOC) at the end of the storage period, because of higher microbial activity. Even though all samples were judged to be unacceptable after 28 days from the sensory point of view, the sample with bio-protective starter under MAP had the best score after 21 days, obtaining a shelf-life extension of about 7 days with respect to control. In conclusion, the combination of MAP and protective starter culture could be an easy way to extend the shelf-life of burrata stored under correct refrigerated conditions.

Keywords: burrata cheese; shelf life; protective culture starter; MAP; sensory analysis; chemical analysis

Citation: Natrella, G.; Gambacorta, G.; Faccia, M. Application of Commercial Biopreservation Starter in Combination with MAP for Shelf-Life Extension of Burrata Cheese. *Foods* **2023**, *12*, 1867. https://doi.org/10.3390/foods12091867

Academic Editors: Barbaros Özer and Bianca Castiglioni

Received: 16 March 2023
Revised: 13 April 2023
Accepted: 27 April 2023
Published: 30 April 2023

1. Introduction

Shelf-life extension of fresh food products is increasingly requested by manufacturers, retailers and consumers. For this challenging task, both conventional (i.e., thermal treatments, cooling, freezing, water activity reduction and use of antimicrobial molecules) and new technologies (i.e., high-pressure processing, pulsed electric fields, ultrasound, membrane filtration) and a combination of them, can be applied [1]. In general, products that suffer the most from short shelf life are those with high pH, water activity and fat content, as are fresh cheeses. For these dairy products, several innovative strategies have been proposed to improve preservability: use of natural bio-active preservatives such as endolysins, lysozyme, lactoferrin or essential oils [2–7]; strategies involving preserving brines [8–11]; non-thermal treatments such as high-pressure processing, pulsed light, ultrasonication and cold plasma [12–16]. Moreover, novel packaging systems (edible films and coatings), modified atmosphere packaging or protective cultures have been tested [17–23].

A fresh cheese that is becoming very popular worldwide is Burrata, an Italian pasta filata cheese with a "double-structured texture". In fact, it is composed of an external pasta filata pouch filled with a cream called "stracciatella," made of double cream mixed with thin strings of mozzarella. The chemical composition is characterized by high fat and moisture content and high pH; consequently, it is highly perishable due to the easy growth of microorganisms and oxidation [24]. For a long time, its consumption has been limited to the

Apulia Region (Southern Italy), where it was developed at the beginning of the last century. The recent improvement in the hygiene conditions of the dairies and the microbiological quality of milk and cream led to better preservability. Still, efforts are continuously made to get further shelf life extension. To this aim, several technological solutions have been proposed, including the application of low temperatures during the whole processing phase [25], the use of protective cultures against spoilage bacteria [26], the use of modified atmosphere packaging (MAP) alone or in combination with lysozyme/EDTA disodium salt [27].

Moreover, a combination of antimicrobial molecules, active coating and MAP was tested by Costa et al. [28]. Other researchers aimed to reduce the fat content to lower the susceptibility to oxidation and, at the same time to improve the nutritional characteristics. Trani et al. [24] partially replaced fat with carob seeds flour and fortified the obtained cheese with polyunsaturated-fatty acids (PUFAs), Costantino et al. [29] used semi-skimmed milk and reduced fat-cream combined with carrageenan xanthan or exopolysaccharides produced by starters. In most of these studies, the results were not applicable in practice due to legal concerns, high costs or excessive variation of the organoleptic characteristics.

The present work aimed to evaluate an innovative and cheap solution applicable at an industrial scale to extend burrata shelf-life without using chemical preservatives. To this purpose, an approach not yet tested on this cheese, which combines a mixed protective starter culture with MAP, was tested. In particular, the investigation focused on the changes in the chemical and sensory characteristics of the cheese during refrigerated storage.

2. Materials and Methods

2.1. Sample Preparation

The cheese samples were prepared in a dairy in Andria (Apulia region, Southern Italy) by dedicated cheesemaking trials (three replicates). As burrata cheese can be sold while wrapped in flexible plastic bags (High-Density Polyethylene with 2 μm of thickness) or not wrapped and immersed in water packaged in trays sealed with plastic film, the experimentation considered both types. As MAP cannot be applied to the packages containing water, the experimental samples were prepared as follows:

- Wrapped burrata with protective starter packaged in MAP 70 N_2% and 30% CO_2 (coded as Ferm-MAP);
- Wrapped burrata packaged in MAP 70 N_2% and 30% CO_2 (coded as MAP);
- Water-immersed burrata with protective starter (coded as Ferm).

A burrata sample without the addition of a starter and immersed in water was used as a control (coded as Ctr).

Details about the experimental design are shown in Figure 1. Briefly, pasteurized milk was pre-acidified with lactic acid to pH 5.8 and then coagulated with 0.2 mL L^{-1} of microbial rennet (153 IMCU, pure chymosin, Sacco srl, Cadorago, Italy); an automatic cutter included in the industrial vat cut the curd (about 1–2 cm cubes) and whey drainage was subsequently done, the acidified curd was mechanically stretched and shaped in hot water (83 °C) to obtain a mozzarella ball. The warm ball was immediately transferred to a "blowing machine" that inflated it to form a pouch and then filled with stracciatella cream (previously prepared to mix UHT cream with freshly prepared mozzarella strands, 1:1 *w*/*w*). After the filling phase, the burrata was tied with a plastic strip and cooled down in chilled water. The protective starter culture was added to the experimental samples by inserting it into the UHT cream used to prepare stracciatella. It was a mixture of *Lactobacillus rhamnosus* (with probiotic function) and *Lactobacillus plantarum* (bacteriocin producer) supplied by Sacco srl (Cadorago, Italy). The samples (125 g weight) were stored in refrigerated conditions as indicated by the sector legislation for this type of dairy product (4 ± 2 °C) and were analyzed in duplicate at 0, 7, 14, 21 and 28 days after the manufacturing process. The sampling time went beyond the expiry date fixed by the manufacturer (14 days from production) to evaluate the effectiveness of the technological solution tested in extending shelf life.

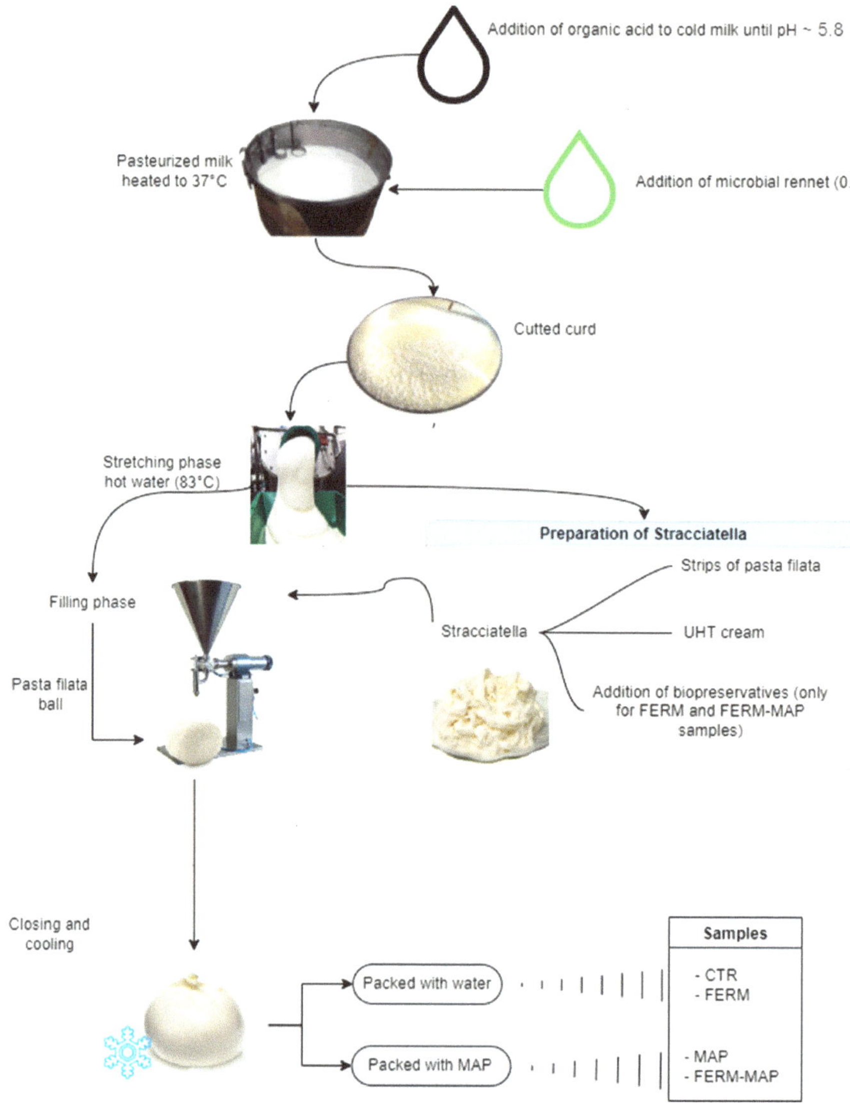

Figure 1. Experimental design of the burrata production.

2.2. Chemical Analyses

The lactose content was determined by following the method reported by Natrella et al. [25]. In brief, ten grams of blended samples were inserted in a 50 mL plastic falcon and added 20 mL of water, shacked for 1 h and then centrifuged at 4 °C × 6000 RCF × 10 min. The supernatant was filtered through a 0.2 µm syringe filter, and 10 µL aliquot was injected into the High-Performance Liquid Chromatography-Refractive Index Detectors system (HPLC-RID) (Agilent Technologies, Palo Alto, CA, USA) equipped with a Rezex RCM-monosaccharide Ca^{2+} column (300 × 7.8 mm, Phenomenex, Torrance, CA, USA) heated at 80 °C. The chromatographic runs were performed in isocratic conditions using deionized water as a mobile phase at 1 mL min^{-1} flow rate; the RID temperature was set at 40 °C. The identification and quantification of the lactose were made by using a reference standard of lactose (SIGMA-ALDRICH, Steinheim, Germany) and an external calibration curve.

Organic acids were separated, as reported by Natrella et al. [3]. The extraction was done as follows: 5 g of blended sample were added with 20 mL of 0.1% orthophosphoric acid in water, shacked for 30 min and then centrifuged at 4 °C × 5000 RCF × 15 min. The supernatant was filtered through a 0.2 μm syringe filter, and 20 μm aliquot was injected into the High-Performance Liquid Chromatography-Diode Array Detector system (HPLC-DAD) (Waters 996, Milford, MA, USA). A Synergi Hydro-RP80 (250 × 4.6 mm, Phenomenex Ltd., Aschaffenburg, Germany) column set to 30 °C was used for the separation. The mobile phase was composed of 0.1% orthophosphoric acid in water (A) and Acetonitrile (B); all solvents were HPLC-GRADE. The gradient was 0–18 min 100% A at 1 mL min^{-1} flow rate, then 18–18.3 min from 100% to 20% A; 18.3–19.5 min increasing flow rate to 1.4 mL min^{-1}, then 19.5–22.5 isocratic and 22.5–23 min from 20% to 100% A held for 20 min in an isocratic condition. Detection was done at λ = 214 nm, and the results were expressed as mg g^{-1} of the sample by using external calibration curves of the analytes considered (SIGMA-ALDRICH, Steinheim, Germany).

Free fatty acids were extracted as reported by McCarthy et al. [30] and separated by a Gas Chromatography-Flame Ionization Detector (GC-FID) system (7890AGC-System, Agilent Technologies, Palo Alto, CA, USA). In brief, 2 μl of the extract was injected into the injector port, and the separation was performed using a HP5 column (30 m × 0.32 mm × 0.25 μm; Agilent Technologies). The oven parameters were: starting temperature 35 °C held for 1 min, then 15 °C min^{-1} until 75 °C; once reached, the temperature was held for 1 min, 3 °C min^{-1} until 90 °C, 20 °C min^{-1} until 180 °C; and finally, held at the isothermal temperature for 5 min. FID temperature was set at 220 °C. Free fatty acids were identified and quantified using standard calibration curves of the analytes considered (SIGMA-ALDRICH, Steinheim, Germany), and results were expressed as mg 100 g^{-1} of the sample.

VOC analysis was done, as reported by Natrella et al. [31]. The extraction technique used was Solid Phase Micro Extraction (SPME) with a divinylbenzene/carboxen/polydimethylsiloxane (DVB/CAR/PDMS) 50/30 mm SPME fiber assembly (Supelco, Bellefonte, PA, USA). The extraction was done by using a Triplus RSH autosampler installed on a GC-MS system (Thermo Fisher Scientific, Milan, Italy). One gram of the sample and 10 μL of the internal standard (3-pentanone, 81.2 ng) were inserted in a 20 mL glass vial and closed by a silicone/PTFE septum and aluminum cap, and then all vials were loaded into the autosampler. The equilibration time was 10 min at 37 °C, then the extraction was carried out at 37 °C for 15 min, during which the fiber was exposed to the vial headspace. The desorption phase was done into the injector port at 220 °C for 2 min by using the helium as carrier gas (99.9995% of purity, Nippon Gasses Industrial srl, Milan, Italy). The molecules were separated by using a VF-WAX MS thermo capillary column (60 m, 0.25 μm i.d., 0.25 mm, Agilent J&W) under the following conditions: oven temperatures 50 °C for 0.1 min, then 13 °C min^{-1} to 180 °C, 18 °C min^{-1} to 220 °C final isothermal for 1.5 min. The detector parameters were detector voltage 1700 V; source temperature 250 °C, ionization energy 70 eV and scan range 33–200 amu. Peak identification was made by means of Xcalibur V2.0 software, in particular Qual Browse (Thermo Scientific, Waltham, MA, USA), by matching with the reference mass spectra of the National Institute of Standards & Technology (NIST) library, while the semi-quantification was done by using the standard internal method.

2.3. Microbiological Analyses

The counts of the most important spoilage and hygiene indicators bacteria were carried out. Ten grams of burrata and 90 mL of Butterfield's phosphate-buffered water (Difco, Sparks, MD, USA) were inserted into a Stomacher bag and homogenized using a BagMixer stomacher (Interscience, St Nom, France). Then, serial dilutions were made of the homogenate and plated on the appropriate media in Petri dishes following the standard methods: *Enterobacteriaceae* [32]; *Pseudomonas* spp. [33]; *Escherichia coli* [34]; coagulase-positive *Staphylococci* [35] and Coliforms [36].

2.4. Sensory Analysis

The analyses were performed by a trained panel composed of 5 experts belonging to the Italian Association of Cheese Tasters (ONAF) with at least five years of experience in the field and selected following ISO 8586:2012 [37]. The TCATA (Temporal Check All That Apply, a qualitative analysis) test is an expansion of the CATA (Check All That Apply) method; it was successfully used in a previous paper with a different approach compared to the typical use of this test, aiming to understand the differences among samples during their shelf-life [25]. The CATA approach (and its expansion) belongs to the flash profile methods used to quickly obtain a sensory profile and is commonly used in consumer science tests involving a high number of consumers. This method was originally used with trained panelists, then its popularity grew with consumers as judges for marketing purposes [38]. However, it has been demonstrated that results provided by the CATA test with a high number of consumers generate similar responses of a classical product characterization (i.e., Quantitative Descriptive Analysis) made by a trained panel [39,40]. For each analysis time (T0, T7, T14, T21 and T28), panelists were provided with a list of descriptors (previously selected in two preliminary sessions by expert panelists) from which they had to select all the words that apply and better describe the product. Sensory evaluations were conducted in a sensory laboratory equipped with individual cabins

2.5. Statistical Analysis

The dataset was used to perform several analyses using XLSTAT (Addinsoft, Paris, France). Analysis of variance (ANOVA) was performed for chemical and microbiological data; Agglomerative Hierarchical Clustering (AHC) was used for VOC profile grouping, while Temporal Check All That Apply (TCATA) was performed for sensory analysis results.

3. Results and Discussion

The variations in the content of lactose, organic acids, FFA and VOC were used as chemical indices to monitor the decay process during cheese storage [3,41]. The lactose and organic acids contents are reported in Figure 2. As expected, lactose decreased during storage, but at different rates: the water-immersed samples underwent a dramatic decrease reaching the final concentrations of 2.59 (Ctr) and 2.72 mg g^{-1} (Ferm), whereas the MAP-packaged samples evidenced a slower drop with the final concentration of 10.63 mg g^{-1} (MAP) and 9.82 mg g^{-1} (Ferm-MAP). The better preservation of lactose in the MAP samples cannot be simply ascribed to an antimicrobial effect of the modified atmosphere since lactose tends to dissolve in the packaging water [41]. The samples packaged in water evidenced already at day 0 lower concentration with respect to the MAP samples (14.0 and 14.4 mg g^{-1} for Ctr and Ferm vs. 15.26 and 15.45 mg g^{-1} for MAP and Ferm-MAP).

Further information is derived from the analysis of the organic acids (Figure 2B), whose formation is strictly related to the metabolism of microorganisms and consequently to the storage time; several authors have tried to classify cheese age by determining the organic acid profile [42–44]. Lactic acid is the main end-product of bacterial activity in all cheeses. It tends to decrease in long-stored/ripened types; other acids (acetic, formic, propionic and others) are secondary fermentation products. Instead, citric acid derives from milk and tends to disappear over time, being used as a fermentation substrate [45–47].

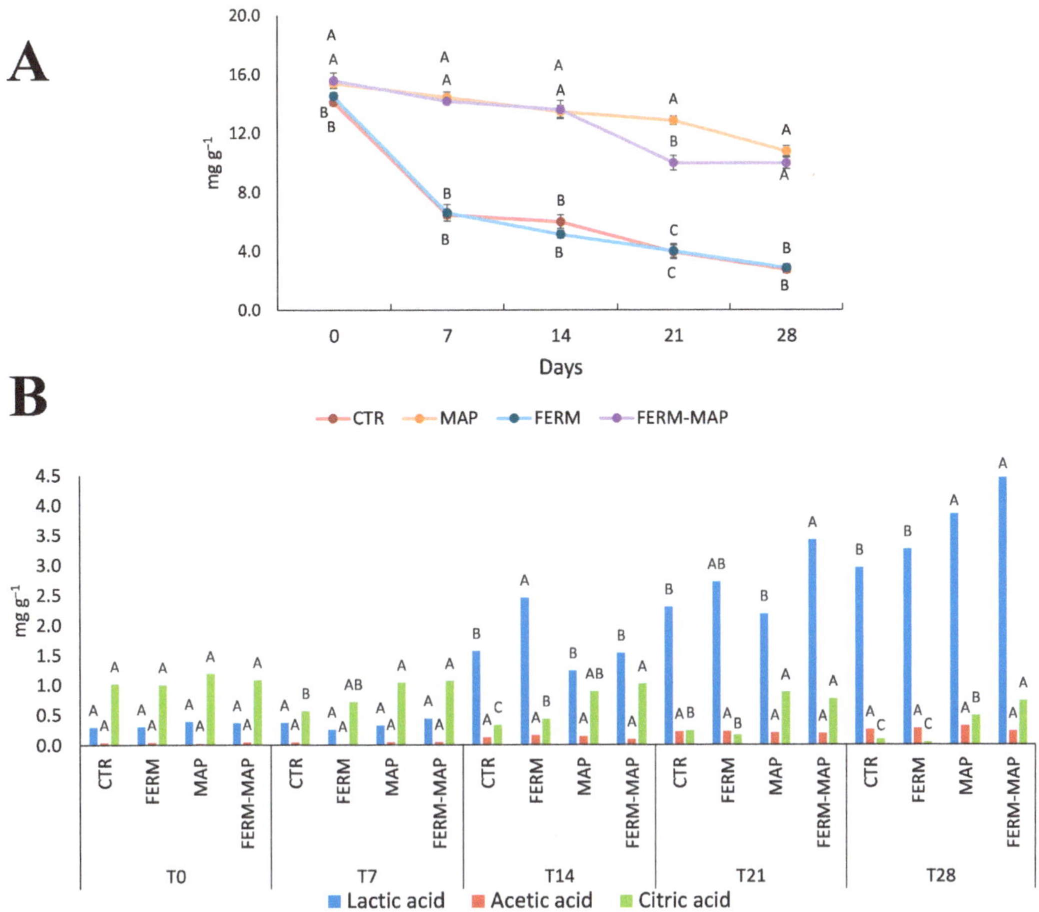

Figure 2. Lactose (**A**) and organic acid (**B**) content in burrata samples, each sampling time was considered separately. Ctr—control burrata; MAP—burrata packaged in the modified atmosphere; Ferm—burrata added microbial protection culture; Ferm-MAP—burrata added microbial protection culture and packaged in modified atmosphere. Results expressed as mg g^{-1} and statistically different at $p < 0.05$. Different letters indicate significant differences among samples considering each time separately (**A**). Different letters indicate significant differences among samples considering both time and acids separately (**B**).

In detail, no differences among samples were observed in the lactic acid content until 2 weeks of storage, as refrigerated storage limited formation. Then, the concentration increased, and differences among samples became significant, with the samples packaged in MAP showing the highest level after 28 days. Concerning acetic acid, the results were similar to that reported by Faccia et al. [48] and Tirloni et al. [49]. Commonly, fresh products like burrata contain a small amount of this acid, and at low concentrations, its role in the sensory characteristics is negligible. The increase of this compound during storage was very slow, reaching the highest amount of 0.32 mg g^{-1} after 28 days in the MAP sample. Finally, citric acid decreased over time; the initial concentration ranged from 1.01 to 1.2 mg g^{-1}, while at the end of the storage, it decreased to a different extent, resembling the behavior of lactose. In fact, the decrease was much more evident in water-immersed than in the MAP samples (0.06 and 0.1 mg g^{-1} in Ctr and Ferm vs. 0.50 and 0.74 mg g^{-1} in

MAP and Ferm-MAP, respectively). Overall, organic acid solubilization in water played a role as they are also water-soluble.

Figure 3 shows the content of free fatty acids (FFA), which is an index of lipolysis that, in cheese, is performed by lipases from milk, rennet paste (when used), and starter and non-starter microorganisms [50,51]. Differently from many other cheeses, lipolysis in Burrata is undesirable since FFA negatively affects the desired mild aroma. For this reason, rennet paste is not used in the manufacturing process, and curd acidification for stretching is not obtained by lactic starter fermentation but by direct milk acidification. Normally, these technological conditions, together with refrigerated storage and the short time from production to consumption, do not allow the significant breakdown of triglycerides into free fatty acids [25]. In the present experimentation, the addition of a protective starter represented a new variable that required monitoring of lipolysis. The obtained results indicated a delay of lipolysis in the three experimental samples compared to the control, with the combination MAP + protective culture showing the best result, probably as a consequence of a synergic effect. The effect was much more evident after 14 days, whereas a fluctuating trend of FFA concentration was observed in the first stages of storage. It was probably due to the concomitant formation and conversion of FFA to other compounds, as reported by several authors [52,53]. The Ferm sample had the highest amount of FFA at day 0, mostly ascribable to the increased presence of free palmitic acid. Then, the MAP sample showed the highest total FFA concentration between 7 and 14 days of storage, also in this case, due to the palmitic acid level. From 21 to 28 days of storage, Ctr evidenced the highest formation of volatile and semivolatile FFA (butyric, caproic, caprylic, capric and myristic) that are known to negatively affect the sensory characteristics of burrata, and almost twice the higher content of total FFA compared to the three experimental samples. Considering the sum of short, medium and long-chain fatty acids (SCFA, MCFA and LCFA), all samples shared the absence of SCFA and the presence of LCFA at day 0; it is worth highlighting that LCFA does not impact any flavor, since they have high odor perception threshold [54]. After 7 days, SCFA became detectable in all samples, even though the concentrations of butyric, caproic and caprylic acids were very low and far from levels able to alter the sensory characteristics. After 14 days of storage, the MAP and Ctr samples had higher amounts of SCFA and LCFA than the samples added with protective cultures, but successively, the situation was reversed, and a significant effect of MAP in delaying lipolysis was observed until the end of the storage.

The observed chemical changes must be interpreted in the light of the storage conditions applied, since refrigeration is known to favor psychrotrophic bacteria that are highly lipolytic [55,56]. Table 1 shows the evolution of the counts of the five bacteria groups considered over time.

In general, the microbial quality of the samples was very good in comparison to the count values reported in the literature [27,28,57]. On fresh burrata samples, the highest counts were found in Ctr and MAP for *Enterobacteriaceae*, Coliforms and *Pseudomonas*, with values ranging from 2.0 to 2.3 log CFU g^{-1}. Probably an immediate control of the starter on these bacteria was observed, limiting their growth in the early storage phase. In addition, the combination of starter and MAP showed the lowest counts (1.5 to 1.9 log CFU g^{-1} for these three microbial species). After 7 days of storage, only the *Pseudomonas* counts showed the highest value in Ctr sample. The differences became more evident after 14 days of storage, then increased over time and regarded all the bacteria groups. Considering the *Enterobacteriaceae* 2.7 log CFU g^{-1} count were found in Ctr at the end of storage, vs. 0.9–1.2 log CFU g^{-1} in the other samples; a similar trend was observed for *Pseudomonas*, ranging from 1 to 1.4 log CFU g^{-1} in the experimental samples vs. 2 log CFU g^{-1} found in the Ctr. The comparison among experimental samples evidenced that the best outcomes were obtained in the case of a combination of the protective starter with MAP; the use of *Lactobacillus plantarum*, as a producer of bacteriocins, may have played a major role, as widely demonstrated in the literature [58–60]; however, a synergic effect with MAP seems to be evident.

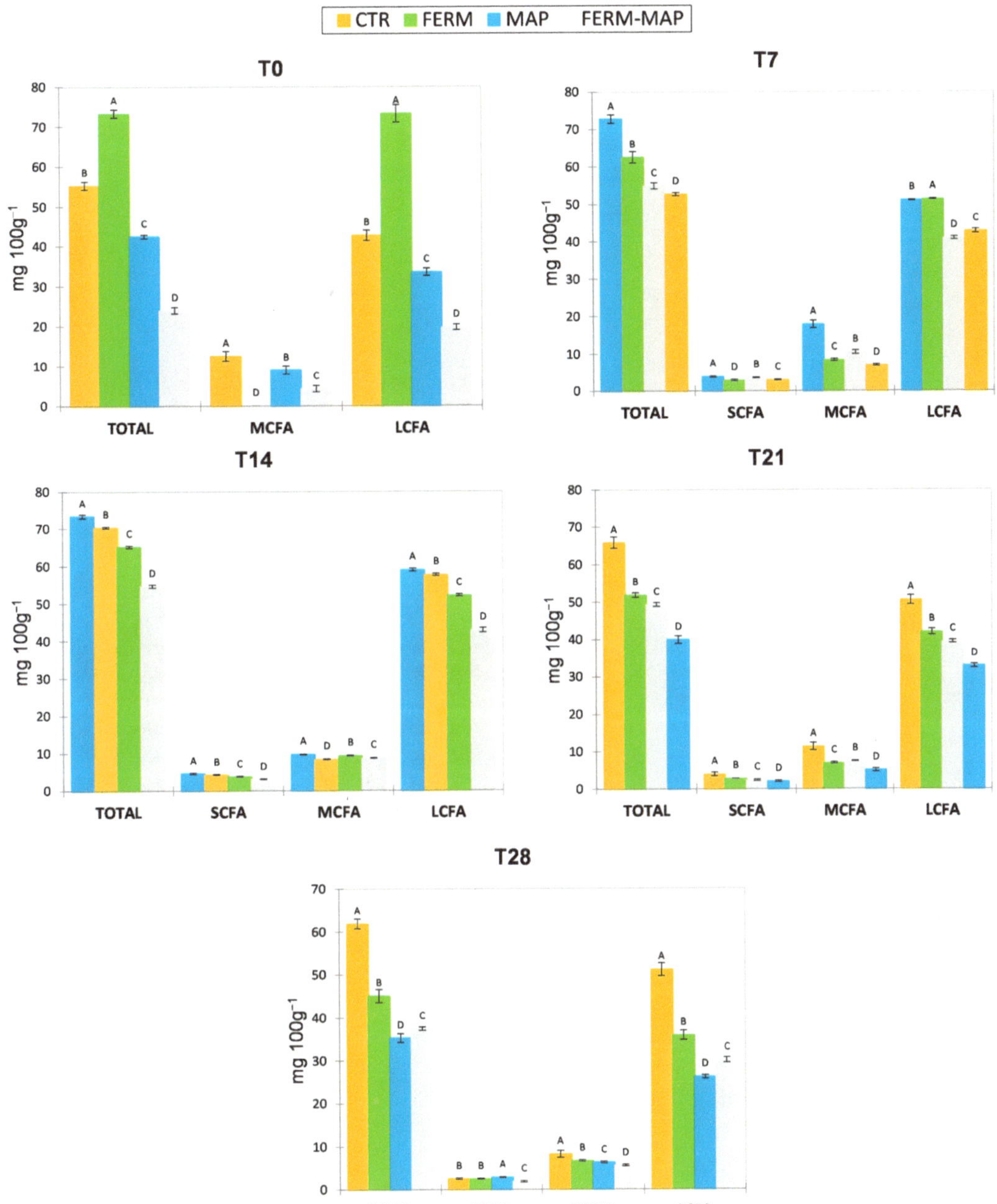

Figure 3. Free fatty acids content grouped by short, medium and long chain fatty acid (SCFA, MCFA and LCFA, respectively) and total amount found in burrata samples from T0 to T28. Ctr—control burrata; MAP—burrata packaged in the modified atmosphere; Ferm—burrata added of microbial protection culture; Ferm-MAP—burrata added of microbial protection culture and packaged in modified atmosphere. Results are expressed as mg 100 g^{-1} ($p < 0.05$). Different letters indicate significant differences among samples considering the total, SCFA, MCFA and LCFA separately at each time.

Table 1. Cell counts (Log CFU g^{-1} ± SD), the average value of three replicates. Different letters indicate the differences among samples (storage times were considered separately). $p < 0.05$.

LOG CFU g^{-1}	Storage (Days)	*Enterobacteriaceae*	Coliforms	*E. Coli*	*Staphylococcus* Coagulase +	*Pseudomonas*
Ctr	T0	2.2 ± 0.1 ^{A}	2.1 ± 0.1 ^{A}	<1 ^{A}	<1 ^{A}	2.3 ± 0.1 ^{A}
Ferm	T0	1.9 ± 0.1 ^{B}	1.9 ± 0.1 ^{B}	<1 ^{A}	<1 ^{A}	2.0 ± 0.1 ^{BC}
MAP	T0	2.0 ± 0.1 ^{AB}	2.0 ± 0.1 ^{AB}	<1 ^{A}	<1 ^{A}	2.1 ± 0.1 ^{AB}
Ferm MAP	T0	1.7 ± 0.2 ^{C}	1.5 ± 0.1 ^{C}	<1 ^{A}	<1 ^{A}	1.9 ± 0.1 ^{C}
Ctr	T7	1.0 ± 0.1 ^{B}	1.3 ± 0.2 ^{C}	<1 ^{A}	<1 ^{A}	3.3 ± 0.1 ^{A}
Ferm	T7	2.7 ± 0.3 ^{A}	2.6 ± 0.2 ^{AB}	<1 ^{A}	<1 ^{A}	3.0 ± 0.1 ^{B}
MAP	T7	2.8 ± 0.2 ^{A}	2.7 ± 0.1 ^{A}	<1 ^{A}	<1 ^{A}	2.7 ± 0.1 ^{C}
Ferm MAP	T7	2.6 ± 0.2 ^{A}	2.5 ± 0.1 ^{B}	<1 ^{A}	<1 ^{A}	2.5 ± 0.1 ^{D}
Ctr	T14	1.0 ± 0.1 ^{A}	1.0 ± 0.1 ^{B}	<2 ^{A}	<2 ^{A}	3.5 ± 0.4 ^{A}
Ferm	T14	0.7 ± 0.1 ^{B}	2.9 ± 0.3 ^{A}	<2 ^{A}	<2 ^{A}	3.0 ± 0.2 ^{A}
MAP	T14	0.7 ± 0.1 ^{B}	2.9 ± 0.3 ^{A}	<2 ^{A}	<2 ^{A}	3.0 ± 0.3 ^{A}
Ferm MAP	T14	0.3 ± 0.1 ^{C}	2.9 ± 0.4 ^{A}	<2 ^{A}	<2 ^{A}	2.9 ± 0.2 ^{A}
Ctr	T21	1.7 ± 0.3 ^{A}	1.7 ± 0.2 ^{A}	<2 ^{A}	<2 ^{A}	3.6 ± 0.2 ^{A}
Ferm	T21	0.9 ± 0.1 ^{B}	0.8 ± 0.1 ^{B}	<2 ^{A}	<2 ^{A}	3.3 ± 0.1 ^{B}
MAP	T21	0.8 ± 0.2 ^{BC}	0.7 ± 0.1 ^{BC}	<2 ^{A}	<2 ^{A}	3.0 ± 0.2 ^{B}
Ferm MAP	T21	0.6 ± 0.1 ^{C}	0.6 ± 0.1 ^{C}	<2 ^{A}	<2 ^{A}	2.9 ± 0.3 ^{B}
Ctr	T28	2.7 ± 0.3 ^{A}	2.6 ± 0.2 ^{A}	<3 ^{A}	<2 ^{A}	2.0 ± 0.2 ^{A}
Ferm	T28	1.2 ± 0.1 ^{B}	1.1 ± 0.1 ^{B}	<2 ^{B}	<2 ^{A}	1.4 ± 0.2 ^{B}
MAP	T28	1.1 ± 0.1 ^{BC}	1.0 ± 0.1 ^{BC}	<2 ^{B}	<2 ^{A}	1.2 ± 0.1 ^{B}
Ferm MAP	T28	0.9 ± 0.1 ^{C}	0.9 ± 0.1 ^{C}	<2 ^{B}	<2 ^{A}	1.0 ± 0.1 ^{C}

The volatile organic compounds data was used to perform Agglomerative Hierarchical Clustering (AHC) based on dissimilarities. It is a clustering (or classification) method that allows to group of samples into clusters by the dendrogram, aiming to better observe the differences in the VOC profile evolution (Figure 4). The dotted line represents the truncation that generates four different clusters, in which burrata samples were separated by the evolution of the VOC profile during storage. The differences among samples' VOC profiles were evident, showing Ctr as having the fastest VOC production process that, overall, penalizes the product.

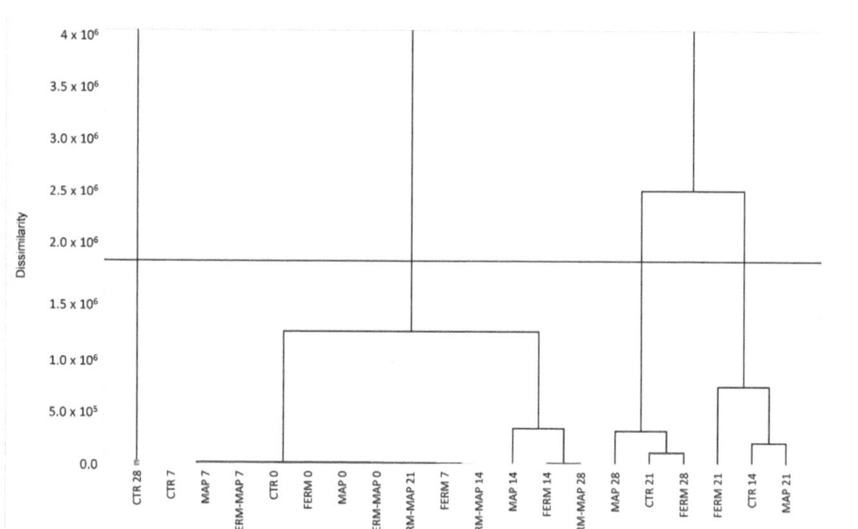

Figure 4. Agglomerative Hierarchical Clustering (AHC) based on dissimilarities of burrata samples volatile organic compounds profile.

The red cluster is more homogeneous than the others being dendrogram flatter; it represents the VOC profile of samples ranging from 0 to 14 days of storage (with some exceptions). This cluster grouped the products until 14 days of storage, but among them are also present the samples having 21 and 28 days of storage, which are Ferm-MAP. The second cluster, the purple one, grouped Ferm and MAP after 21 days of storage and Ctr after 14 days. This meant that the VOC profile of the latter was ascribable to the older experimental samples. Then, the same results were found within the green cluster showing Ctr after 21 days which clustered with MAP and Ferm after 28 days of storage. Finally, the fourth cluster (black one) is represented only by Ctr T28, the sample with the highest amount of VOC. Therefore, it is clear that among samples, Ctr was the richest in volatile compounds and the one that perishes more rapidly, being grouped with older experimental samples. The high VOC production is usually related to microbial growth [61].

On the contrary, the most amazing results were ascribable to the Ferm-MAP sample, which showed reduced VOC production compared to Ctr. Moreover, such a sample at days 21 and 28 clustered with samples within the Ctr producers' shelf-life, theoretically highlighting a more delicate aroma than the Ctr sample. This result suggests a strong inhibition of volatile production by combining protective culture starter and MAP, leading to a less pronounced off-flavor. To sum up, the VOC profile of the Ctr sample after 14 and 21 days of storage was grouped with older experimental samples, while Ctr after 28 days was the worst sample, the richest in VOC compounds. The total VOC amount of Ctr at the end of the storage had a 3-fold higher concentration than the other samples (6641 $\mu g\ kg^{-1}$ vs. 1467, 1925 and 653 $\mu g\ kg^{-1}$ of Ferm, MAP and Ferm-MAP at T28, respectively). These results were better than those reported by Natrella et al. [25], in which samples after 21 days had a higher total VOC than that found in this work after 28 days of storage, which could reflect a higher off-flavor perception. After 28 days, the highest chemical classes found were alcohols and acids for all samples except for Ferm-MAP, in which the ketones class was the most representative.

The sensory results were subjected to multivariate analysis to summarize the evolution during the storage of the samples (Figure 5). The two components (PCs) of the plot explained 71.82% of the total variance. In general, the first PC explained most of the variance with 59.82%, in which it is possible to observe the separation among samples based on the storage time as a discrimination variable. In fact, sample worsening is visible from the left (fresh products) to the right (stored products). On the other hand, dissimilarity among the type of sample is shown along the second principal component. Better results were obtained by combining protective culture starter with MAP, which followed the negative side of the second PC in the bottom right quadrant, where better sensory descriptors were placed compared to the ones on the upper-right side. During the first week of storage, no differences were found among the samples—all remained fresh and similar to each other. Then, starting from days 14 to 28, the samples separate into space, with Ctr and Ferm upward on the positive side of PC2 and MAP and Ferm-MAP downward through the negative side. The differences among samples were based on the main sensory characteristic perceived by panelists, mostly after the two weeks of storage. Firstly Ctr, then Ferm at day 28 were described as having sulfuric and sour milk odor notes, while MAP and Ferm-MAP samples were characterized by notes of yogurt and cooked milk. Thus, milder descriptors were used to refer to the latter samples, even if, after 28 days, all samples were considered inedible. On the contrary, after 21 days, the MAP and Ferm-MAP samples were considered by the panelists to be very similar to the same sample after 14 days of storage.

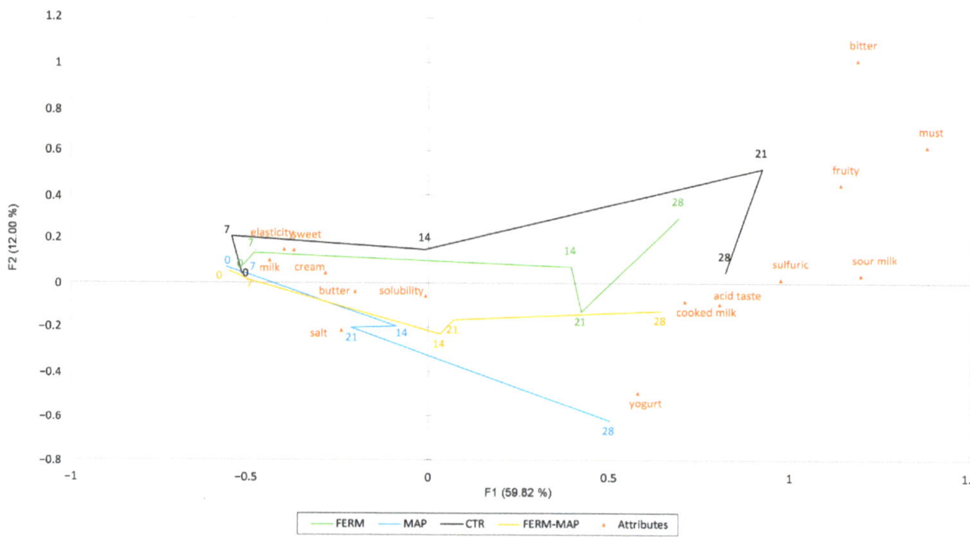

Figure 5. Multivariate analysis, Temporal Check All That Apply (TCATA) based on sensory results. Ctr—Control burrata; MAP—burrata packaged in the modified atmosphere; Ferm—burrata added of microbial protection culture; Ferm-MAP—burrata added of microbial protection culture and packaged in modified atmosphere.

In sum, the overall characteristics decrease over time for almost all products (from the left to the right side of the plot); the panelists highlighted the worst quality of samples after 28 days of storage, while after 21 days, some samples still were acceptable, i.e., MAP packaged samples were similar to the same samples at T14, which still had good characteristics, and were very different from Ctr. These outcomes revealed that MAP alone or in combination with protective starter, whose bacteriocins have been demonstrated to have antimicrobial activity against a wide range of bacteria [58–60], better preserved the sensory characteristics of burrata in agreement with VOC and the other chemicals results.

4. Conclusions

The use of MAP in conjunction with bioprotective starter allowed for better preserve the chemical and sensory characteristics of burrata, which gained a few days of shelf life compared to the control (from 14 to 21 days). The combination of a modified atmosphere and protective starter allowed, during storage, a significant delay in lactose and citric acid degradation, which are directly linked to microbial growth. Accordingly, the microbial counts of undesired bacteria (*Enterobacteriaceae*, Coliforms, *Staphylococcus*, *Pseudomonas* spp.) were lower than control burrata as well as microbial metabolites such as organic acids, free fatty acids and volatile organic compounds, among which are more probable to find molecules responsible of off-flavors in this cheese. These results were confirmed by the sensory evaluation. In conclusion, the synergy between the modified atmosphere and bioprotective starter, in conjunction with the good quality of raw matter and good manufacturing practices, can significantly improve the microbiological stability of burrata without using chemical additives. Further investigation is needed to assess the effect of the application of modified atmosphere packaging in the presence of the governing liquid when it is technically possible.

Author Contributions: G.N. contributed to the investigation, writing, review and editing of the original draft. G.G. contributed to writing, investigation, supervision, and conceptualization. M.F. contributed to conceptualization, investigation, writing, review and editing. All authors have read and agreed to the published version of the manuscript.

Funding: This research received no external funding.

Institutional Review Board Statement: Not applicable.

Informed Consent Statement: Not applicable.

Data Availability Statement: Data are contained within the article.

Acknowledgments: This work was supported by the project PSR P.S.R. Puglia 2014/2020—Misura 16—Cooperazione—Sottomisura 16.2 "Sostegno a progetti pilota e allo sviluppo di nuovi prodotti, pratiche, processi e tecnologie" project BURRATA (CUP: B89J20000130007; DDS 94250034439).

Conflicts of Interest: The authors declare no conflict of interest.

References

1. Barba, F.J.; Ahrné, L.; Xanthakis, E.; Landerslev, M.G.; Orlien, V. Innovative Technologies for Food Preservation. In *Innovative Technologies for Food Preservation*; Elsevier: Amsterdam, The Netherlands, 2018; Volume 2, pp. 25–51. ISBN 9780128110324.
2. Ibarra-Sánchez, L.A.; Van Tassell, M.L.; Miller, M.J. Antimicrobial behavior of phage endolysin PlyP100 and its synergy with nisin to control Listeria monocytogenes in Queso Fresco. *Food Microbiol.* **2018**, *72*, 128–134. [CrossRef] [PubMed]
3. Natrella, G.; Difonzo, G.; Calasso, M.; Costantino, G.; Caponio, F.; Faccia, M. Evolution of VOC and Sensory Characteristics of Stracciatella Cheese as Affected by Different Preservatives. *Foods* **2020**, *9*, 1446. [CrossRef] [PubMed]
4. Noori, N.; Yahyaraeyat, R.; Khosravi, A.; Atefi, P.; Basti, A.A.; Akrami, F.; Bahonar, A.; Misaghi, A. Effect of *Zataria multiflora* Boiss. Essential Oil on Growth and Citrinin Production by *Penicillium citrinum* in Culture Media and Mozzarella Cheese. *J. Food Saf.* **2012**, *32*, 445–451. [CrossRef]
5. Sara, G.; Nasrin, H.S.J.; Mehran, M.; Kianoush, K.D. Application of zein antimicrobial edible film incorporating Zataria multiflora boiss essential oil for preservation of Iranian ultrafiltered Feta cheese. *Afr. J. Biotechnol.* **2015**, *14*, 2014–2021. [CrossRef]
6. Barbara, S.; Milena, S.; Daniela, C.; Rosaria, C.M.; Antonio, B. Modelling the Amounts of Carbon Dioxide in the Headspace to Assess the Coliforms of Mozzarella Cheese. *Ann. Microbiol. Res.* **2017**, *1*, 4–8. [CrossRef]
7. Caputo, L.; Quintieri, L.; Bianchi, D.M.; Decastelli, L.; Monaci, L.; Visconti, A.; Baruzzi, F. Pepsin-digested bovine lactoferrin prevents Mozzarella cheese blue discoloration caused by Pseudomonas fluorescens. *Food Microbiol.* **2015**, *46*, 15–24. [CrossRef]
8. Huang, X.; Nzekoue, F.K.; Renzi, S.; Alesi, A.; Coman, M.M.; Pucciarelli, S.; Sagratini, G.; Silvi, S. Influence of modified governing liquid on shelf-life parameters of high-moisture mozzarella cheese. *Food Res. Int.* **2022**, *159*, 111627. [CrossRef]
9. Faccia, M.; Gambacorta, G.; Natrella, G.; Caponio, F. Shelf life extension of Italian mozzarella by use of calcium lactate buffered brine. *Food Control* **2019**, *100*, 287–291. [CrossRef]
10. Öztürk, M.; Güncü, B.G. Effect of Brine Calcium Concentration on the Surface Solubilization and Texture of Fresh Perline Mozzarella Cheese. *Turk. J. Agric. Food Sci. Technol.* **2021**, *9*, 650–654. [CrossRef]
11. Mizuno, R.; Abe, T.; Koishihara, H.; Okawa, T. The Effect of Preservative Liquid Composition on Physicochemical Properties of Mozzarella Cheese. *Food Sci. Technol. Res.* **2016**, *22*, 261–266. [CrossRef]
12. Ricciardi, F.E.; Plazzotta, S.; Conte, A.; Manzocco, L. Effect of pulsed light on microbial inactivation, sensory properties and protein structure of fresh ricotta cheese. *LWT* **2021**, *139*, 110556. [CrossRef]
13. Wan, Z.; Misra, N.; Li, G.; Keener, K.M. High voltage atmospheric cold plasma treatment of Listeria innocua and Escherichia coli K-12 on Queso Fresco (fresh cheese). *LWT* **2021**, *146*, 111406. [CrossRef]
14. Coutinho, N.M.; Silveira, M.R.; Rocha, R.S.; Moraes, J.; Ferreira, M.V.S.; Pimentel, T.C.; Freitas, M.Q.; Silva, M.C.; Raices, R.S.; Ranadheera, C.S.; et al. Cold plasma processing of milk and dairy products. *Trends Food Sci. Technol.* **2018**, *74*, 56–68. [CrossRef]
15. Hnosko, J.; Gonzalez, M.S.-M.; Clark, S. High-pressure processing inactivates Listeria innocua yet compromises Queso Fresco crumbling properties. *J. Dairy Sci.* **2012**, *95*, 4851–4862. [CrossRef]
16. Evert-Arriagada, K.; Hernández-Herrero, M.; Guamis, B.; Trujillo, A. Commercial application of high-pressure processing for increasing starter-free fresh cheese shelf-life. *LWT Food Sci. Technol.* **2014**, *55*, 498–505. [CrossRef]
17. Göksen, G.; Fabra, M.J.; Ekiz, H.I.; López-Rubio, A. Phytochemical-loaded electrospun nanofibers as novel active edible films: Characterization and antibacterial efficiency in cheese slices. *Food Control* **2020**, *112*, 107133. [CrossRef]
18. Ordoñez, R.; Contreras, C.; González-Martínez, C.; Chiralt, A. Edible coatings controlling mass loss and Penicillium roqueforti growth during cheese ripening. *J. Food Eng.* **2021**, *290*, 110174. [CrossRef]
19. Zhang, L.; Zhang, Z.; Chen, Y.; Ma, X.; Xia, M. Chitosan and procyanidin composite films with high antioxidant activity and pH responsivity for cheese packaging. *Food Chem.* **2020**, *338*, 128013. [CrossRef]
20. Mastromatteo, M.; Conte, A.; Faccia, M.; Del Nobile, M.A.; Zambrini, A.V. Combined effect of active coating and modified atmosphere packaging on prolonging the shelf life of low-moisture Mozzarella cheese. *J. Dairy Sci.* **2014**, *97*, 36–45. [CrossRef]
21. Atallah, A.A.; Ismail, E.A.; Yehia, H.M.; Elkhadragy, M.F.; Khater, E.-S.G. Proteolytic Development and Volatile Compounds Profile of Domiati Cheese under Modified Atmosphere Packaging. *Fermentation* **2022**, *8*, 358. [CrossRef]
22. Spanu, C.; Piras, F.; Mocci, A.; Nieddu, G.; De Santis, E.; Scarano, C. Use of Carnobacterium spp protective culture in MAP packed Ricotta fresca cheese to control *Pseudomonas* spp. *Food Microbiol.* **2018**, *74*, 50–56. [CrossRef]

23. Cosentino, S.; Viale, S.; Deplano, M.; Fadda, M.E.; Pisano, M.B. Application of Autochthonous *Lactobacillus* Strains as Biopreservatives to Control Fungal Spoilage in Caciotta Cheese. *BioMed. Res. Int.* **2018**, *2018*, 3915615. [CrossRef] [PubMed]
24. Trani, A.; Gambacorta, G.; Gomes, T.F.; Loizzo, P.; Cassone, A.; Faccia, M. Production and characterisation of reduced-fat and PUFA-enriched Burrata cheese. *J. Dairy Res.* **2016**, *83*, 236–241. [CrossRef] [PubMed]
25. Natrella, G.; Gambacorta, G.; Faccia, M. Use of dry ice as innovative technology to preserve the chemical and microbial characteristics of burrata cheese. *J. Food Process. Preserv.* **2022**, *46*, e16908. [CrossRef]
26. Minervini, F.; Conte, A.; Del Nobile, M.A.; Gobbetti, M.; De Angelis, M. Dietary Fibers and Protective Lactobacilli Drive Burrata Cheese Microbiome. *Appl. Environ. Microbiol.* **2017**, *83*, e01494-17. [CrossRef]
27. Conte, A.; Brescia, I.; Del Nobile, M. Lysozyme/EDTA disodium salt and modified-atmosphere packaging to prolong the shelf life of burrata cheese. *J. Dairy Sci.* **2011**, *94*, 5289–5297. [CrossRef]
28. Costa, C.; Lucera, A.; Conte, A.; Zambrini, A.V.; Del Nobile, M.A. Technological Strategies to Preserve Burrata Cheese Quality. *Coatings* **2017**, *7*, 97. [CrossRef]
29. Costantino, G.; Calasso, M.; Minervini, F.; De Angelis, M. Use of Exopolysaccharide-Synthesizing Lactic Acid Bacteria and Fat Replacers for Manufacturing Reduced-Fat Burrata Cheese: Microbiological Aspects and Sensory Evaluation. *Microorganisms* **2020**, *8*, 1618. [CrossRef]
30. McCarthy, C.M.; Kelly, P.M.; Wilkinson, M.G.; Guinee, T.P. Effect of fat and salt reduction on the changes in the concentrations of free amino acids and free fatty acids in Cheddar-style cheeses during maturation. *J. Food Compos. Anal.* **2017**, *59*, 132–140. [CrossRef]
31. Natrella, G.; Gambacorta, G.; Faccia, M. Volatile organic compounds throughout the manufacturing process of Mozzarella di Gioia del Colle PDO cheese. *Czech J. Food Sci.* **2020**, *38*, 215–222. [CrossRef]
32. *ISO 21528-2:2017*; Microbiology of the Food Chain Horizontal Method for the Detection and Enumeration of Enterobacteriaceae Part 2: Colonycount Technique. ISO (International Organization for Standardization): Geneva, Switzerland, 2017.
33. *ISO 11059:2009 (IDF/RM 225:2009)*; Milk and Milk Products. Method for the Enumeration of *Pseudomonas* spp. ISO (International Organization for Standardization): Geneva, Switzerland, 2017.
34. *ISO 16649-3:2015*; Microbiology of the Food Chain-Horizontal Method for the Enumeration of Beta-Glucuronidase-Positive Escherichia Coli—Part 3: Detection and Most Probable Number Technique Using 5-bromo-4-chloro-3-indolyl-ß-D-glucuronide. ISO (International Organization for Standardization): Geneva, Switzerland, 2015.
35. *ISO 6888-1:2018*; Microbiology of the Food Chain—Horizontal Method for the Enumeration of Coagulase-Positive Staphylococci (Staphylococcus aureus and Other Species)—Part 1: Method Using Baird-Par. ISO (International Organization for Standardization): Geneva, Switzerland, 2018.
36. *ISO 4832:2007*; Microbiology of Food and Animal Feeding Stuffs: Horizontal Method for the Enumeration of Coliforms: Colony-Count Technique. ISO (International Organization for Standardization): Geneva, Switzerland, 2007.
37. *ISO 8586:2012*; Sensory Analysis—General Guidelines for the Selection, Training and Monitoring of Selected Assessors and Expert Sensory Assessors. ISO (International Organization for Standardization): Geneva, Switzerland, 2012.
38. Ares, G.; Jaeger, S. Check-all-that-apply (CATA) questions with consumers in practice: Experimental considerations and impact on outcome. *Rapid Sens. Profiling Tech.* **2023**, 257–280. [CrossRef]
39. Ares, G.; Barreiro, C.; Deliza, R.; Giménez, A.; Gámbaro, A. Application of a check-all-that-apply question to the development of chocolate milk desserts. *J. Sens. Stud.* **2010**, *25*, 67–86. [CrossRef]
40. Dooley, L.; Lee, Y.-S.; Meullenet, J.-F. The application of check-all-that-apply (CATA) consumer profiling to preference mapping of vanilla ice cream and its comparison to classical external preference mapping. *Food Qual. Prefer.* **2010**, *21*, 394–401. [CrossRef]
41. Adda, J.; Gripon, J.; Vassal, L. The chemistry of flavour and texture generation in cheese. *Food Chem.* **1982**, *9*, 115–129. [CrossRef]
42. Lombardi, A.; Bevilacqua, A.; Califano, A. Variation in organic acids content during ripening of Reggianito cheese in air-tight sealed bags. *Food Chem.* **1994**, *51*, 221–226. [CrossRef]
43. Lues, J.; Botha, W. Relationships amongst South African processed, young and matured Cheddar cheese pertaining to organic acid content and non-starter population. *Food Res. Int.* **1998**, *31*, 449–457. [CrossRef]
44. Califano, A.N.; Bevilacqua, A.E. Freezing low moisture Mozzarella cheese: Changes in organic acid content. *Food Chem.* **1999**, *64*, 193–198. [CrossRef]
45. Tirloni, E.; Bernardi, C.; Ghelardi, E.; Celandroni, F.; Andrighetto, C.; Rota, N.; Stella, S. Biopreservation as a potential hurdle for Bacillus cereus growth in fresh cheese. *J. Dairy Sci.* **2020**, *103*, 150–160. [CrossRef]
46. Passerini, D.; Laroute, V.; Coddeville, M.; Le Bourgeois, P.; Loubière, P.; Ritzenthaler, P.; Cocaign-Bousquet, M.; Daveran-Mingot, M.-L. New insights into *Lactococcus lactis* diacetyl- and acetoin-producing strains isolated from diverse origins. *Int. J. Food Microbiol.* **2013**, *160*, 329–336. [CrossRef]
47. Alegría, Á.; González, P.; Delgado, S.; Flórez, A.B.; Hernández-Barranco, A.M.; Rodríguez, A.; Mayo, B. Characterisation of the technological behaviour of mixtures of mesophilic lactic acid bacteria isolated from traditional cheeses made of raw milk without added starters. *Int. J. Dairy Technol.* **2016**, *69*, 507–519. [CrossRef]
48. Faccia, M.; Natrella, G.; Gambacorta, G. Analysis of the water-soluble compounds as a tool for discriminating traditional and industrial high moisture mozzarella made with citric acid. *Int. J. Food Sci. Technol.* **2021**, *56*, 5352–5361. [CrossRef]
49. Tirloni, E.; Bernardi, C.; Rosshaug, P.; Stella, S. Potential growth of Listeria monocytogenes in Italian mozzarella cheese as affected by microbiological and chemical-physical environment. *J. Dairy Sci.* **2019**, *102*, 4913–4924. [CrossRef] [PubMed]

50. Thierry, A.; Collins, Y.F.; Mukdsi, M.C.A.; McSweeney, P.L.H.; Wilkinson, M.G.; Spinnler, H.E. Lipolysis and Metabolism of Fatty Acids in Cheese. In *Cheese*; Elsevier BV: Amsterdam, The Netherlands, 2017; pp. 423–444. ISBN 9780124170124.
51. Cadwallader, K.R.; Singh, T.K. Flavours and Off-Flavours in Milk and Dairy Products. In *Advanced Dairy Chemistry: Volume 3: Lactose, Water, Salts and Minor Constituents*; McSweeney, P., Fox, P.F., Eds.; Springer: New York, NY, USA, 2009; Volume 3, pp. 634–690. ISBN 978-0-387-84864-8.
52. Tarakci, Z.; Temiz, H.; Aykut, U.; Turhan, S. Influence of Wild Garlic on Color, Free Fatty Acids, and Chemical and Sensory Properties of Herby Pickled Cheese. *Int. J. Food Prop.* **2011**, *14*, 287–299. [CrossRef]
53. Akın, N.; Aydemir, S.; Koçak, C.; Yıldız, M.A. Changes of free fatty acid contents and sensory properties of white pickled cheese during ripening. *Food Chem.* **2003**, *80*, 77–83. [CrossRef]
54. Kaminarides, S.; Stamou, P.; Massouras, T. Changes of organic acids, volatile aroma compounds and sensory characteristics of Halloumi cheese kept in brine. *Food Chem.* **2007**, *100*, 219–225. [CrossRef]
55. Faccia, M.; Gambacorta, G.; Pasqualone, A.; Summo, C.; Caponio, F. Quality Characteristics and Consumer Acceptance of High-Moisture Mozzarella Obtained from Heat-Treated Goat Milk. *Foods* **2021**, *10*, 833. [CrossRef] [PubMed]
56. Fuentes, L.; Mateo, J.; Quinto, E.J.; Caro, I. Changes in quality of nonaged pasta filata Mexican cheese during refrigerated vacuum storage. *J. Dairy Sci.* **2015**, *98*, 2833–2842. [CrossRef] [PubMed]
57. Rea, S.; Marino, L.; Stocchi, R.; Branciari, R.; Loschi, A.R.; Miraglia, D.; Ranucci, D. Differences in chemical, physical and microbiological characteristics of Italian Burrata cheeses made in artisanal and industrial plants of the Apulia Region. *Ital. J. Food Saf.* **2016**, *5*, 5879. [CrossRef] [PubMed]
58. Abee, T.; Krockel, L.; Hill, C. Bacteriocins: Modes of Action and Potentials in Food Preservation and Control of Food Poisoning. *Int. J. Food Microbiol.* **1995**, *28*, 169–185. [CrossRef]
59. Akhtar, S.; Nawaz, S.K. Antimicrobial efficacy of Lactobacillus plantarum strain against the B. cereus, B. subtilis, S. aureus and E.coli strains. *Biosci. J.* **2023**, *39*, e39061. [CrossRef]
60. Milioni, C.; Martínez, B.; Degl'innocenti, S.; Turchi, B.; Fratini, F.; Cerri, D.; Fischetti, R. A novel bacteriocin produced by Lactobacillus plantarum LpU4 as a valuable candidate for biopreservation in artisanal raw milk cheese. *Dairy Sci. Technol.* **2015**, *95*, 479–494. [CrossRef]
61. Natrella, G.; Gambacorta, G.; De Palo, P.; Maggiolino, A.; Faccia, M. Volatile organic compounds in milk and mozzarella: Comparison between two different farming systems. *Int. J. Food Sci. Technol.* **2020**, *55*, 3403–3411. [CrossRef]

foods

MDPI

Article

Impact of Milk Thermization on the Quality Characteristics of P.D.O. "Canestrato Pugliese" Ovine Hard Cheese

Giuseppe Natrella, Giuseppe Gambacorta, Giacomo Squeo and Michele Faccia *

Department of Soil, Plant and Food Science, University of Bari, Via Amendola 165/A, 70126 Bari, Italy
* Correspondence: michele.faccia@uniba.it; Tel.: +39-080-544-2939

Abstract: The use of raw milk is compulsory in the manufacturing process of most of the European protected designation of origin (PDO) cheeses but, for ovine products, it is often responsible for faulty productions. Since pasteurization is hardly compatible with the PDO concept, a milder treatment (thermization) is allowed in some cases. An investigation was undertaken to assess the effect of thermization on the overall quality of Canestrato Pugliese, a PDO ovine hard cheese of Southern Italy that can be manufactured exclusively from raw milk. Three types of cheese were produced using raw, mild-thermized and high-thermized milk inoculated with a thermophilic commercial starter. The results demonstrated that the heat treatment did not cause remarkable differences in the gross composition, but the microbiological profiles had some differences despite the use of the selected starter. The raw milk cheese contained higher levels (0.5–1 log units) of mesophilic lactobacilli, total viables, total coliforms and enterococci with respect to the thermized counterparts, with the high-thermized cheese showing the lowest levels; these microbiological differences fitted well with the higher content and the different High Performance Liquid Chromatography (HPLC) pattern of soluble nitrogen. The sensory analysis revealed that the thermized cheeses lost some typical sensory characteristics, probably as a consequence of the reduced indigenous microbiota populations. It was concluded that milk thermization could be applied to Canestrato Pugliese manufacturing only together with the development and use of an autochthonous starter.

Keywords: ovine cheese; PDO Canestrato Pugliese; milk thermization; proteolysis; sensory analysis

Citation: Natrella, G.; Gambacorta, G.; Squeo, G.; Faccia, M. Impact of Milk Thermization on the Quality Characteristics of P.D.O. "Canestrato Pugliese" Ovine Hard Cheese. *Foods* **2023**, *12*, 1080. https://doi.org/10.3390/foods12051080

Academic Editor: Maurice O'Sullivan

Received: 27 January 2023
Revised: 20 February 2023
Accepted: 28 February 2023
Published: 3 March 2023

1. Introduction

Protected designation of origin (PDO) and protected geographical indication (PGI) are two instruments created by the European Union to protect typical food products of specific geographical areas within the European framework [1]. These products are manufactured in a given area using recognized manufacturing practices that are described in the official production protocol forming an integral part of the law establishing the acknowledgement [2]. For both PDO and PGI products, protocol updating is a laborious process, requiring time and attention. That is problematic because, overtime, some specific technical aspects of the manufacturing process can become incompatible with the changes of the food legislation and the consumers' expectations. This is the case for the use of unpasteurized milk in cheesemaking, whose safety is highly debated worldwide, as well as its unsuitability for obtaining a constant quality [3]. Even though most PDO cheeses continue to be manufactured from raw milk, the possibility of applying thermal treatments is increasingly being required from the producers [4,5] in order to reduce the risk of defects. Milk pasteurization causes the elimination, along with pathogenic microorganisms, of indigenous microbiota, which make the use of a starter in cheesemaking necessary. In addition to this, other modifications can play a role in cheese quality, such as the denaturation of whey proteins, inactivation/activation of native milk enzymes and changes in the mineral equilibrium. All these modifications, whose intensity depends on the parameters of the heat treatment, often impact proteolysis and lipolysis, leading to a different flavor with respect to the cheese obtained from raw milk [6–10]. For all these reasons, pasteurization is considered a problem for PDO/PGI cheeses, as the modifications of the milk characteristics tend to weaken the linkage with the geographical area. In order to minimize the adverse effects, milk thermization could be an alternative to pasteurization. According to the EU regulations, thermization is a sub-pasteurization treatment, performed under mild conditions (i.e., 57–68 °C for no less than 15 s in a heat exchanger), so as to guarantee the preservation of phosphatase activity (CE Dir 46/92) and, consequently, part of the indigenous microbiota. The treatment always causes the elimination of psychrotrophic bacteria and the reduction of total bacterial count, whereas a marked reduction of *Enterobacteriaceae* and coliforms is obtained only adopting suitable time/temperature combinations [11–13]. Of course, for thermization it is also necessary to add a starter before cheesemaking, in order to fortify the indigenous microflora surviving the heat treatment that is not sufficient for carrying out the suitable fermentations. This thermal treatment can also be performed directly in the vat if, as often happens at a small farm level, a heat-exchanger is not available. Besides preserving part of the indigenous microbiota, thermization presents other advantages with respect to pasteurization, such as less damage to the milk thermolabile compounds, the better preservation of coagulability and lower energy consumption for heating and cooling.

The possibility of applying milk thermization before cheesemaking is much more requested by the manufacturers of PDO ovine cheeses than the bovine counterpart, since sheep milk quality is often affected by high microbial load. Italy is the European Union (EU) country with the highest number of PDO ovine cheeses (seventeen), all belonging to the class of semi-hard or hard cheeses. Out of seventeen, eight have received the authorization to apply a thermal treatment to milk, while the remaining nine must be manufactured from raw milk [14], as reported in Table 1.

Table 1. PDO and PGI Italian cheeses made from sheep milk (alone or in combination) and heat treatments allowed.

Cheese	Type of Milk	Acknowledgement	Treatment Allowed
1. Canestrato di Moliterno	sheep + goat	PGI	thermization
2. Canestrato Pugliese	sheep	PDO	none
3. Casciotta d'Urbino	sheep + cow	PDO	pasteurization
4. Fiore Sardo	sheep	PDO	none
5. Formaggio di Fossa di Sogliano	cow + sheep	PDO	pasteurization
6. Murazzano	sheep + cow	PDO	any heat treatment
7. Pecorino Crotonese	sheep	PDO	any heat treatment
8. Pecorino delle Balze Volterrane	sheep	PDO	none
9. Pecorino del Monte Poro	sheep	PDO	none
10. Pecorino di Filiano	sheep	PDO	none
11. Pecorino di Picinisco	sheep	PDO	none
12. Pecorino Romano	sheep	PDO	thermization
13. Pecorino Sardo	sheep	PDO	any heat treatment
14. Pecorino Siciliano	sheep	PDO	none
15. Pecorino Toscano	sheep	PDO	any heat treatment
16. Piacentinu Ennese	sheep	PDO	none
17. Vastedda del Belice	sheep	PDO	none

For PDO/PGI cheeses, the ban on the heat treatment could only be removed by demonstrating that the overall quality does not significantly change and the linkage with the territory is not lost. Unfortunately, information on the effect of thermization on cheese quality is scarce. Xanthopoulos et al. [15] investigated Anevato, a Greek PDO spreadable cheese made from goat milk, and concluded that thermization allowed for the better preservation of part of the milk indigenous microbiota with respect to pasteurization (about one-half log unit of difference for the most important microorganisms groups in terms of colony forming units). However, both heat-treated milk cheeses lacked the characteristic cheesy flavor of raw milk cheese. In cheddar cheese, both pasteurization and thermization (65 °C × 15 s) had no effect on primary proteolysis but reduced the levels of free amino acids and the intensity of lipolysis during ripening [8]. Pirisi et al. [16] investigated the effect of milk thermization on the quality of Fiore Sardo and found that the heat treatment did not significantly influence the composition and the secondary proteolysis, whereas significant differences were observed in the rheological properties and in the lypolitic pattern. Caboni et al. [17] investigated the same cheese and concluded that while the effects of milk thermization on macro-compositional parameters and free fatty acid levels were not evident, strong differences between cheese produced from raw and thermized milk were detectable by using a Gas Chromatography- Mass Spectrometry (GC-MS) metabolomics approach. Unfortunately, the study did not include a sensory evaluation. Finally, a recent work of Dedola et al. [18] proposed the dosage of enzyme α-l-fucosidase as a tool for discriminating Fiore Sardo PDO cheeses obtained from raw or thermized milk, but in this case the sensory characteristics were also not investigated.

Canestrato Pugliese is one of the nine ovine Italian PDO cheeses for which the heat treatment of milk is not allowed. It is a semi-hard to hard cheese made in the Apulia Region (Southern Italy) from milk of sheep of local flocks, whose designation of origin is connected to ancient sheep transhumance. Transhumance is a very old form of pastoralism consisting of the seasonal movement of livestock and herders between higher pastures in the summer, and lower pastures during the winter. The north-west part of Apulia has not too cold winters and presents vast lowland areas with natural pastures; for this reason, it has been the summer destination for sheepherders from the surrounding mountain areas since Roman times [19]. As sheep lactation mostly took place from February to May, the Apulian lowlands were crawling with shepherds making cheeses, the most important of which was a hard type that could be easily stored. It was called Canestrato from the name of the rush basket ("Canestro") in which it was molded and kept for the first period of ripening that lasted from a few months to about one year. When transhumance ceased to be practiced (1950–1960s) and sheep breeding became sedentary, the cheese continued to be manufactured in dairy industries and farmers cooperatives. The uniqueness linked to history, the particular environment in which the flocks graze, the manufacturing procedure and typical organoleptic characteristics led to the acknowledgement of the designation of origin, which was first obtained in 1985 at the national level and then confirmed in 1996 at the EU level. Unfortunately, the recent changes in both livestock management and consumer habits has caused the total production volume of Canestrato Pugliese to dramatically decline in the last two decades. The main problem caused by the changes in consumer habits was the increased demand of "perfect" food products, which led to the depreciation of non-standardized cheeses which often are those obtained from raw milk. In addition to this, the use of raw sheep milk brings with it a high risk of faulty production that cannot be tolerated by the manufacturers. The consequence of this framework is that the dairy industry has abandoned the production of Canestrato Pugliese, which is currently manufactured only in two small artisanal dairies. The absence of interest from the dairy industry, together with difficulties in finding animal care workers, is continuing to discourage the sheep farmers and, if the situation does not change, the Apulian breeding sheep sector will disappear. The application of milk thermization could increase the interest of dairy enterprises to restart the production of this cheese, but no research has been conducted about its effect on cheese quality. Albenzio et al. [20] investigated the effect of milk pasteurization and of heating the curd in hot whey (80 °C for 30 s) on proteolysis and lipolysis. Both treatments led to lower counts of lactic acid bacteria in cheese and slower proteolysis and lipolysis than the control cheese made from raw milk without any thermal treatment. Unfortunately, the cheeses were not subjected to sensory analysis, even though the authors speculated that it should be expected that cheeses obtained from raw milk reach their optimum sensory quality earlier than those from pasteurized milk. Piombino et al. [21] investigated the sensory characteristics of Canestrato Pugliese made from raw or pasteurized milk, and found important differences, which were confirmed by the gas chromatography-mass-spectrometry-olfactometry analysis of volatile compounds. The aim of the present research was to assess the effect of milk thermization on the chemical, microbiological and sensory characteristics of Canestrato Pugliese cheese, in order to establish the extent by which the heat treatment influences the overall quality.

2. Materials and Methods

2.1. Cheese Manufacturing

The milk used in the experimentation derived from a batch of ovine bulk milk (about 4000 L) collected in two days from 17 sheep farms located in the province of Bari and Foggia (Apulia region, Italy). From this batch, three 1000 L aliquots were taken, one of which was used raw for preparing the control cheese (CR) while the other two were subjected to thermization in a heat exchanger for preparing thermized milk cheeses. Two different heating conditions were tested: 62 °C × 20 s for preparing mild-thermized cheese (CMT) and 68 °C × 20 s for preparing high-thermized cheese (CHT). Two cheesemaking replicates were performed for each type of milk by applying the official production protocol [22], except for the addition of a selected lactic acid bacteria (LAB) starter (a mixed culture of *Streptococcus thermophilus* and *Lactobacillus delbrueckii* subsp. *bulgaricus*, Sacco Srl, Cadorago, Italy) for reintegrating the microbiota damaged by the heat treatment. The culture was also added to the raw milk for manufacturing the control cheese, in order to normalize the technological conditions. The experimental design is summarized in Figure 1, whereas the cheesemaking protocol adopted is shown in Figure 2.

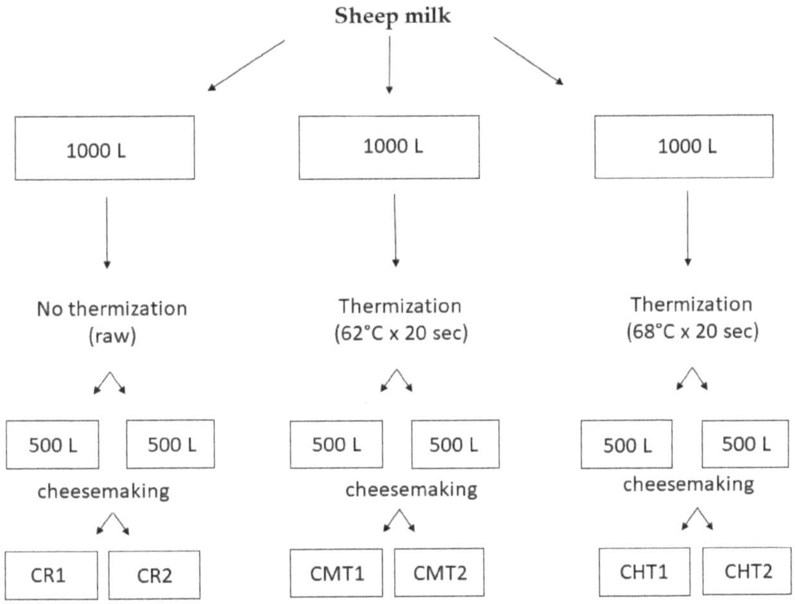

Figure 1. Experimental design applied in the investigation.

2.2. Chemical Analyses

The milk used in the experimentation was subjected to the analysis of the gross composition by infrared analyzer (Milko-Scan, Foss Electric, Hillerød, Denmark). The cheeses were sampled at 0, 7, 14, 21, 28, 42, 56, 70, 82, 150, 180, 240 and 300 days of ripening. Representative samples were prepared by eliminating the rind (about 0.5 cm) and cutting a triangular wedge weighting 0.5 kg. The wedge was then grated, and, after thorough mixing, portions were taken to perform the analyses. The following determinations were carried out: moisture (oven drying), pH (pH meter equipped with a penetration probe, Hanna Instruments, Woonsocket, RI, USA), NaCl (chloride analyzer, Sherwood Scientific Ltd., Cambridge, UK), fat (Soxhlet method), total nitrogen (Kjeldahl method), water-soluble nitrogen (WSN) according to Kuchroo and Fox [23]. All analyses were carried out in triplicate. Proteolysis was investigated by urea-poly acrylamide gel electrophoresis (PAGE) according to the method of Andrews [24]. The main casein fractions were identified by comparison with a milk sample taken from the vat and with the data from the scientific literature; the protein bands in the gel were quantified by densitometry. Proteolysis was also investigated by Reverse-Phase High Performance Liquid Chromatography (RP-HPLC) analysis of water soluble nitrogen on an Agilent Technologies apparatus with Ultraviolet (UV) detection (Palo Alto, CA, USA), under the operating conditions reported in a previous paper [25].

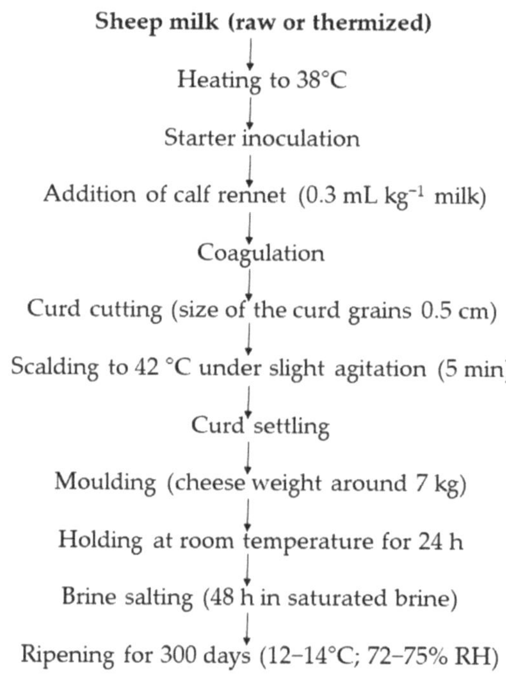

Figure 2. Cheesemaking process applied in the experimentation.

2.3. Microbiological Analyses

Analyses were carried out on raw milk, vat milk (i.e., after starter addition), curd and cheese at 0, 7, 14, 28, 56, 150 and 300 days of ripening. On raw milk, the count of somatic cell and total bacteria was performed by flow cytometry (Fossomatic and Bactoscan, Foss Electric, Denmark). For vat milk, curd and cheese, a 20 mL or 20 g sample was diluted in 180 mL of 2% (w/v) sodium citrate solution and homogenized in a Stomacher Lab-Blender. Serial dilutions were made in quarter strength Ringer's solution and plated on specific media for viable counts. The following groups were enumerated: total viable (TV, 32 °C, plate count agar, Oxoid, Basingstoke, UK), total coliforms (TC, 37 °C, VRB agar, Biolife, Milan, Italy); presumptive mesophilic lactococci (MLc, 30 °C, M17 agar supplemented with 0.1% cycloheximide, Merck, Darmstad, Germany); presumptive mesophilic lactobacilli (MLb, 30 °C, MRS agar supplemented with 0.1%, cycloheximide Merck, Darmstad, Germany); presumptive thermophilic lactobacilli (TLb, 45 °C, MRS agar supplemented with 0.1%, cycloheximide Merck, Darmstad, Germany); presumptive thermophilic streptococci (TLs, 45 °C, lactose M17 agar supplemented with 0.1%, cy-cloheximide Merck, Darmstad, Germany); presumptive enterococci (Ec, 37 °C, Slanetz and Bartley agar, Oxoid, Basingstoke, UK); yeasts and molds (Y&M, 25 °C yeast glucose chloramphenicol media, Biolife, Milan, Italy).

2.4. Sensory Analysis

The cheeses were evaluated by a panel composed of 7 trained assessors, selected by following ISO standard 8586-1 and certified by the Italian Association of Cheese Tasters (ONAF, Cuneo, Italy) after attending a 20-h course for cheese evaluation (description and quantification of aroma, taste and texture). The sensory evaluation involved the use of two different tests: a discrimination test (paired comparison) and a descriptive test (quantitative descriptive analysis, QDA). The panel activity started with two attribute-generation sessions performed on commercial samples of PDO Canestrato Pugliese resulting in the definition of 13 attributes (5 regarding aroma, 4 taste and 4 texture). Then, the paired comparison was performed by offering the cheese samples in white plastic dishes identified by a 4-digit code in all possible pairwise options, with balanced presentation. The panelists were asked to assess the presence of a difference between the sample pairs and, if present, to indicate on a scorecard which sample had higher overall flavor intensity. Successively, the assessors performed the QDA analysis by tasting the samples one by one and judging them on a form in which the 13 established attributes were quantified on a 6-point scale. After statistical elaboration, the results of the analyses were discussed in a final open session.

2.5. Statistical Analysis

The data were statistically processed by XLSTAT software (version 2020.1.3, Addinsoft Inc., New York, NY, USA). Discrete variables were described by their mode values and continuous variables by their means. For microbiological analyses performed on milk, standard deviation (SD) was calculated. For chemical and microbiological analyses performed on cheese, the results were subjected to one-way ANOVA followed by Tukey's honestly significant difference test at a critical value for significance of $p < 0.05$; as regards the sensory analysis, the nonparametric variables were compared by using the Kruskal–Wallis test.

3. Results and Discussion

3.1. Chemical and Microbiological Analyses

The milk used to produce the cheeses contained 7.11% fat, 5.61% protein, and 4.64% lactose and had a pH of 6.55; somatic cell and total bacteria counts were 730,000 mL^{-1} and 540,000 mL^{-1}, respectively. Thermization had a slight but significant effect on the cheese gross composition (Figure 3a,b). At the end of ripening, CMT and CHT were less humid and contained less protein and more fat than CR. On the other hand, the NaCl concentration was almost the same, ranging from a minimum of 3.10% in CMT to a maximum of 3.22% in CR. Overall, the compositional differences between raw and thermized cheeses were rather small, and were almost absent in the two thermized cheeses. It is worth highlighting that thermization had a scarce impact on moisture retention, differently from pasteurization that causes more water retention because of the greater denaturation of whey proteins, part of which tends to be entrapped into the curd [26,27].

In comparison with the compositional parameters, the differences in pH were much more relevant (see Figure 3b). Except for the samples analyzed at 7 days, the values were always lower in the thermized samples than in the raw one. The pH curves presented two "acidification peaks", the first one after 7 days and the second one after 56 days. The first peak corresponded to the acidification phase of the starter, with the three samples behaving very similarly. The second acidification peak should be ascribable to other fermenting bacteria groups since the starters are known to undergo autolysis after the first stages of ripening [28]. In this phase, CR showed a significantly higher pH than CMT and CHT, which could be explained by the higher formation of low molecular weight alkaline compounds deriving from secondary proteolysis. This aspect will be further discussed later.

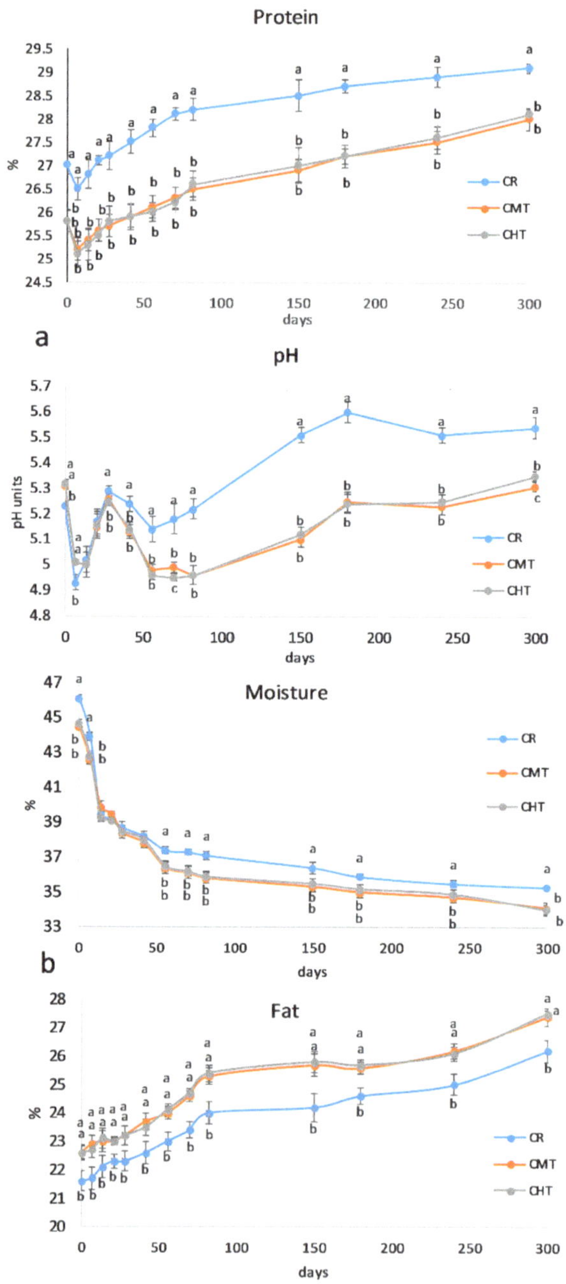

Figure 3. (**a**) Evolution of the moisture and fat contents in Canestrato Pugliese cheeses made from raw (CR), mild-thermized (CMT) and high-thermized (CHT) milk during ripening (0–300 days). Mean values of three analyses performed on two cheese wheels for each type of cheese. Different letters for the same ripening time indicate different values at $p > 0.05$; no letters means no difference. (**b**) Evolution of the protein content and pH in Canestrato Pugliese cheeses made from raw (CR), mild-thermized (CMT) and high-thermized (CHT) milk during ripening (0–300 days). Mean values of three analyses performed on two cheese wheels for each type of cheese. Different letters for the same ripening time indicate different values at $p > 0.05$.

Useful information about this point derived from the microbiological analysis of the vat milks (i.e., after the addition of the starter) and of the curds, whose results are shown in Table 2. As expected, the raw milk had higher counts than the two thermized milks, with mesophilic lactobacilli (MLb) and mesophilic lactococci (MLc) being the most represented groups, followed by enterococci (Ec). The same three groups were the most abundant in the two thermized milks, with the difference that Ec exceeded the other two. Enterococci are non-starter lactic acid bacteria that play an important role during ripening of artisanal hard and semi-hard cheeses [29]. According to Mc Auley et al. [30], their heat resistances greatly vary, depending on the species, and thermoduric enterococci can survive pasteurization. For TLb and TSt, which included the LAB species added with the starter, the higher counts in the raw milk indicated the presence of indigenous species that combined with the starter. It is worth mentioning that these two groups were not present at the level expected: that is not surprising, considering that the starter was added as a lyophilized culture, which needs time to grow, and that the milk samples were taken from the vat a few minutes after inoculation. The count value of total coliforms (TC) in the raw milk was rather high (5.47 cfu/g) but after thermization decreased by about 1.5–2.0 log units. Coliforms are commonly considered as an index of milk hygiene and their presence depends on the technological level of the farms and the efficiency of refrigeration throughout the whole production chain. In sheep milk, the populations are commonly higher than in cow milk: the survey conducted by de Garnica et al. [31] reported count values ranging from a minimum of 1.30 to a maximum of 6.64 cfu/g. The high TC count in the raw milk used in the present experimentation probably depended on the fact that, at the moment of the cheesemaking trials, the bulk milk used was the sum of 17 milk sub-batches having different quality and different levels of freshness (from a minimum of 18 h to a maximum of about 70 h). Indeed, that is the normal situation occurring in this geographical area, where the farms are scattered along a long route and, considering the low milk yield of sheep, milk collection takes a long time.

Table 2. Counts * of the major microorganism groups (cfu/g) in the vat milk (after starter addition) and in the curd. RM = raw milk; MTM = mild-thermized milk; HTM = high-thermized milk. TV = total viable; TC = total coliforms; Ec = enterococci; MLb = mesophilic lactobacilli; TLb = thermophilic lactobacilli; MLc = mesophilic lactococci; TSt = thermophilic streptococci; Y&M = yeasts and molds. Mean values ± standard deviation from two analyses on each type of milk and curd.

Group	RM	MTM	HTM	Curd R	Curd MT	Curd HT
TV	6.69 ± 0.42	5.17 ± 0.34	5.85 ± 0.21	9.17 ± 0.34	8.87 ± 0.21	8.91 ± 0.27
TC	5.53 ± 0.34	3.93 ± 0.04	2.55 ± 0.15	5.47 ± 0.24	4.05 ± 0.03	3.40 ± 0.14
Ec	5.89 ± 0.29	4.82 ± 0.11	4.40 ± 0.64	6.28 ± 0.11	5.35 ± 0.10	4.57 ± 0.15
MLb	6.67 ± 0.34	4.68 ± 0.28	4.26 ± 0.21	7.05 ± 0.09	5.98 ± 0.25	5.36 ± 0.12
TLb	5.32 ± 0.09	3.41 ± 0.31	3.63 ± 0.22	7.28 ± 0.24	6.15 ± 0.19	6.28 ± 0.04
MLc	6.21 ± 0.30	4.30 ± 0.43	3.52 ± 0.22	6.91 ± 0.18	6.00 ± 0.08	5.82 ± 0.25
TSt	2.71 ± 0.20	1.57 ± 0.11	2.09 ± 0.15	5.49 ± 0.11	5.56 ± 0.15	5.51 ± 0.06
Y&M	5.40 ± 0.08	3.53 ± 0.05	2.07 ± 0.56	5.40 ± 0.08	3.77 ± 0.10	3.55 ± 0.20

As expected, the microbiological profile deeply changed at the end of the in-vat cheesemaking process, with a marked increase in all microbial groups in the curd. In particular, the differences in the LAB counts observed in the milks tended to disappear or decrease, whereas those regarding the other microbial groups remained relevant.

Figure 4 shows the evolution of LAB counts in the cheeses throughout the entire ripening process. The curves of MLc and MLb were rather similar, with differences among the three cheeses corresponding at most to half a log unit. In the first weeks, these microorganisms reached a maximum count value of between 8.4 and 9.00 log cfu/g, and then constantly decreased over time. Albenzio et al. [20] reported the same trend for MLb in Canestrato Pugliese manufactured from raw and pasteurized milk, but the difference in the cell density between two cheeses was much higher with respect to our experimentation. This finding confirms that milk thermization allows for the preservation of a part of the indigenous LAB, maintaining a linkage with the territory. Differently from mesophilic, the two thermophilic LAB groups evidenced a dramatic drop at 28 days, except for the raw milk cheese that evidenced the drop of TLb at 56 days. These drops should correspond to the autolysis phase of the starter cultures, which in the CR cheeses was probably delayed (or made less evident) due to the presence of viable indigenous species. As for the non-LAB groups (Figure 5), the most relevant differences regarded total coliforms. In particular, CR had higher counts than the two thermized cheeses for the whole time in which this bacteria group was present. It is known that coliforms in hard cheeses tend to disappear rapidly in connection with the decrease in moisture content and increase in NaCl concentration.

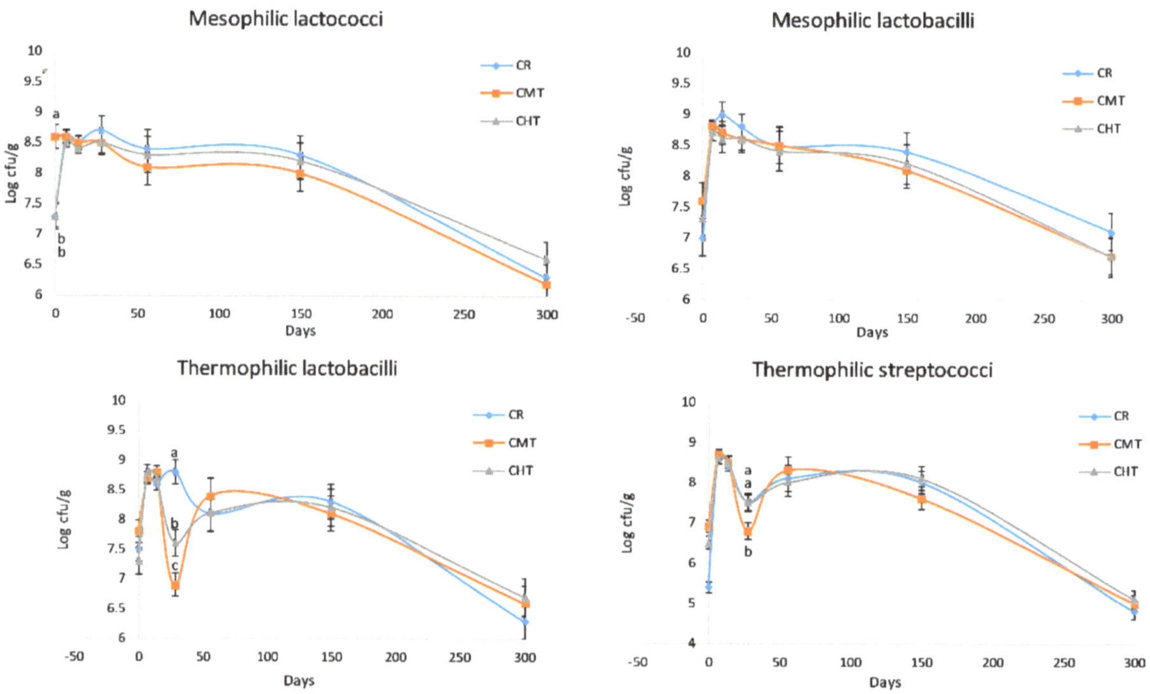

Figure 4. Counts * of the major lactic acid bacteria groups (cfu/g) in Canestrato Pugliese cheeses made from raw (CR), mild-thermized (CMT) and high-thermized (CHT) milk during ripening (0–300 days). Mean values of two analyses on two cheese wheels for each type of cheese.

From the figure, it can be observed that the value of <1 log cfu/g was reached after about 3 months in CMT and CHT and about 4 months in CR. Another interesting feature was the evolution of enterococci. After an early drop, much more evident in thermized cheeses, the populations sharply increased and reached a maximum around 2 months, successively they started to decrease again but remained at high levels at the end of ripening. Interestingly, the trend in CR and CMT was similar from 4 months onwards, with count values at 300 days about 1 log unit higher than in CHT.

Finally, yeasts and molds evidenced a peak growth in correspondence with the second month, with count values that exceeded 7 log units cfu/g in all cheeses, with a maximum of 7.7 in CR. These values are very high, more than those reported in the same cheese by Corbo et al. [32], in which the maximum level approached 6 log units cfu/g at 19 days of ripening. However, the comparison is not reliable, as the authors did not perform the investigation on real PDO Canestrato Pugliese, whose weight must be of 7 or 14 kg, but on mini cheeses weighing only 1 kg. Another factor could be the ripening environments, which are known to play a major role in the contamination and growth of these microorganisms. In our experimentation, the cheese ripened on wooden boards in a natural warehouse with controlled temperature and relative humidity; in this room, after 1 month, the rind was totally covered by a lay of mold that underwent the first washing at 56 days, in perfect correspondence with the count peak.

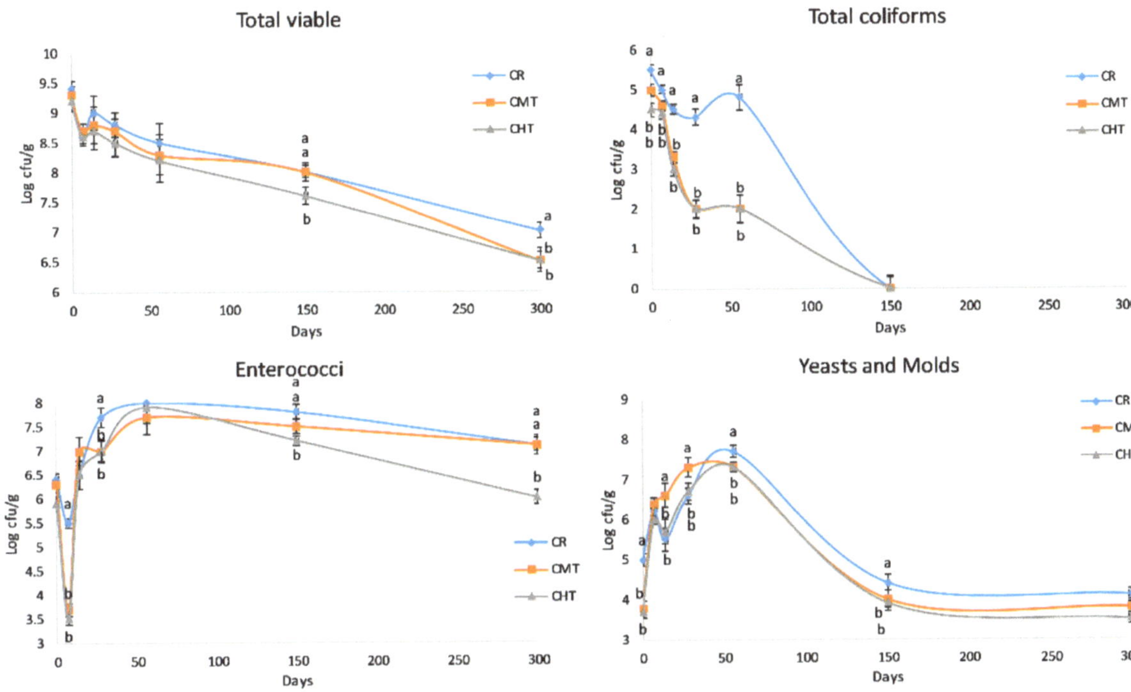

Figure 5. Counts of the non- lactic acid bacteria groups (cfu/g) in Canestrato Pugliese cheeses made from raw (CR), mild-thermized (CMT) and high-thermized (CHT) milk during ripening (0–300 days). Mean values of two analyses on two cheese wheels for each type of cheese.

Figure 6 shows the urea-PAGE electropherograms of the samples. The patterns were rather similar, indicating a relatively slow rate of proteolysis and faster degradation of α-S1 casein than β casein in all cheeses, in good agreement with the findings reported in a previous paper [33]. The main difference detectable between raw and thermized cheeses regarded the bands corresponding to the α-S1-I fraction and to the proteolytic products, which appeared to accumulate slightly faster in CR. This finding suggests a possible different rate of primary proteolysis but, considering that the electrophoretic techniques are semi-quantitative, this hypothesis needed confirmation. A further possible difference was observed in the zone between β and αS1 casein, where a diffuse unknown band is present. Even though the patterns in that zone suffer from smearing, it seems that this band was absent in milk and formed more intensely in the mild-thermized cheese. Pirisi et al. [34] reported a band positioned in the same area in Fiore Sardo cheese, and considered it as an unidentified compound, even though it was recognized by polyclonal antibodies against αS1-casein. In contrast, Sousa and Malcata reported a band in that area that was identified as a primary proteolytic product of β casein released by the activity of residual rennet [35].

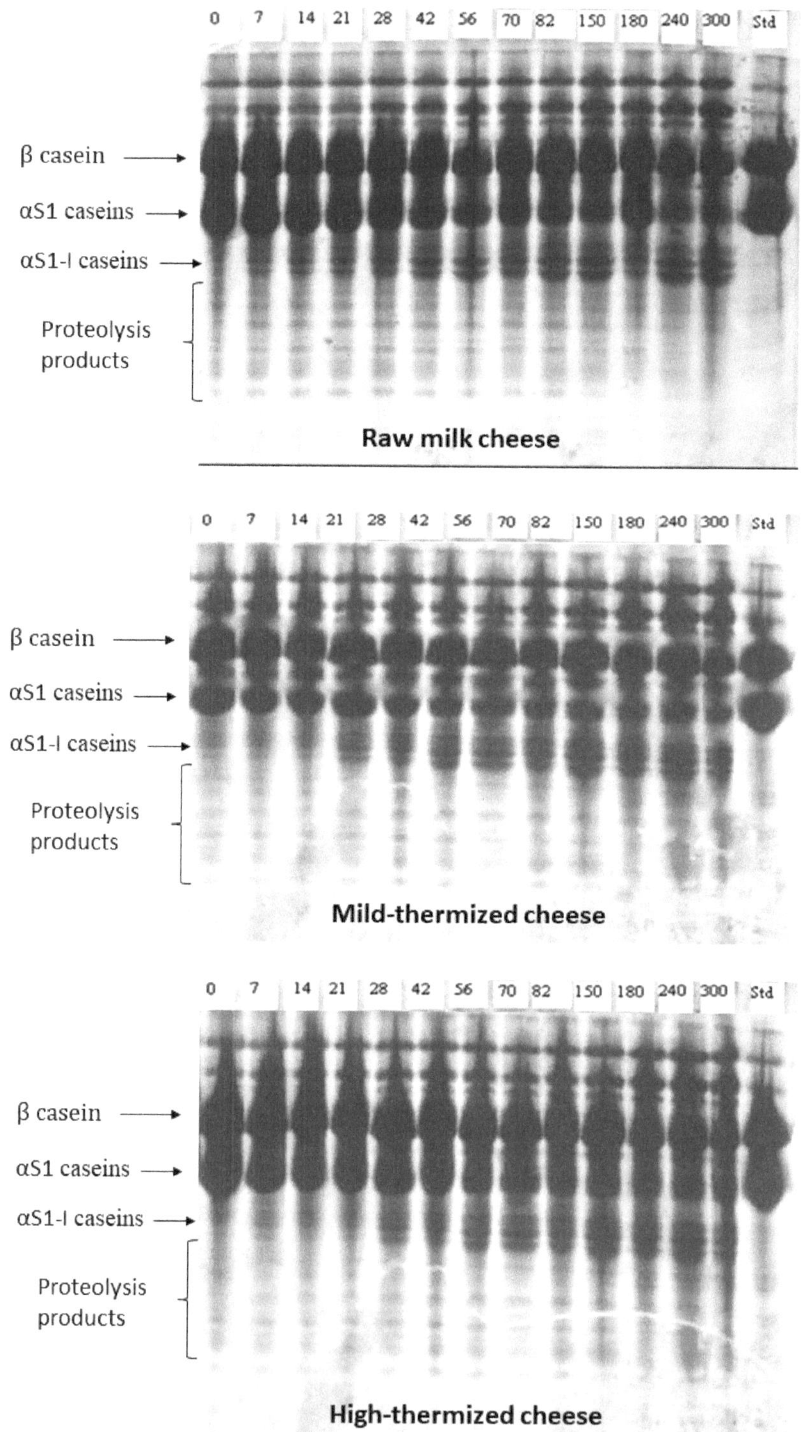

Figure 6. Urea-PAGE of Canestrato Pugliese cheese made from raw and thermized milk. Std = standard (sheep milk); from 0 to 300 = days of ripening.

Indeed, a confirmation was derived from the study of the soluble nitrogen fraction. In fact, the rate of its formation was almost the same until the first month of ripening; successively, it started to increase more rapidly in CR than in CMT and CHT and the final value in CR was about 20% higher (Figure 7). Consequently, the faster proteolysis in the raw milk cheese was confirmed. For a possible explanation, it must be considered that this biochemical event depends on the type of rennet used and the amount resituated in the curd, on the activity of the major milk protease (plasmin) and on the presence of proteolytic microorganisms [36]. In our experimental conditions: (i) the type of rennet was the same; (ii) the similar compositional characteristics of the curds should not have determined a different extent of rennet retention; (iii) the heat treatments applied were too weak to impact plasmin activity, since the whole plasmin system is affected only by higher time-temperature conditions [37]. Consequently, it is likely that the differences in proteolysis are attributable to the different activity of proteolytic microorganisms. As reported above, the raw milk cheese had higher counts of several potentially proteolytic microorganism groups (mesophilic lactic acid bacteria, enterococci, coliforms, and yeasts and molds) than the thermized ones. Given the high counts level observed for enterococci and their remarkable proteolytic activity [38,39], these microorganisms might have played a primary role. Nevertheless, taken together, the contribution to the peptidasic activity from the other groups should not be neglected, since the counts at some stages of ripening were relevant, as reported above for yeasts and molds.

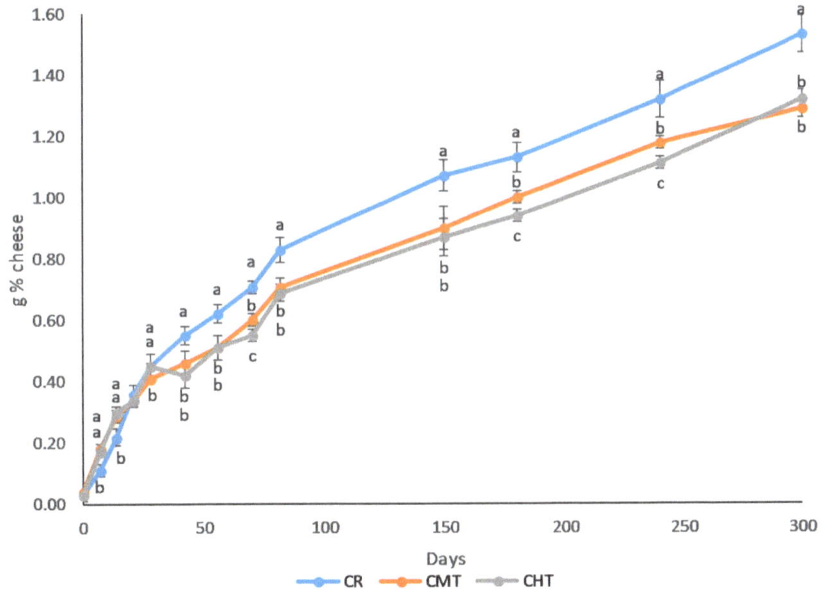

Figure 7. Evolution of the soluble nitrogen fraction in Canestrato Pugliese cheese made from raw and thermized milk. Different letters for the same ripening time indicate different values at $p > 0.05$; no letters means no difference.

The RP-HPLC study of WSN at 300 days of ripening revealed both quantitative and qualitative differences, supplying further information about proteolysis (Figure 8). The raw milk cheese had a higher total peak area and a less complex profile with respect to the other two cheeses, which were rather similar to each other. For a better interpretation of the qualitative differences, the chromatograms were divided into three parts: hydrophilic, intermediate and hydrophobic. It is known that the hydrophilic peaks mostly correspond to small nitrogen compounds with polar characteristics such as hydrophilic free amino acids and small peptides, resulting from the activity of bacterial peptidases [40,41]. In CR, the hydrophilic part was characterized by three major peaks (the first one corresponding to non-retained compounds), whose sum represented more than 60% of the total peak area; in CMT and CHT, a high number of peaks was present, with different retention times and a much lower area with respect to CR. The intermediate part of the chromatogram also showed significant differences between raw and thermized cheeses, but in this zone, the differences were mostly quantitative. The huge total area of hydrophilic peaks in the raw milk cheese fitted well with the above hypothesis that proteolytic microorganisms were mainly responsible for faster proteolysis.

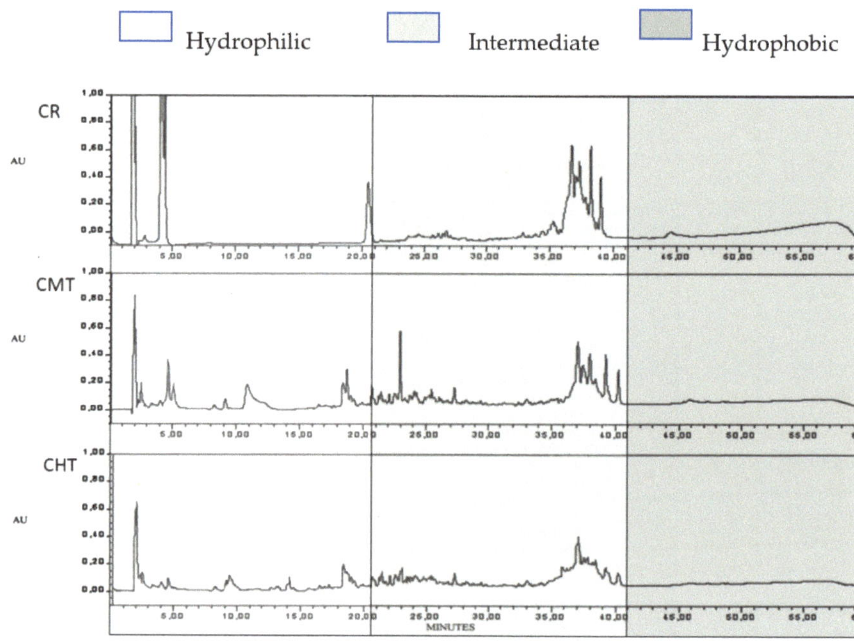

Figure 8. RP-HPLC of the soluble nitrogen fraction of Canestrato Pugliese cheese made from raw and thermized milk at 300 days of ripening. CR = raw milk cheese; CMT = mild-thermized cheese; CHT = high-thermized cheese.

3.2. Sensory Analysis

During the paired comparison tests, all panelists were able to discriminate the raw milk cheese from the thermized ones that, in turn, were not judged as different from each other. QDA analysis allowed for deepening the reasons for the differences perceived, as shown in Table 3. CR cheese received significantly higher scores for three aroma (cheesy-sweat, sheep barn and dirty socks), one taste (spicy) and one texture (eyes) attributes. Overall, during the final discussion, the panel defined it as "rough" and "typical", whereas the terms used to define the two thermized cheeses were "fragrant" and "correct". Information is available in the literature on the origin of the three aroma descriptors that discriminated the samples. The cheesy-sweat odor is mostly connected to short chain free fatty acids, such as butanoic and hexanoic [42,43]: the former can both derive from microbial fermentation and lipolysis, the latter only from lipolysis. Unfortunately, we did not investigate the volatile compounds, nor lipolysis. The sheep barn odor has been reported to have strong correlation with *p*-cresol, a phenolic compound directly deriving from milk and closely dependent on animal feeding, with milk from grass-fed animals presenting higher concentration with respect to milk from total mixed ratio-fed animals [44]. As in the present investigation, the milk used was the same in all cheesemaking trials, so this sensory attribute should have a different origin. According to the literature, *p*-cresol in cheese can also derive from the microbial catabolism of tyrosine [45]. The volatile compound responsible for the dirty socks odor in cheese (3-methylbutanoic acid) also derives from the microbial catabolism of an amino acid (leucine) [46,47]. In the present experimentation, the higher secondary proteolysis observed in the raw milk cheese in connection with the higher counts of proteolytic microorganisms, could fit these pathways. Finally, the higher scores for the spicy and eyes descriptors can also be attributed to the different microbiological profile. In hard cheese, the spicy taste is a typical consequence of advanced lipolysis with the formation of short-chain free fatty acids, mostly butanoic and hexanoic, the same that are responsible of the cheesy-sweat odor. Even though we did not investigated lipolysis, it is worth mentioning that many strains belonging to both coliforms and enterococci groups can be strongly lipolytic [48]. The higher coliform counts can also be a good candidate to explain the presence of small eyes in the cheese, which were already evident after a few weeks of ripening.

Table 3. Sensory attributes (modal values) for Canestrato cheese made from raw and thermized milk at 300 days of ripening. CR = raw milk cheese; CMT = mild-thermized cheese; CHT = high-thermized cheese. Sig = statistical significance (* = different at $p < 0.05$).

Attributes	CR	CMT	CHT	Min–Max	Sig
		AROMA			
Cheesy-sweat	4	3	3	3–5	*
Sheep barn	2	0	0	0–2	*
Butter	1	1	1	0–1	
Toasted	1	1	1	0–1	
Dirty socks	2	0	0	0–3	*
		TASTE			
Salty	2	2	2	2–3	
Bitter	0	0	1	0–1	
Umami	2	2	2	1–2	
Spicy	2	1	1	1–2	*
		TEXTURE			
Eyes	2	1	0	0–2	*
Hard	4	4	4	3–4	
Crumbly	3	3	3	2–3	
Greasy	3	3	3	2–3	
Soluble	3	2	2	2–3	

4. Conclusions

The application of milk thermization in the manufacturing process of Canestrato Pugliese PDO cheese affected the overall quality. Even though the gross composition underwent only minor changes, the sensory characteristics of the raw milk cheese were more intense with respect to the thermized ones, despite of the use of the same starter in the manufacturing process. The result of the investigation clearly addresses a relevant role of the indigenous microbiota, which is probably responsible of faster secondary proteolysis with the related aroma-active molecules. The most important information that can be drawn from the present work is that heat treatment alone cannot be proposed as a tool to minimize defective productions, since the original characteristics of the cheese tend to fade. The only possible solution might be the use of an autochthonous whey starter, able to maintain the linkage with the territory and provide a complex microbiota, resembling the main microbiological characteristics of raw milk as much as possible, including the presence of enterococci. Although they are not yet considered GRAS (generally recognized as safe) microorganisms, their presence in a long-ripened hard cheese such as Canestrato Pugliese does not raise safety concerns.

Author Contributions: Conceptualization, resources, writing—original draft preparation, M.F.; methodology, software, investigation, G.N.; validation, formal analysis, writing—review and editing G.G. and G.S. All authors have read and agreed to the published version of the manuscript.

Funding: This research received no external funding.

Data Availability Statement: Data is contained within the article.

Conflicts of Interest: The authors declare no conflict of interest.

References

1. Melini, V.; Melini, F. Recent Advances in Food Protected Designations of Origin. In *Comprehensive Foodomics*; Cifuentes, A., Ed.; Elsevier: Amsterdam, The Netherlands, 2021; pp. 417–437.
2. Martelo-Vidal, M.J.; Vázquez, M. Advances in Ultraviolet and Visible Light Spectroscopy for Food Authenticity Testing. In *Advances in Food Authenticity Testing*; Elsevier: Amsterdam, The Netherlands, 2016; pp. 35–70.
3. West, H.G. Food fears and raw-milk cheese. *Appetite* **2008**, *51*, 25–29. [CrossRef]
4. de Sainte Marie, C.; Mariani, M.; Millet, M.; Cerdan, C.; Casabianca, F. Can raw milk cheese and pasteurised milk cheese coexist? Unthinkable or never really considered? *Rev. Agric. Food Environ. Stud.* **2020**, *101*, 287–309. [CrossRef]
5. Formaggi Europei: Dop e Latte Crudo. Available online: https://www.fondazioneslowfood.com/it/i-nostri-temi/biodiversita/osservatorio-sulla-biodiversita/formaggi-europei-dop-latte-crudo/ (accessed on 5 July 2022).
6. Benfeldt, C.; Sørensen, J.; Ellegård, K.H.; Petersen, T.E. Heat Treatment of Cheese Milk: Effect on Plasmin Activity and Proteolysis During Cheese Ripening. *Int. Dairy J.* **1997**, *7*, 723–731. [CrossRef]
7. Benfeldt, C.; Sørensen, J. Heat treatment of cheese milk: Effect on proteolysis during cheese ripening. *Int. Dairy J.* **2001**, *11*, 567–574. [CrossRef]
8. Hickey, D.K.; Kilcawley, K.N.; Beresford, T.P.; Wilkinson, M.G. Lipolysis in Cheddar Cheese Made from Raw, Thermized, and Pasteurized Milks. *J. Dairy Sci.* **2007**, *90*, 47–56. [CrossRef]
9. Pappa, E.C.; Bontinis, T.G.; Samelis, J.; Sotirakoglou, K. Assessment of the Microbiological Quality and Biochemical Parameters of Traditional Hard Xinotyri Cheese Made from Raw or Pasteurized Goat Milk. *Fermentation* **2022**, *8*, 20. [CrossRef]
10. Chambers, D.H.; Esteve, E.; Retiveau, A. Effect of milk pasteurization on flavor properties of seven commercially available French cheese types. *J. Sens. Stud.* **2010**, *25*, 494–511. [CrossRef]
11. Matselis, E.; Roussis, I.G. Proteinase and lipase production by Pseudomonas fluorescens. Proteolysis and lipolysis in thermized ewe's milk. *Food Control.* **1998**, *9*, 251–259. [CrossRef]
12. Peng, S.; Hummerjohann, J.; Stephan, R.; Hammer, P. Short communication: Heat resistance of Escherichia coli strains in raw milk at different subpasteurization conditions. *J. Dairy Sci.* **2013**, *96*, 3543–3546. [CrossRef]
13. Tilocca, B.; Costanzo, N.; Morittu, V.M.; Spina, A.A.; Soggiu, A.; Britti, D.; Roncada, P.; Piras, C. Milk microbiota: Characterization methods and role in cheese production. *J. Proteomics* **2020**, *210*, 103534. [CrossRef]
14. Italian Ministry of Agriculture, Food and Forestry Policies. Available online: https://www.politicheagricole.it/flex/cm/pages/ServeBLOB.php/L/IT/IDPagina/7469 (accessed on 25 December 2022).
15. Xanthopoulos, V.; Polychroniadou, A.; Litopoulou-Tzanetaki, E.; Tzanetakis, N. Characteristics of Anevato Cheese made from Raw or Heat-treated Goat Milk Inoculated with a Lactic Starter. *LWT Food Sci. Technol.* **2000**, *33*, 483–488. [CrossRef]
16. Pirisi, A.; Pinna, G.; Papoff, C.M. Effect of milk thermisation on Fiore Sardo PDO cheese: 1. Physicochemical characteristics. *Sci. Tecn. Latt. Cas.* **1999**, *50*, 353–366.
17. Caboni, P.; Maxia, D.; Scano, P.; Addis, M.; Dedola, A.; Pes, M.; Murgia, A.; Casula, M.; Profumo, A.; Pirisi, A. A gas chromatography-mass spectrometry untargeted metabolomics approach to discriminate Fiore Sardo cheese produced from raw or thermized ovine milk. *J. Dairy Sci.* **2019**, *102*, 5005–5018. [CrossRef] [PubMed]
18. Dedola, A.S.; Piras, L.; Addis, M.; Pirisi, A.; Piredda, G.; Mara, A.; Sanna, G. New Analytical Tools for Unmasking Frauds in Raw Milk-Based Dairy Products: Assessment, Validation and Application to Fiore Sardo PDO Cheese of a RP-HPLC Method for the Evaluation of the α-l-Fucosidase Activity. *Separations* **2020**, *7*, 40. [CrossRef]
19. Cammerino, A.R.B.; Biscotti, S.; De Iulio, R.; Monteleone, M. The sheep tracks of transhumance in the Apulia region (South Italy): Steps to a strategy of agricultural landscape conservation. *Appl. Ecol. Env. Res.* **2018**, *16*, 6977–7000. [CrossRef]
20. Albenzio, M.; Corbo, M.; Rehman, S.; Fox, P.; De Angelis, M.; Corsetti, A.; Sevi, A.; Gobbetti, M. Microbiological and biochemical characteristics of Canestrato Pugliese cheese made from raw milk, pasteurized milk or by heating the curd in hot whey. *Int. J. Food Microbiol.* **2001**, *67*, 35–48. [CrossRef]
21. Piombino, P.; Pessina, R.; Genovese, A.; Lisanti, M.T.; Moio, L. Sensory profiling, volatiles and odor-active compounds of Canestrato pugliese PDO cheese made from raw and pasteurized ewes' milk. *Ital. J. Food Sci.* **2008**, *20*, 225–237.
22. European Commission (EC). Regulation No 1107/96 of 12 June 1996 On the registration of geographical indications and designations of origin under the procedure laid down in Article 17 of Council Regulation (EEC) No 2081/92. *Off. J. Eur. Communities* **1996**, *L148*, 1–10.
23. Kuchroo, C.N.; Fox, P.F. Soluble nitrogen in Cheddar cheese: Comparison of extraction procedures. *Milchwiss. Milk Sci. Int.* **1982**, *37*, 331–335.
24. Andrews, A.T. Proteinases in normal bovine milk and their action on caseins. *J. Dairy Res.* **1983**, *50*, 45–55. [CrossRef]
25. Faccia, M.; Gambacorta, G.; Caponio, F.; Pati, S.; Di Luccia, A. Influence of type of milk and ripening time on proteolysis and lipolysis in a cheese made from overheated milk. *Int. J. Food Sci. Technol.* **2007**, *42*, 427–433. [CrossRef]

26. Bachmann, H.P.; Banks, J.; Beresford, T.; Bütikofer, U.; Grappin, R.; Lavanchy, P.; Lindblad, O.; McNulty, D.; McSweeney, P.L.; Skeie, S. Interlaboratory Comparison of Cheese Making Trials: Model Cheeses Made From Raw, Pasteurized and Microfiltered Milks. *LWT Food Sci. Technol.* **1998**, *31*, 585–593. [CrossRef]
27. Awad, S. Texture and flavour development in Ras cheese made from raw and pasteurised milk. *Food Chem.* **2006**, *97*, 394–400. [CrossRef]
28. Wilkinson, M.G.; LaPointe, G. Invited review: Starter lactic acid bacteria survival in cheese: New perspectives on cheese microbiology. *J. Dairy Sci.* **2020**, *103*, 10963–10985. [CrossRef]
29. Giraffa, G. Functionality of enterococci in dairy products. *Int. J. Food Microbiol.* **2003**, *88*, 215–222. [CrossRef] [PubMed]
30. McAuley, C.M.; Gobius, K.S.; Britz, M.L.; Craven, H.M. Heat resistance of thermoduric enterococci isolated from milk. *Int. J. Food Microbiol.* **2012**, *154*, 162–168. [CrossRef]
31. de Garnica, M.L.; Santos, J.A.; Gonzalo, C. Short communication: Influence of storage and preservation on microbiological quality of silo ovine milk. *J. Dairy Sci.* **2011**, *94*, 1922–1927. [CrossRef]
32. Corbo, M.R.; Albenzio, M.; De Angelis, M.; Sevi, A.; Gobbetti, M. Microbiological and biochemical properties of Canestrato Pugliese hard cheese supplemented with bifidobacteria. *J. Dairy Sci.* **2001**, *84*, 551–561. [CrossRef] [PubMed]
33. Santoro, M.; Faccia, M. Influence of mould size and rennet on proteolysis and composition of Canestrato Pugliese cheese. *Ital. J. Food Sci.* **1998**, *10*, 217–228.
34. Pirisi, A.; Pinna, G.; Addis, M.; Piredda, G.; Mauriello, R.; De Pascale, S.; Caira, S.; Mamone, G.; Ferranti, P.; Addeo, F.; et al. Relationship between the enzymatic composition of lamb rennet paste and proteolytic, lipolytic pattern and texture of PDO Fiore Sardo ovine cheese. *Int. Dairy J.* **2007**, *17*, 143–156. [CrossRef]
35. Sousa, M.J.; Malcata, F.X. Comparative biochemical evolution during ripening of bovine, ovine and caprine cheeses manufactured with extracts of flowers of Cynara cardunculus. *Z. Lebensm. Unters. Forsch. A* **1997**, *205*, 97–103. [CrossRef]
36. Faccia, M.; Trani, A.; Loizzo, P.; Gagliardi, R.; La Gatta, B.; Di Luccia, A. Detection of αs1-I casein in mozzarella Fiordilatte: A possible tool to reveal the use of stored curd in cheesemaking. *Food Control.* **2014**, *42*, 101–108. [CrossRef]
37. Ismail, B.; Nielsen, S.S. Invited review: Plasmin protease in milk: Current knowledge and relevance to dairy industry. *J. Dairy Sci.* **2010**, *93*, 4999–5009. [CrossRef]
38. Sarantinopoulos, P.; Kalantzopoulos, G.; Tsakalidou, E. Effect of Enterococcus faecium on microbiological, physicochemical and sensory characteristics of Greek Feta cheese. *Int. J. Food Microbiol.* **2002**, *76*, 93–105. [CrossRef]
39. Franz, C.M.A.P.; Holzapfel, W.H.; Stiles, M.E. Enterococci at the crossroads of food safety? *Int. J. Food Microbiol.* **1999**, *47*, 1–24. [CrossRef]
40. Faccia, M.; Gambacorta, G.; Liuzzi, V.A.; Alviti, G.; Di Luccia, A. Influence of cheese weight and type of rennet on composition and proteolysis of Canestrato Pugliese cheese II. Chromatographic characterization of soluble nitrogen. *Ital. J. Food Sci.* **2003**, *15*, 75–84.
41. Sousa, M.; Ardö, Y.; McSweeney, P.L. Advances in the study of proteolysis during cheese ripening. *Int. Dairy J.* **2001**, *11*, 327–345. [CrossRef]
42. Delgado, F.J.; González-Crespo, J.; Cava, R.; Ramírez, R. Formation of the aroma of a raw goat milk cheese during maturation analysed by SPME-GC-MS. *Food Chem.* **2011**, *129*, 1156–1163. [CrossRef] [PubMed]
43. Moio, L.; Addeo, F. Grana Padano cheese aroma. *J. Dairy Res.* **1998**, *65*, 317–333. [CrossRef]
44. Kilcawley, K.; Faulkner, H.; Clarke, H.; O'Sullivan, M.; Kerry, J. Factors Influencing the Flavour of Bovine Milk and Cheese from Grass Based versus Non-Grass Based Milk Production Systems. *Foods* **2018**, *7*, 37. [CrossRef] [PubMed]
45. Curtin, Á.C.; McSweeney, P.L.H. Catabolism of Amino Acids in Cheese during Ripening. In *Cheese: Chemistry, Physics and Microbiology and Microbiology. Volume 1*; Fox, P.F., McSweeney, P.L.H., Cogan, T.M., Guinee, T.P., Eds.; Academic Press: Cambridge, MA, USA, 2004; pp. 435–454.
46. Yvon, M.; Rijnen, L. Cheese flavour formation by amino acid catabolism. *Int. Dairy J.* **2001**, *11*, 185–201. [CrossRef]
47. Curioni, P.M.G.; Bosset, J.O. Key odorants in various cheese types as determined by gas chromatography-olfactometry. *Int. Dairy J.* **2002**, *12*, 959–984. [CrossRef]
48. Foulquié Moreno, M.R.; Sarantinopoulos, P.; Tsakalidou, E.; De Vuyst, L. The role and application of enterococci in food and health. *Int. J. Food Microbiol.* **2006**, *106*, 1–24. [CrossRef] [PubMed]

foods

MDPI

Article

Optimization of Modified Atmosphere Packaging for Sheep's Milk Semi-Hard Cheese Wedges during Refrigerated Storage: Physicochemical and Sensory Properties

Marta Albisu [1], Sonia Nieto [2], Olaia Martínez [3], María Ángeles Bustamante [1], Luis Javier R. Barron [1] and Ana Isabel Nájera [1,*]

[1] Lactiker Research Group, Faculty of Pharmacy, Universidad del País Vasco/Euskal Herriko Unibertsitatea, 01006 Vitoria-Gasteiz, Álava, Spain
[2] Efficient and Sustainable Processes Department, Bizkaia Technology Park, AZTI, P.O. Box 609, 48160 Derio, Bizkaia, Spain
[3] Texture Analysis Laboratory, G3S Research Group, Faculty of Pharmacy, Universidad del País Vasco/Euskal Herriko Unibertsitatea, 01006 Vitoria-Gasteiz, Álava, Spain
* Correspondence: anaisabel.najera@ehu.eus; Tel.: +34-945-013-077

Abstract: Modified atmosphere packaging (MAP) has become a good potential strategy to retain quality throughout the shelf life of perishable foods. The aim of this work was to evaluate different packaging atmospheres on semi-hard protected designation of origin Idiazabal cheese wedges. Six different packaging treatments (air, vacuum, and CO_2/N_2 gas mixtures in the ratio of 20/80, 50/50, 80/20, and 100/0% *v/v*, respectively) were studied. Changes in gas headspace composition, cheese gross composition, weight loss, pH, acidity, colour, and textural and sensory properties were investigated during 56 days of refrigerated storage at 5 ± 1 °C. MAP was the most effective preserving technique compared to air- and vacuum-packaging treatments. The cheese characteristics with the greatest discriminating weight in the preservation techniques were paste appearance, holes, flavour, a* (redness) and b* (yellowness) colour parameters, and slope to hardness. Air-packaged cheeses, on 35 day, presented a mouldy flavour. Vacuum packaging affected paste appearance (greasy, plastic marks, and non-homogeneous colour) and holes (occluded and unnatural appearance) starting after 14 packaging days. MAP mixtures with CO_2 concentration between 50/50 and 80/20% CO_2/N_2 (*v/v*) are recommended to ensure sensory quality and stability in the distribution of these raw sheep-milk cheese wedges.

Keywords: ripened cheese preservation; modified atmosphere packaging; vacuum; cheese wedges; sensory properties

Citation: Albisu, M.; Nieto, S.; Martínez, O.; Bustamante, M.Á.; Barron, L.J.R.; Nájera, A.I. Optimization of Modified Atmosphere Packaging for Sheep's Milk Semi-Hard Cheese Wedges during Refrigerated Storage: Physicochemical and Sensory Properties. *Foods* **2023**, *12*, 849. https://doi.org/10.3390/foods12040849

Academic Editors: Michele Faccia and Giuseppe Natrella

Received: 2 December 2022
Revised: 31 January 2023
Accepted: 13 February 2023
Published: 16 February 2023

1. Introduction

World cheese production in 2022 was approximately 22.08 million tonnes [1], where the European Union (EU) is the world's most important cheese manufacturer, with a production of 10.4 million tonnes in 2021 [2]. The most widely produced cheeses are from cow's milk, with continuous processing throughout the year, while small-ruminant cheese production is smaller and seasonal. The production of sheep's milk cheeses has increased substantially in the Mediterranean countries. Spain is one of the largest producers in the EU, with 72.2 thousand tonnes in 2021 [3].

Given the seasonality of this cheese production, the herds spend a large part of the time grazing, which influences the quality of the milk and the regularity of cheese manufacture. Sheep-milk cheeses are highly appreciated by consumers, and this is partly because many of them are sold under quality brands such as protected designation of origin (PDO) or protected geographical indication (PGI) certifications [4]. Idiazabal PDO is a seasonal, pressed hard, and semi-hard raw-milk sheep's cheese produced in the Basque Country (northern Spain). The milk-producing sheep must be of Latxa breed. The weight of Idiazabal PDO cheese is between 1–3 kg, with a minimum ripening time of two months [5]. These cheeses maintain acceptable sensory quality for a short period according to their brief manufacture season [6]. Therefore, it is extremely necessary to improve the preservation method in order to increase availability over a longer period without changing the sensory features. This traditionally made cheese is usually sold as a whole piece. However, there is an increasing consumer demand for portioned cheeses, and packaging is of great importance to ensure their quality and safety [7].

Shelf-life extension by packaging optimization has become a powerful strategy to improve marketing needs and reduce food waste or loss [8]. Different packaging systems have proven their usefulness to prolong the shelf life of cheeses and specifically to prevent several faults that may come with portioning. When combined with low-temperature storage, vacuum and modified atmosphere packaging (MAP) have effectively proven to be useful in cheese preservation [9], but the results depend on the cheese variety [10,11].

Vacuum packaging reduces oxidative damage and inhibits aerobic microorganisms' growth. It might prevent dehydration and weight loss in cheeses as well as the incorporation of undesirable odours [12]. This preservation method has been successfully tested for hard and semi-hard cheeses [12–14]. However, negative effects have also been described, as this packaging provides anaerobic conditions and favours pathogens and moulds growth [15,16] as well as changes in cheese appearance [17,18].

MAP has been reported to extend the shelf life of cheeses, based on microbiological and sensory parameters, with certain gas mixtures (CO_2 and N_2) and for some types of non-pasteurized cheeses such as Crottin de Chavignol, Pasta Filata, and some mould-surface cheeses [19–21]. Carbon dioxide is the most important gas from a microbiological perspective because it inhibits the growth of spoilage bacteria such as aerobic Gram (−) and moulds and to a lesser extent Gram (+) bacteria and yeasts. On the other hand, N_2 is used as a filler preventing package collapse. The gaseous composition of the atmosphere can change during storage time due to cheese breathing, biochemical reactions, and the diffusion of gases through the packaging material. A CO_2 concentration between 20 and 60% (*v/v*) in the atmosphere is required for antimicrobial effect [9]. Usually, high concentrations of CO_2 are used to suppress undesirable microbial growth, particularly in hard and semi-hard cheeses [21,22]. However, atmosphere design is more complex for raw-milk cheeses or those with starters added [23]. For these cheeses, CO_2 concentration must be carefully adjusted and controlled; otherwise, the bacterial metabolism, enzymatic activities, lipid oxidation, and proteolysis involved in the development of cheese flavour may be modified, and off-flavours may appear during storage [10,13,21,24]. Moreover, this approach is of great interest for semi-hard cheeses presented in portion packs, as inner faces are exposed to atmospheres, and the best packaging options must be chosen in order to offer products with a longer durability to consumers. Preservation studies play a fundamental role in increasing the conservation time of portioned hard and semi-hard cheeses and can be of significant importance for the profitability and sustainability of the dairy sector [9].

Therefore, the aim of this work was to evaluate the influence of different types of packaging conditions—air, vacuum, and four different modified atmospheres—to see how they affect the quality of a sheep's milk cheese. Physicochemical, colour, and textural properties and sensory quality of Idiazabal semi-hard cheese wedges with a maturity degree of three months were studied for a period of two months. The ultimate goal is to select which of the packaging options best preserves quality during the distribution stage of this type of cheese.

2. Materials and Methods

2.1. Cheese Collection, Packaging, and Sampling

A total of 60 raw sheep's milk cheeses with three ripening months were purchased at a local dairy farm registered at the Idiazabal PDO Council. Cheeses were produced in different days, 24 h from each other, which constituted two batches. Idiazabal cheese, as it is produced with raw milk, needs a minimum of two ripening months to meet food safety and sensory requirements. Three ripening months were chosen, as it is then when the cheeses have already developed their typical sensory characteristics [5].

From the whole pieces of cheese, 360 wedges (175–200 g) were obtained. Cheese wedges were cut and immediately packed in commercial polyamide/polyethylene (20/70) pouches of 90 μ thickness (Merkapack, Vitoria, Spain). Pouches presented oxygen permeability of $\leq$80 $cm^3/m^2 \times$ bar $\times$ 24 h (75% HR), carbon dioxide permeability of $\leq$174 $cm^3/m^2 \times$ bar $\times$ 24 h (0% HR), and $\leq$2 $g/m^2 \times$ bar $\times$ 24 h (85% HR) for water vapour. Cheese wedges were packaged under air, vacuum, and four different CO_2/N_2 gas mixtures: 20/80, 50/50, 80/20, and 100/0% *v/v*, (MAP1, MAP2, MAP3, and MAP4, respectively). Pouches were evacuated, flushed, and sealed using MAP equipment (Model EVT-450/20, Irimar, Lesaka, Spain) with gas injection. Carburos Metálicos-Grupo Air Products (Cornellá de Llobregat, Spain) supplied food-grade gases, and a binary gas hand-operated mixer model MM-2K N_2/CO_2 (Witt, Witten, Germany) was used. Two sets of samples were prepared and stored in a refrigerator at 5 ± 1 °C for 56 days.

Headspace gas composition, physicochemical, colour, and instrumental textural and sensory analyses were conducted on the cheese samples at 14, 21, 28, 35, 42, 49, and 56 storage days; the same analyses were carried out on the cheeses prior to packaging.

2.2. *Headspace Gas Composition of Packed Cheese Wedges*

Immediately after packaging in the laboratory, pouches were subjected to a visual inspection of the sealing area to check for possible failures. Gas composition was verified by means of a gas analyser, Oxybaby (Witten, Germany), at every starting point when supplying a new ratio of gases. On each sampling day, O_2 and CO_2 concentrations were checked again.

2.3. *Physicochemical Analysis of Cheese*

Cheese wedges were weighed before packaging and on each sampling day on an Adam balance (Milton Keynes, UK). Before the analyses, the samples were equilibrated at 17 ± 2 °C. Sample temperature was controlled with a penetration thermometer Testo model 104-IR in triplicate (Barcelona, Spain).

pH was determined, in quadruplicate, at different points along each wedge by means of a pH meter with a penetration electrode (Crison, Barcelona, Spain).

Titratable acidity was measured in duplicate according to ISO/TS 11869 [25] method for fermented milks, adapted according to the AOAC 920 method for cheeses [26], in which the volume of filtered aliquot was modified and expressed as g lactic acid/100 g cheese.

A Zeutec model 110-A100-1 infrared spectral analyser 2.0 (Rendsburg, Germany) was used to determine dry matter concentrations, total protein, and fat, previously calibrated using the application worxG2 software with a multiple linear regression (MLR) model. Measurements were performed in duplicate for each batch from a homogeneous fraction obtained from grating each cheese wedge after the removal of approximately 1 cm of rind.

2.4. *Cheese Colour Measurement*

A Minolta Chroma Meter CR-200 (Madrid, Spain) was used for colour measurement on one side of the cheese wedges in triplicate. CIELab values (lightness, L*; redness, a*; and yellowness, b*) were measured with the standard illuminant D65 and a visual angle of 10°.

The yellow index (Yi) colour expression [27] used by Romani et al. [13] and Favati et al. [17] was calculated. In addition, the yellowness index (Zi) parameter [28] reported by del Caro et al. [29] was also calculated.

$$\mathrm{Yi} = \frac{142.86b^*}{L^*} \qquad \mathrm{Zi} = 100\left(\frac{L^* + 16}{116} - \frac{b^*}{200}\right)^3$$

2.5. Texture Profile Analysis of Cheese

Sample cubes were prepared as follows from each wedge: a 1 cm thick slice was removed from the rind in one lateral side of each wedge, and then, using a guide, three consecutive 1 cm thick slices were cut. Five cubes (1 cm × 1.25 cm × 1.25 cm) were obtained from each slice (Figure 1). These cubes were used as repetitions for texture profile analysis (TPA). Texture was assessed with a TA.XT2plus texture analyser (Stable Micro System, Surrey, UK) by means of TPA [30], with a 5 kg load cell. Two consecutive compression cycles at 15% were performed on cheese cubes, always with the narrower side upwards and using an aluminium cylindrical probe (diameter = 2.5 cm) at 1 mm/s. Cheeses were at 17 ± 2 °C during the assay. Texture Expert Exceed software was used for data processing.

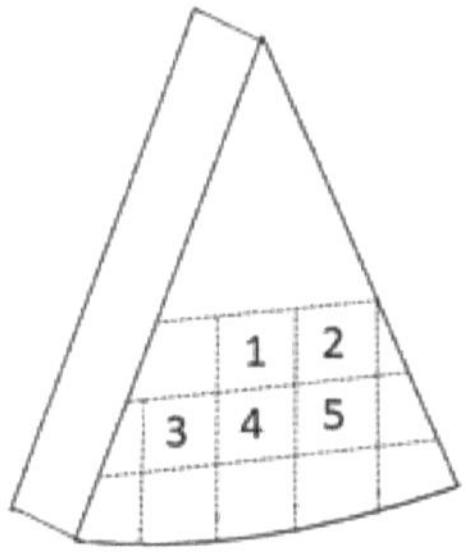

Figure 1. Scheme of the cube shaped samples used as repetitions for texture profile analysis within each slice obtained from cheese wedges.

The texture parameters studied were the following [31,32]. Hardness: the highest force peak of the first compression cycle (N). Slope: expressed as N/s, from the start of the curve to the maximum peak at the first compression cycle. It can be referred to as slope to hardness; as the slope becomes higher, the material might have less tendency to deform before fracture. Springiness: ratio between the distances the sample was compressed in the second downstroke divided by the first downstroke. Cohesiveness: ratio of the areas under the curve of the second compression cycle to the first compression cycle. Chewiness is calculated as cohesiveness × hardness × springiness (N). Resilience: ratio between the area under the curve of the withdrawal divided by the area under the curve in the downstroke, both in the first compression cycle.

2.6. Cheese Sensory Analysis

Sensory evaluation was carried out with seven trained assessors aged between 35 and 60 years (three men and four women). Informed consent was obtained from all subjects involved in the study. A discontinuous seven-point scale was used for texture, flavour, paste appearance, and paste holes, where 1 was the lower score, and 7 was the best score, as required for quality control of Idiazabal PDO cheese. Scores lower than 4 indicated that cheeses presented defects, and assessors were asked to identify which defects were perceived. Scores from 4 to 6 were marked when there were not defects, but the sensory characteristics were not totally appropriate [33].

Nine training sessions were conducted (around 90 min each). Four panellists belonged to the PDO Idiazabal official sensory panel, and three assessors had previous experience in sensory analysis in sheep-milk cheeses [34]. The first three training sessions were addressed to the assessors who did not belong to the PDO Idiazabal official sensory panel. All the assessors attended to the onward sessions. In the next four sessions, references were presented together with cheese samples. In the last two training sessions, the assessors evaluated cheese samples without references in order to harmonize results within the panel [33].

Sensory assessments were conducted in individual booths at the sensory laboratory, which complied with ISO 8589 standard [35]. Cheeses for each session were tempered at

17 ± 2 °C and presented rind-free and cut into parallelograms of 1.5 cm × 1.5 cm × 5 cm, and samples were randomly presented coded with a three-digit number obtained from Fizz software 2.40H (Biosystemes, Couternon, France). Low-mineralization water and Granny Smith apples were used to remove aftertaste between samples. Subsequently, whole wedges were randomly presented and identified with different three-digit numbers from Fizz software to score paste appearance and holes. In each session, assessors analysed wedge samples packaged in the six different treatments. A replicate of each session was conducted within the same day with a half-hour break and randomly presented with different coded numbers.

2.7. Data Treatment and Statistical Analysis

SPSS IBM Statistics software version 26.0 (New York, NY, USA) was used for statistical analysis (SPSS INC., Chicago, IL, USA). Two-way analysis of variance (ANOVA) was used to determine the significant differences in headspace and physicochemical and colour parameters from the different packaging treatments over the study period using packaging treatment and storage time as fixed factors. Subsequently, the Tukey's test was applied to pairwise comparisons between cheeses packaged under the different treatments and on each sampling day separately. Kruskal–Wallis H test was used to check for possible significant differences between storage condition and storage time, regarding instrumental texture and sensory parameters. A stepwise discriminant analysis was applied to physicochemical, instrumental colour, texture profile, and sensory parameters to classify cheese samples from the different packaging methods considering all MAP samples as a unique group. Statistical significance was declared at $p \leq 0.05$.

3. Results and Discussion

3.1. Headspace Gas Composition

The headspace of cheese wedges was analysed for O_2 and CO_2 concentration except for the vacuum-packed samples. On day 56, MAP4 pouches were totally collapsed, and the measurement was not carried out.

3.1.1. O_2 Concentration

In the air-packaging treatment, O_2 concentration decreased from day 0 (20.3%) to day 56 (0.7%), resulting in a total reduction of 96.4%. From day 42, it remained constant with 0.5–0.7% (Figure 2). Film permeability, aerobic microorganisms' metabolism, oxidative and enzymatic reactions involving oxygen, and cheese respiration could cause a progressive decrease in O_2 concentration in the air-packaged wedges [10,36,37]. A decrease in the O_2 concentration occurred in Domiati cheese packaged in air during cold storage, with levels from 19.9% to 0.2%, although very high barrier film was used for the samples packaging [38].

In the MAP cheese wedges, O_2 concentration at the beginning of the storage time was residual (±0.4%), and it did not change during storage. Steady-state conditions between microbial respiration rate and O_2 permeation through packaging material could explain this result [36], indicating non-presence of failures in the packaging [39]. According to Garabal et al. [12], there were no differences ($p > 0.05$) in O_2 content among different MAP atmospheres of packaged cheeses, with mean values close to 0.2%. These concentrations were similar for Samso cheese [10] and Havarti cheese stored with 20 to 100% CO_2 modified atmospheres [40].

3.1.2. CO_2 Concentration

In air-packaging treatment, there was a progressive and very pronounced increase of CO_2 (2.9 times more) in the first 14 days (Figure 2; Table 1). After 56 days, the CO_2 concentration increased 5.6 times. This progressive increase could be due to the gas permeability through the packaging material and the microbial growth in the cheese matrix.

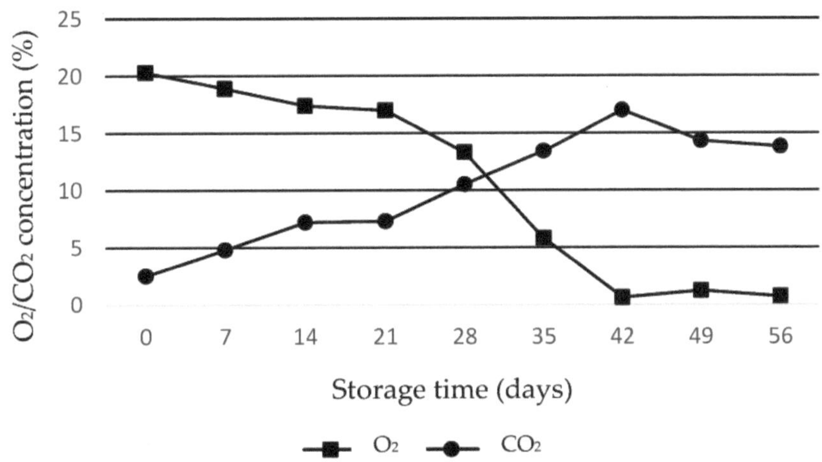

Figure 2. Evolution of O_2 and CO_2 concentrations of cheese wedges stored for eight weeks in air-packaging treatment.

Table 1. Mean, standard deviation, and significance level of ANOVA for CO_2 concentration (%) in the headspace of packaged cheese wedges stored for eight weeks at different atmosphere treatments.

Day	Air	MAP1	MAP2	MAP3	MAP4
0	2.45 ± 0.78 d	23.98 ± 0.67 a	52.90 ± 0.78 a	83.80 ± 0.35 a	100.08 ± 0.04 a
14	7.20 ± 1.56 cd	19.20 ± 0.99 ab	44.75 ± 0.78 ab	76.80 ± 0.99 ab	100.00 ± 0.00 a
21	7.25 ± 0.78 cd	17.70 ± 0.57 b	40.60 ± 1.41 bc	73.65 ± 2.62 ab	99.95 ± 0.07 a
28	10.49 ± 0.69 bc	18.30 ± 1.56 ab	40.50 ± 1.98 bc	70.40 ± 6.36 ab	97.95 ± 2.90 a
35	13.35 ± 4.03 abc	17.05 ± 0.21 b	37.53 ± 2.02 bcd	64.20 ± 8.34 ab	94.65 ± 7.42 a
42	17.00 ± 0.57 a	17.95 ± 2.05 ab	38.35 ± 4.45 bcd	66.00 ± 9.90 ab	93.80 ± 8.77 a
49	14.25 ± 0.49 ab	17.15 ± 1.34 b	33.90 ± 2.40 cd	59.80 ± 14.85 ab	99.10 ± 0.00 a
56	13.80 ± 0.57 ab	16.70 ± 2.97 b	29.50 ± 4.38 d	47.65 ± 14.07 b	-

MAP1, 20/80% CO_2/N_2 (*v*/*v*); MAP2, 50/50% CO_2/N_2 (*v*/*v*); MAP3, 80/20% CO_2/N_2 (*v*/*v*); MAP4, 100/0% CO_2/N_2 (*v*/*v*). Different letters (a–d) in the same column indicate significant differences ($p \leq 0.05$) during storage for each packaging treatment.

Increases in CO_2 concentration were reported for semi-hard and hard cheeses packed under MAP [11,40]. Several authors related this effect to microbial growth and cheese ripening, together with some oxidative and enzymatic reactions [11,37,41]. In the present study, in the air-packaged wedges after 28 storage days, CO_2 concentration was increased 4.3 times, whereas that of O_2 decreased 1.5 times (Figure 2). These data agreed with those reported for MAP-packaged Domiati cheeses. The increase of CO_2 concentration in the gas headspace might be mainly associated with O_2 consumption by microorganisms [38].

In MAP1, MAP2, and MAP3 treatments (Table 1), a progressive decrease of the CO_2 concentration reaching a mean value of 39% (MAP1: 30.36%, MAP2: 44.23%, and MAP3: 43.14%) was observed at the end of the storage period. These results could be explained by the gas dissolution in the cheese matrix. Several authors pointed out that the low CO_2 concentrations detected in headspaces may be attributed to its dissolution in the cheese matrix, its consumption by anaerobic microorganisms, or by CO_2 loss through the barrier film [10,11,37,41]. Solomakos et al. [42] observed no important changes on CO_2 content in the headspace of cheese stored during 85 days under 50/50% CO_2/N_2 (*v*/*v*) MAP conditions. For MAP4, CO_2 concentration did not change until day 49 (Table 1). However, from day 28, some pouches started to collapse and on day 56 were totally collapsed, so the measurement could not be carried out. The absence of variation in the proportion of CO_2 in the latter treatment is due to the fact that it is the only gas present in the package. The progressive decrease of the volume inside the pouches until the collapse at the end of storage may be related to the fact that it is a raw-milk cheese with a higher number of lactic acid bacteria (LAB) and/or anaerobic microorganisms that can consume this gas. Both the consumption of gas by LAB and/or the dissolution of CO_2 in the cheese can increase the acidity, which was reflected in the results of the sensory analysis with the appearance of acid off-flavours on day 56.

3.2. Physicochemical Analysis of Cheeses

None of the physicochemical parameters measured in cheese wedges showed significant differences ($p \leq 0.05$) for both the packaging treatment and storage time or the interaction between them. Thus, cheese wedges had mean values of 0.39 ± 0.43% for weight loss, 5.00 ± 0.06 for pH, 1.22 ± 0.07 g lactic acid/100 g cheese for titratable acidity, 66.13 ± 0.54% for dry matter, 24.68 ± 0.34% for protein, and 35.50 ± 0.53% for fat.

The percentage of weight loss percentage of cheese wedges during storage was not significant ($p > 0.05$) for all packaging treatments, with the mean value being 0.39 ± 0.43%. The plastic material used for packaging prevented dehydration and weight loss of cheese wedges [12,22]. This was consistent with the findings of different MAP gas mixtures that did not significantly ($p > 0.05$) influence weight losses in ripened cheeses packaged under MAP [11,37,43]. Favati et al. [17] reported weight losses of 0.15% in Provolone cheese.

The possible dissolution of CO_2 in MAP-packaged cheeses matrix mentioned before did not affect pH value, and it remained stable throughout the storage. Other studies observed a similar behaviour for pH values in the case of other cheese types packaged under MAP conditions during refrigerated storage (ranges 4.7–4.8 for aged white cheese) [41]. Solomakos et al. [42] observed a pH decrease of air (from 5.52 to 5.10) and MAP (from 5.52 to 4.95) cheeses during storage at 10 °C, probably due to further activity and growth of LAB as compared to a lower microbial growth at 4 °C. The presence of CO_2 in the headspaces was expected to cause a decrease in pH and an increase in the acidity of the cheese samples because of CO_2 dissolution occurring at low temperature in cheese and the formation of carbonic acid [22,24]. Provolone samples packaged with 100% CO_2 at 4 °C presented higher acidity, free fatty acids, and free amino acids contents than other gas mixtures [17]. However, other authors indicated that CO_2 is mainly absorbed on food surface, which may lead to acidification of some spots along the surface rather than in the matrix [44]. Pintado and Malcata [45] reported that storing cheese above refrigeration temperatures (12 or 18 °C) resulted in a very strong pH decrease in the cheeses. These results confirm that temperature control during storage is important. On the other hand, an increase in pH (from 5.25 to 5.40 in 50% CO_2 MAP) during storage can be due to proteolysis and associated formation of amines and ammonium [12]. It has been reported that after 45 storage days, the total amount of free amino acids in cheeses was approximately two times higher than that observed at the beginning of the process, suggesting a high rate of proteolysis during storing. At the same time, microbial degradation could be favoured by low concentrations of O_2 in MAP packaging. A lack of any effect of the CO_2 on cheese proteolysis during storage was reported by Alves et al. [46]. Gonzalez-Fandos et al. [22] associated increased proteolysis in vacuum-packaged Cameros cheese to the higher counts of microorganisms.

3.3. Instrumental Colour Parameters

None of the colour parameters showed significant differences ($p > 0.05$) for either storage time or the interaction between packaging treatment and storage time. The colour parameters L* and Zi did not show significant differences ($p > 0.05$) over time in any of the packaging treatments, while a*, b*, and Yi parameters showed differences ($p \leq 0.01$) for the packaging treatment (Table 2). In particular, a* parameter was able to significantly distinguish air-packaged cheeses (mean value −3.52) from the rest of the treatments (mean value −2.68).

Table 2. Mean, standard deviation, and significance level of ANOVA for packaging treatment, storage time, and interaction between them for the colour parameters L*, a*, b*, Yi, and Zi.

	L*	a*	b*	Yi	Zi
Packaging treatment	ns	**	**	**	ns
Storage time	ns	ns	ns	ns	ns
Interaction	ns	ns	ns	ns	ns
x ± SD	81.27 ± 2.22	−2.80 ± 0.47	12.27 ± 0.93	21.54 ± 1.56	47.04 ± 3.37

L*, lightness; a*, redness; b*, yellowness; yellow index Yi, 142.86 b*/L*; yellowness index Zi, 100(L* + 16/116)–(b*/200)^3; ** $p \leq 0.01$; ns, not-significant differences.

The only significant difference ($p \leq 0.05$) over the storage period was for a* in the air-packaging treatment, going from −2.96 to −3.16, from 0 to day 56. This change was sharper the first 14 days (−3.82), and then, this value was practically maintained until the end of the storage period. The rest of the packaging treatments showed no significant differences ($p > 0.05$) over time.

Colour parameters b* and Yi showed significant differences ($p \leq 0.05$) over time only in MAP1 treatment. The rest of the packaging treatments showed no differences ($p > 0.05$) over time. Comparing the treatments with each other, the air- and vacuum-packed cheeses (mean values for both treatments were 13.19 for b* and 22.98 for Yi) were similar, while MAP treatments (mean values for all four treatments were 12.08 for b* and 21.11 for Yi) were grouped together. The parameters Yi and Zi did not seem to discriminate more than L*, a*, and b*. A similar trend was found for the colour parameter, L*, during the first two storing months in a blue cheese [47] and Crottin-de-Chavignol-type goat cheese [19]. In portioned Canestrato pugliese cheese, vacuum and MAP might stabilize cheese colour during storage [48]. Low O_2 transmission rates and low residual O_2 levels in the headspace and dissolved in the cheese can help to avoid photo-oxidation of the food matrix. Trobetas et al. [11] observed a gradual discoloration, L* and b* values decreased, and a* value increased in Graviera hard cheeses packed under MAP and exposed to light at 4 °C, which was related in part to riboflavin degradation induced by light. The values for a* and b* of the samples stored in dark remained constant. Retinol and xanthophyll have been detected in low concentrations in sheep and goat milk but not β-carotene [49].

Favati et al. [17] did not detect differences in the colour parameter Yi for cow's milk cheese packaged in portions at different CO_2 concentrations, as reported Romani et al. [50]. Avila et al. [51] described that the increase in parameters a* and b* might be mainly due to the cheese concentration components coming from dehydration throughout ripening.

3.4. Texture Profile Analysis

Storage brought a significant initial increase ($p \leq 0.05$) in some instrumental texture parameters within the first two weeks of storage (Figure 3). This was observed for hardness, slope, and chewiness in packaging treatments of air, vacuum, and MAP1 (Table S1). For hardness, this difference was also observed in MAP2-packaged cheese wedges. For further CO_2 concentrations, this change was not noticeable. Along the eight-week storage, values tended to decrease, getting closer to those registered initially at day 0 when approaching the last storage stages (Figure 3).

Statistical analysis showed significant ($p \leq 0.05$) changes along time in all the packaging treatments for hardness, slope, chewiness, and resilience. Atallah et al. [38] described an initial increase in cohesiveness, springiness, and chewiness that decreased from day 30 onwards. They concluded that this depended on the type of milk, production methods, and other processing conditions. Costa et al. [37] reported an increase in hardness during the first 20 storage days for ripened cheeses and related it to the change in moisture con-tent. From that day on, they observed a reduction in instrumental hardness as well, which they attributed to detrimental phenomena and moulds growth, according to the literature. Changes in moisture content were not detected, as happened in most of the physicochemical parameters studied. However, initial changes were reported in gas balances and colour a* parameters, especially for air-storage samples, which could be related to initial increase in the values of some texture attributes already described. CO_2 dissolution in the cheese matrix might have prevented this effect happening from 20/80% CO_2/N_2 (v/v) on, as CO_2 has proven to maintain physical, nutritional, and organoleptic features and to improve cheese microstructure through component interactions [52]. Nevertheless, microbial growth in two weeks' time might have been enough to cause a decrease in texture values after the initial increase in air, vacuum, and MP1 samples and during storage for the other MAP treatments.

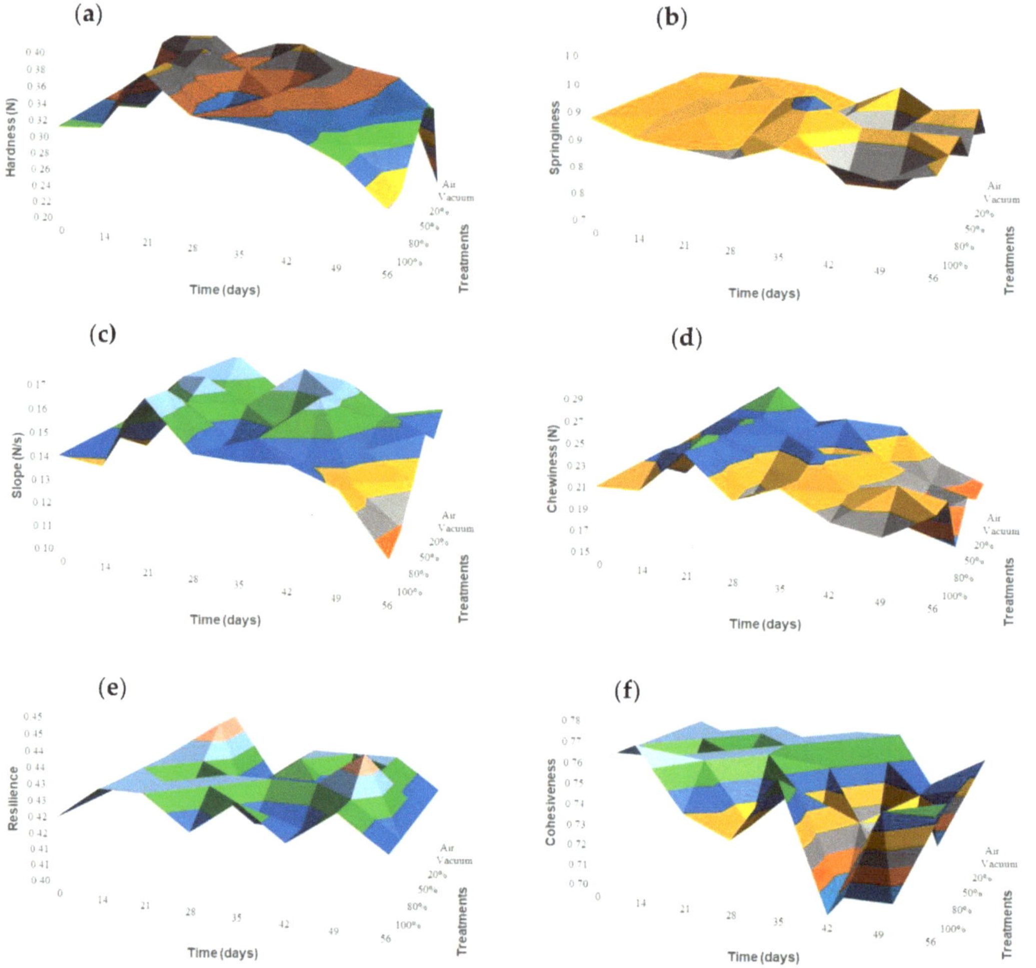

Figure 3. Three-dimensional surface chart for mean values of each parameters obtained in the texture profile analysis (TPA). Each graph represents a texture parameter from the TPA: (**a**) hardness, (**b**) springiness, (**c**) slope to hardness, (**d**) chewiness, (**e**) resilience, and (**f**) cohesiveness. Texture parameters are represented with respect to time (0 to 56 days) for each storage condition (air, vacuum, and increasing CO_2 percentages from 20% to 100%).

Generalized significant ($p \leq 0.05$) changes in cheese texture between packaging treatments were only perceived for hardness and chewiness. Hardness showed a significant decrease along time, with the lowest values from 42 days on. This decrease was higher for those MAP treatments with highest CO_2 concentration (>50%). Significant changes among treatments were also registered for chewiness in all the storage time points except for 21 and 35 days. Chewiness values tended to decrease along time for air-packaged, vacuum, MAP1, and MAP2 treatments, while the last storage stages showed similar values for MAP3 compared to storage at 14 days (0.237 ± 0.049 N to 0.217 ± 0.060 N) and higher for MAP4 (0.212 ± 0.053 N to 0.218 ± 0.032 N) compared to storage at 14 days. Some previous studies had different results, probably due to differences in milk characteristics, ripening time, and storage conditions. Kirkin et al. [41] showed that hardness was higher in 75/25% CO_2/N_2 MAP (*v/v*) compared with the vacuum packaging considering the overall mean during the entire storage period. Favati et al. [17] reported the lowest shear force values for vacuum-packed Provolone cheeses compared to CO_2-containing atmospheres (10–100%).

3.5. Sensory Properties of Cheeses

On days 49 and 56, it was not possible to analyse texture and flavour of air-packaged cheese wedges due to the growth of mould spots in the paste (Table 3). The texture of air-packaged cheeses was dry and lumpy after 21 storage days, and flavour scores were below acceptance from day 35 onwards with mouldy notes.

Table 3. Sensory texture and flavour: mean, standard deviation, and significance level of Kruskal–Wallis H statistics treatment of packaged cheese wedges stored for eight weeks at different atmosphere treatments.

Day	Air	Vacuum	MAP1	MAP2	MAP3	MAP4
			Texture			
0	5.36 ± 0.63 a1	5.36 ± 0.63 a1	5.36 ± 0.63 a1	5.36 ± 0.63 a1	5.36 ± 0.63 a1	5.36 ± 0.63 a1
14	5.07 ± 0.92 a12	5.29 ± 0.9 a12	5.14 ± 0.86 a12	5.18 ± 0.77 a1	5.07 ± 0.83 a1	4.75 ± 0.75 a12
21	4.84 ± 0.89 a12	5.00 ± 0.68 a12	5.07 ± 1.00 a12	4.79 ± 0.98 a1	4.43 ± 0.51 a1	4.71 ± 0.83 a12
28	4.50 ± 0.76 a12	4.79 ± 0.80 a12	4.71 ± 0.73 a12	4.86 ± 0.77 a1	5.00 ± 0.88 a1	4.64 ± 0.74 a12
35	4.21 ± 0.43 a2	4.64 ± 0.74 a12	4.50 ± 0.65 a12	4.77 ± 0.70 a1	4.86 ± 0.86 a1	4.14 ± 0.66 a2
42	4.50 ± 0.52 a12	4.92 ± 0.83 a12	4.79 ± 0.58 a12	4.93 ± 0.88 a1	5.14 ± 0.77 a1	4.50 ± 0.66 a12
49	-	4.79 ± 0.80 a12	4.50 ± 0.85 a12	5.21 ± 0.89 a1	4.79 ± 0.56 a1	4.21 ± 0.43 a12
56	-	4.36 ± 0.50 a2	4.43 ± 0.51 ab2	4.64 ± 0.74 ab1	4.86 ± 0.66 a1	3.86 ± 0.77 b2
			Flavour			
0	5.50 ± 0.52 a1	5.50 ± 0.5 a1	5.50 ± 0.52 a1	5.50 ± 0.52 a1	5.50 ± 0.52 a1	5.50 ± 0.52 a1
14	5.07 ± 0.83 a12	5.14 ± 0.66 a12	5.07 ± 0.83 a1	5.21 ± 0.58 a1	5.07 ± 0.73 a1	5.04 ± 0.80 a12
21	4.43 ± 0.65 a234	4.93 ± 0.9 a23	4.93 ± 0.73 a1	5.00 ± 0.78 a1	4.86 ± 0.86 a1	4.29 ± 0.73 a23
28	4.23 ± 0.73 a234	4.29 ± 0.61 a3	4.64 ± 0.84 a123	4.86 ± 0.77 a1	4.79 ± 0.89 a1	4.43 ± 0.76 a23
35	3.93 ± 0.47 bc34	4.50 ± 0.6 ab23	4.50 ±0.85 ab123	4.57 ± 0.7 ab1	4.71 ± 0.73 ab1	4.14 ± 0.6 bc23
42	3.36 ± 1.00 c4	5.21 ± 0.70 ab12	4.86 ± 0.86 ab12	4.93 ± 0.83 ab1	5.36 ± 0.63 a1	4.29 ± 0.92 bc23
49	-	4.79 ± 0.80 a12	4.50 ± 0.85 a12	5.21 ± 0.89 a1	4.79 ± 0.56 a1	4.21 ± 0.43 a12
56	-	4.50 ± 0.94 ab23	3.71 ± 0.71 b3	4.57 ± 0.76 a1	4.79 ± 0.43 a1	3.71 ± 0.91 ab3

MAP1, 20/80% CO_2/N_2 (*v/v*); MAP2, 50/50% CO_2/N_2 (*v/v*); MAP3, 80/20% CO_2/N_2 (*v/v*); MAP4, 100/0% CO_2/N_2 (*v/v*). Different letters (a–c) in the same row indicate significant differences ($p \leq 0.05$) between the different packaging conditions on that day. Different numbers (1–4) in the same column indicate significant differences ($p \leq 0.05$) during storage for each packaging condition.

Short storage times have been described for cheeses kept in air on account of mould growth [19,22,44,53]. Vacuum-packaged cheese wedges were acceptable at all storage times, with slight differences in texture and flavour. Other vacuum-packaged cheeses were softer and more elastic due to possible fat migration. Garabal et al. [12] and Romani et al. [13,50] detected an increase in acidity with this preservation method.

Among the MAP treatments, only MAP1 samples showed differences ($p \leq 0.05$) between day 0 and 56 for texture, but on day 49, assessors signalled wet-mouldy flavours, and cheeses had an assessment below the limit in flavour parameter on the last storage day. Esmer et al. [19] reported that cheese packaged at low CO_2 concentration quality was randomly affected on 42 days, and Garabal et al. [12] described the cheeses as friable and grainy. From a safety pointy of view, this low CO_2 concentration was at the limit for microorganisms' inhibition, which corresponds to the presence of mouldy flavours in this study [9].

Sensory texture and flavour scores for MAP4 were below the minimum acceptance mark at 56 days (Table 3). Assessors indicated texture defects as lumpy and fracturable, and crystals were perceived on chewing. Juric et al. [10] found dry and crumbly texture in cheeses packed with high CO_2 concentration. Agarwal et al. [54] described calcium lactate crystals in Cheddar cheese packages with 100/0% CO_2/N_2 (v/v) after four storage weeks. A possible reason for crystal formation is the cheese's superficial drying, which may favour the onset of calcium lactate crystallization. Assessors highlighted acid, rancid, and pungent notes as off-flavours. These off-flavours may be the result of CO_2 solubilisation in the cheese matrix since the pouches, as discussed above in Section 3.1.2, were collapsed. Romani et al. [50] and Gonzalez-Fandos et al. [22] observed that high CO_2 atmosphere produced great flavour variations, and they attributed this to the solubilisation of CO_2 in the cheese matrix, which produced acidity, and the storage was shortened compared to other MAP conditions [53]. MAP2 and MAP3 showed no significant changes for texture and flavour ($p > 0.05$) during 56 storage days. Several authors pointed out that atmosphere close to 50/50% CO_2/N_2 (v/v) is the best for preserving cheese flavour [11,13,22,41,44,50,53,55].

For paste appearance, the panel scored vacuum- and air-packaged cheeses as unacceptable from day 14 and 28 onwards, respectively (Table 4). Vacuum packaging on day 14 showed a greasy, plastic-like paste appearance, with very small white spots (Figure S1). At the end of the storage period, non-homogeneous colourings appeared with white areas and pronounced marks caused by packaging shrinkage. This anomalous look was also observed in Parmegiano Reggiano cheese, and it was described as the oil-dropping phenomena, in which there is a migration of fat to the surface due to lipid hydrolysis [13,50]. From day 28 onwards, air-packaged samples exhibited a non-homogeneous appearance and obtained a score below 4. On day 42, batches presented spot moulds and small crystals (well-defined, round, white marks with relief) on the paste. During the last two storage days, moulds were more noticeable. Costa et al. [37] and Atallah et al. [38] indicated that appearance was the limiting factor in air-packaged cheeses. MAP4 scored below the limit on days 21 and 49. On day 21, they presented a whiter appearance and small crystals and, on day 49, small black spots in the paste, non-homogeneous colour, and small crystals. The presence of crystals corresponded to the perception described as a defect in texture at the end of storage. At high CO_2 concentration, free ionic calcium combines with lactate through a mechanisms involving carbonic acid, resulting in calcium lactate crystals [56]. Costa et al. [37] found a variation in surface colour with crystals at these conditions, also possibly due to calcium lactate formation that may result from growth of non-starter LAB. MAP1 and MAP2 samples remained above the acceptance limit value at all times, and no differences ($p > 0.05$) were found for paste appearance in MAP3 cheeses throughout storage (Table 4).

According to Idiazabal PDO specifications, cheese must present few small irregular holes homogeneously distributed throughout the cheese paste [5]. Vacuum-packed wedges showed a significant ($p \leq 0.05$) degradation from day 14 onwards (Table 4), and holes started to occlude due to packaging pressure. This fact is aggravated over time, and after two storage months, the cheese paste had sinkholes where holes were initially located. The holes of air-packaged cheeses were scored lower than the MAP-preserved cheeses. This could be due to the intrinsic variation in the cheese holes distribution. Cheeses stored under MAP conditions showed small variations and were always above the limit of the disqualification score. The best-rated cheeses were MAP2 and MAP3. In the scientific literature, there are no remarks on the behaviour of the natural paste holes in MAP-packaged and vacuum-packaged cheese.

3.6. Discriminant Analysis

Discriminant analysis was applied to cheese physicochemical variables, instrumental colour, and texture and sensory parameters to classify cheese samples according to the packaging treatment applied considering MAP conditions as a unique group and irrespective of the storage time. Figure 4 shows the cheese sample distribution in the graph displaying the two canonical discriminant functions.

Table 4. Sensory paste appearance and holes: mean, standard deviation, and significance level of Kruskal–Wallis H statistics treatment of packaged cheese wedges stored for eight weeks at different atmosphere treatments.

Paste Appearance							
Day	**Air**	**Vacuum**	**MAP1**	**MAP2**	**MAP3**	**MAP4**
0	5.43 ± 0.65^{a1}	5.43 ± 0.65^{a1}	5.43 ± 0.65^{a1}	5.43 ± 0.65^{a1}	5.43 ± 0.65^{a1}	5.43 ± 0.65^{a1}
14	4.71 ± 1.07^{ab12}	3.86 ± 0.66^{b12}	5.14 ± 0.66^{a12}	5.21 ± 0.89^{a12}	4.64 ± 0.63^{ab1}	5.27 ± 0.83^{a12}
21	4.21 ± 0.70^{ab12}	3.36 ± 0.74^{b2}	4.36 ± 1.01^{ab2}	4.64 ± 0.84^{a123}	4.50 ± 1.02^{a1}	3.93 ± 0.73^{ab23}
28	3.71 ± 0.73^{bc23}	3.14 ± 0.36^{c23}	4.43 ± 0.85^{ab12}	4.36 ± 0.74^{ab23}	4.93 ± 0.83^{a1}	4.21 ± 0.98^{ab12}
35	3.36 ± 0.50^{b234}	3.21 ± 0.58^{b23}	4.36 ± 0.84^{a2}	4.21 ± 0.70^{a23}	4.79 ± 0.80^{a1}	4.00 ± 0.78^{ab23}
42	1.79 ± 1.05^{b4}	2.78 ± 0.43^{b234}	4.71 ± 0.91^{a12}	4.57 ± 0.75^{a123}	5.29 ± 0.73^{a1}	4.29 ± 0.73^{a12}
49	2.14 ± 0.66^{b34}	2.21 ± 0.43^{b34}	4.86 ± 0.53^{a12}	4.50 ± 0.52^{a123}	4.93 ± 0.73^{a1}	2.86 ± 0.53^{b3}
56	1.64 ± 0.74^{b4}	1.64 ± 0.63^{b4}	4.29 ± 0.61^{a2}	4.00 ± 0.78^{a3}	4.50 ± 0.65^{a1}	4.22 ± 0.80^{a12}
Holes						
0	5.64 ± 0.50^{a1}	5.64 ± 0.50^{a1}	5.64 ± 0.50^{a1}	5.64 ± 0.50^{a1}	5.64 ± 0.50^{a1}	5.64 ± 0.50^{a1}
14	4.36 ± 1.08^{ab2}	4.00 ± 0.78^{b12}	5.21 ± 1.31^{ab12}	5.36 ± 0.74^{ab1}	5.36 ± 0.50^{ab12}	5.79 ± 0.70^{a1}
21	4.14 ± 0.36^{ab2}	3.35 ± 0.63^{b23}	4.14 ± 0.66^{ab3}	4.43 ± 0.51^{a3}	4.71 ± 0.47^{a2}	4.21 ± 0.43^{a2}
28	4.29 ± 0.72^{bc2}	3.50 ± 0.65^{c23}	4.64 ± 0.84^{ab123}	5.50 ± 0.52^{a1}	5.5 ± 0.52^{a1}	4.86 ± 0.77^{ab12}
35	4.00 ± 0.55^{bc2}	3.28 ± 0.61^{c23}	4.64 ± 0.84^{ab123}	5.21 ± 0.70^{a123}	5.21 ± 0.80^{a12}	4.50 ± 0.52^{ab2}
42	3.93 ± 1.54^{bc23}	2.71 ± 0.47^{bc34}	5.43 ± 0.76^{a1}	5.50 ± 0.52^{a1}	5.64 ± 0.63^{a1}	4.64 ± 0.75^{ab2}
49	2.29 ± 0.47^{c3}	2.70 ± 0.83^{bc34}	4.36 ± 0.50^{ab23}	4.50 ± 0.52^{a23}	5.14 ± 0.66^{a12}	4.07 ± 0.73^{ab2}
56	4.14 ± 1.02^{b2}	1.00 ± 0.00^{c4}	5.14 ± 0.66^{ab12}	5.36 ± 0.50^{a12}	5.00 ± 0.55^{ab12}	4.64 ± 0.63^{ab2}

MAP1, 20/80% CO_2/N_2 (*v/v*); MAP2, 50/50% CO_2/N_2 (*v/v*); MAP3, 80/20% CO_2/N_2 (*v/v*); MAP4, 100/0% CO_2/N_2 (*v/v*). Different letters (a–c) in the same row indicate significant differences ($p \leq 0.05$) between the different packaging conditions on that day. Different numbers (1–4) in the same column indicate significant differences ($p \leq 0.05$) during storage for each packaging condition.

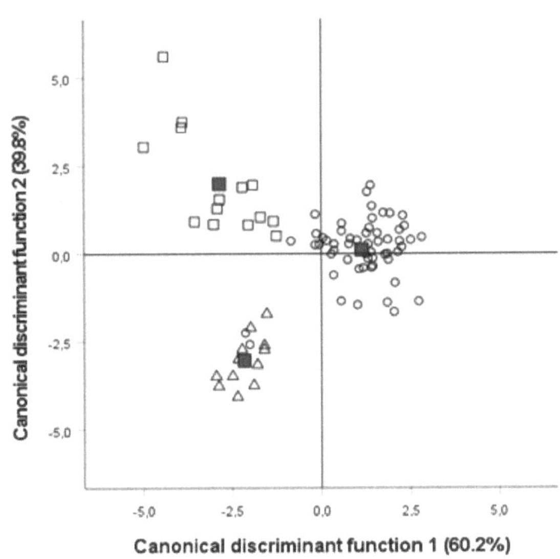

Figure 4. Graph for the two canonical discriminant functions corresponding to the stepwise discriminant analysis on compositional, instrumental colour, and texture and sensory parameters of packaged cheese wedges stored for eight weeks at different atmosphere treatments. Air, Δ; vacuum, □; modified atmosphere packaging, ○.

In general, results showed that 97.6% of the samples were correctly classified into their corresponding treatment group (air-packaged, vacuum-packaged, and MAP). The discriminant variables with higher correlation with canonical functions in the structure matrix were the sensory parameters paste appearance and holes, flavour, the colour parameters b* and a*, and the instrumental texture parameter slope. The cross-validation method used for sample classification reported that all air-packaged samples were correctly grouped and that the 96.4% and 92.9% of the MAP and vacuum-cheese samples, respectively, were correctly assigned.

4. Conclusions

In this study of semi-hard raw-milk cheeses, different gas-packaging treatments were tested (air, vacuum, and MAP). Experimental results highlighted that none of the treatments changed the physicochemical composition. There were no significant physicochemical changes during the storage period studied. The characteristics with the greatest weight in the packaging treatment differentiation were paste appearance and holes, flavour, a*and b* colour parameters, and slope texture parameter. Sensory analysis based on Idiazabal cheeses quality control requirements was decisive to select the best packaging conditions.

MAP-preserved cheese wedges' quality was better than that of air or vacuum packaging. The absence of oxygen in MAP and vacuum conditions contributed to colour stability in view of the changes observed in the parameter a* in air-packaged cheese wedges. The air-packed atmosphere was not the best option for storing cheese wedges since they had a short shelf life caused by gas changes in the atmosphere. Together with the colour, the limiting factor was the presence of moulds, giving mouldy flavours. The traditional vacuum-packaging was not a valid option either; although many parameters were not affected, sensory appearance was low-rated in very early stages of storage, rendering these cheeses unacceptable.

Regarding MAP treatments, very low concentrations of CO_2 were not sufficient to inhibit the growth of microorganisms, as mouldy flavours were observed, and texture was compromised. For $\geq$50/50% CO_2/N_2 (*v/v*) atmosphere, texture parameters (hardness, chewiness, and slope) remained stable, while changes were significant for wedges packaged under lower CO_2 concentration. However, at 100% CO_2 concentration, the pouches collapsed, sensory texture declined, and off-flavours appeared. For a storage period of two months, mixtures between 50/50 and 80/20% CO_2/N_2 (*v/v*) resulted as the most useful techniques to ensure sensory quality for these cheeses.

These results can be of great interest for dairy farms, cheese industries, distribution chains, retail points, and consumers, as these preservation techniques can improve the cheese storage period. Given the inherent interest in the sector, the behaviour of sustainable materials (recyclable or biodegradable) in selected packaging options could be further explored.

Supplementary Materials: The following supporting information can be downloaded at: https://www.mdpi.com/article/10.3390/foods12040849/s1, Table S1: Mean, standard deviation and significance level of Kruskal–Wallis H for texture analysis profile of packaged cheese wedges stored for eight weeks at different atmosphere treatments. Figure S1: Photograph of cheese paste with examples of white spot area in red and small crystals in blue.

Author Contributions: Experimental design, A.I.N. and M.A.; physicochemical analysis, S.N., A.I.N. and M.Á.B.; texture profile analysis, S.N. and O.M.; sensory analysis, S.N. and M.A.; writing—original draft preparation, A.I.N., M.A. and O.M.; writing and editing, A.I.N. and M.A.; statistical analysis, A.I.N., M.A., O.M. and L.J.R.B.; supervision, L.J.R.B.; funding acquisition, L.J.R.B. All authors have read and agreed to the published version of the manuscript.

Funding: The Basque Government (Consolidated Research Group IT944-16 and IT1568-22) provided financial support. Vitoria-Gasteiz, Spain.

Institutional Review Board Statement: Not applicable.

Informed Consent Statement: Informed consent was obtained from all panellists involved in the study.

Data Availability Statement: Data available on request from the corresponding author.

Acknowledgments: Members of the sensory panel are acknowledged for their indispensable help with the quality and descriptive analysis.

Conflicts of Interest: The authors declare no conflict of interest.

References

1. Stadista. Produccion Anual de Queso a Nivel Mundial de 2015 a 2022. Available online: https://es.statista.com/estadisticas/1311313/produccion-de-queso-en-el-mundo/ (accessed on 24 January 2023).
2. Eurostat. Milk and Milk Products Statistics. Available online: https://ec.europa.eu/eurostat/statistics-explained/index.php?title=Milk_and_milk_product_statistics#Milk_products. (accessed on 18 January 2023).
3. Ministerio de Agricultura, Pesca y Alimentacion (MAPA). Estadisticas Agrarias de 2021. Available online: https://www.mapa.gob.es/es/estadistica/temas/estadisticas-agrarias/cuadro_b_2021 (accessed on 19 January 2023).
4. Figueroa Sanchez, A.; Perea Muñoz, J.; Caballero-Villalobos, J.; Arias Sanchez, R.; Garzon, A.; Angon Sanchez de Pedro, E. Coagulation process in Manchega sheep milk from Spain: A path analysis approach. *J. Dairy Sci.* **2021**, *104*, 7544–7554. [CrossRef] [PubMed]
5. Idiazabal PDO Cheese. Available online: https://www.quesoidiazabal.eus (accessed on 2 November 2022).
6. Zabaleta, L.; Albisu, M.; Ojeda, M.; Gil, P.F.; Etaio, I.; Perez-Elortondo, F.J.; de Renobales, M.; Barron, L. Occurrence of sensory defects in semi-hard ewe's raw milk cheeses. *Dairy Sci. Technol.* **2016**, *96*, 53–65. [CrossRef]
7. Poças, M.F.; Pintado, M. Packaging and Shelf Life of Cheese. In *Food Packaging and Shelf Life*; Robertson, G.L., Ed.; CRC Press, Taylor and Francis Group: Boca Ratón, FL, USA, 2009; pp. 103–125.
8. Ciccullo, F.; Cagliano, R.; Bartezzaghi, G.; Perego, A. Implementing the circular economy paradigm in the agri-food supply chain: The role of food waste prevention technologies. *Resour. Conserv. Recycl.* **2021**, *164*, 105114. [CrossRef]
9. Najera, A.I.; Nieto, S.; Barron, L.J.R.; Albisu, M. A review of the preservation of hard and semi-hard cheeses: Quality and safety. *Int. J. Environ. Res. Public Health* **2021**, *18*, 9789. [CrossRef] [PubMed]
10. Juric, M.; Bertelsen, G.; Mortensen, G.; Petersen, M.A. Light-induced colour and aroma changes in sliced, modified atmosphere packaged semi-hard cheeses. *Int. Dairy J.* **2003**, *13*, 239–249. [CrossRef]
11. Trobetas, A.; Badeka, A.; Kontominas, M.G. Light-induced changes in grated Graviera hard cheese packaged under modified atmospheres. *Int. Dairy J.* **2008**, *18*, 1133–1139. [CrossRef]
12. Garabal, J.I.; Rodriguez-Alonso, P.; Franco, D.; Centeno, J.A. Chemical and biochemical study of industrially produced San Simón da Costa smoked semi-hard cow's milk cheeses: Effects of storage under vacuum and different modified atmospheres. *J. Dairy Sci.* **2010**, *93*, 1868–1881. [CrossRef]
13. Romani, S.; Sacchetti, G.; Pittia, P.; Pinnavaia, G.G.; Dalla Rosa, M. Physical, Chemical, Textural and Sensorial Changes of Portioned Parmigiano Reggiano Cheese Packed under Different Conditions. *Food Sci. Technol. Int.* **2002**, *8*, 203–211. [CrossRef]
14. Piscopo, A.; Zappia, A.; de Bruno, A.; Poiana, M. Qualitative variations on Calabrian Provola cheeses stored under different packaging conditions. *J. Dairy Res.* **2015**, *82*, 499–505. [CrossRef]
15. Hocking, A.D.; Faedo, M. Fungi causing thread mould spoilage of vacuum packaged Cheddar cheese during maturation. *Int. J. Food Microbiol.* **1992**, *16*, 123–130. [CrossRef]
16. Bellio, A.; Astegiano, S.; Traversa, A.; Bianchi, D.M.; Gallina, S.; Vitale, N.; Zuccon, F.; Decastelli, L. Behaviour of Listeria monocytogenes and Staphylococcus aureus in sliced, vacuum-packaged raw milk cheese stored at two different temperatures and time periods. *Int. Dairy J.* **2016**, *57*, 15–19. [CrossRef]
17. Favati, F.; Galgano, F.; Pace, A.M. Shelf-life evaluation of portioned Provolone cheese packaged in protective atmosphere. *LWT Food Sci. Technol.* **2007**, *40*, 480–488. [CrossRef]
18. Tansman, G.F.; Kindstedt, P.S.; Hughes, J.M. Crystal fingerprinting: Elucidating the crystals of Cheddar, Parmigiano-Reggiano, Gouda, and soft washed-rind cheeses using powder x-ray diffractometry. *Dairy Sci. Technol.* **2015**, *95*, 651–664. [CrossRef] [PubMed]
19. Esmer, O.K.; Balkir, P.; Seckin, A.K.; Irkin, R. The Effect of Modified Atmosphere and Vacuum Packaging on the Physicochemical, Microbiological, Sensory and Textural Properties of Crottin de Chavignol Cheese. *Food Sci. Technol. Res.* **2009**, *15*, 367–376. [CrossRef]
20. Rodriguez-Aguilera, R.; Oliveira, J.C.; Montanez, J.C.; Mahajan, P.V. Effect of modified atmosphere packaging on quality factors and shelf-life of mould surface-ripened cheese: Part II varying storage temperature. *LWT Food Sci. Technol.* **2011**, *44*, 337–342. [CrossRef]
21. Todaro, M.; Palmeri, M.; Cardamone, C.; Settanni, L.; Mancuso, I.; Mazza, F.; Scatassa, M.L.; Corona, O. Impact of packaging on the microbiological, physicochemical and sensory characteristics of a "pasta filata" cheese. *Food Package Shelf Life* **2018**, *17*, 85–90. [CrossRef]
22. Gonzalez-Fandos, E.; Sanz, S.; Olarte, C. Microbiological, physicochemical and sensory characteristics of Cameros cheese packaged under modified atmospheres. *Food Microbiol.* **2000**, *17*, 407–414. [CrossRef]

23. Cosentino, C.; Paolino, R.; Valentini, V.; Musto, M.; Ricciardi, A.; Adduci, F.; D'Adamo, C.; Pecora, G.; Freschi, P. Effect of jenny milk addition on the inhibition of late blowing in semihard cheese. *J. Dairy Sci.* **2015**, *98*, 5133–5142. [CrossRef]
24. Colchin, L.M.; Owens, S.L.; Lyubachevskaya, G.; Boyle-Roden, E.; Russek-Cohen, E.; Rankin, S.A. Modified Atmosphere Packaged Cheddar Cheese Shreds: Influence of Fluorescent Light Exposure and Gas Type on Color and Production of Volatile Compounds. *J. Agric. Food Chem.* **2001**, *49*, 2277–2282. [CrossRef]
25. *ISO/TS 11869*; Fermented Milks—Determination of Titratable Acidity—Potentiometric Method. International Organization for Standardization: Geneva, Switzerland, 2012.
26. AOAC. *Acidity of Milk, Titrimetric Method AOAC Official Method 920.124*; Association of Official Agricultural Chemists: Rockville, MD, USA, 2005.
27. Rohm, H.; Jaros, D. Colour of hard cheese. *Z. Für Lebensm.-Unters. Und Forsch.* **1996**, *203*, 241–244. [CrossRef]
28. Frau, M.; Simal, S.; Femenia, A.; Sanjuan, E.; Rossello, C. Use of principal component analysis to evaluate the physical properties of Mahon cheese. *Eur. Food Res. Technol.* **1999**, *210*, 73–76. [CrossRef]
29. del Caro, A.; Fadda, C.; Sanguinetti, A.M.; Carboni, M.G.; Pinna, G.; Naes, T.; Menichelli, E.; Piga, A. Influence of Technology and Ripening on Textural and Sensory Properties of Vacuum Packaged Ewe's Cheese. *Czech J. Food Sci.* **2016**, *34*, 456–462. [CrossRef]
30. Szczesniak, A.S. Texture is a sensory property. *Food Qual. Prefer.* **2002**, *13*, 215–225. [CrossRef]
31. Bourne, M.C. Texture profile analysis. *Food Technol.* **1978**, *32*, 62–66.
32. Wee, M.S.M.; Goh, A.T.; Stieger, M.; Forde, C.G. Correlation of instrumental texture properties from textural profile analysis (TPA) with eating behaviours and macronutrient composition for a wide range of solid foods. *Food Funct.* **2018**, *9*, 5301–5312. [CrossRef] [PubMed]
33. Ojeda, M.; Etaio, I.; Fernandez-Gil, M.P.; Albisu, M.; Salmeron, J.; Perez-Elortondo, F.J. Sensory quality control of cheese: Going beyond the absence of defects. *Food Control* **2015**, *51*, 371–380. [CrossRef]
34. Valdivielso, I.; Albisu, M.; de Renobales, M.; Barron, L.J.R. Changes in the volatile composition and sensory properties of cheeses made with milk from commercial sheep flocks managed indoors, part-time grazing in valley, and extensive mountain grazing. *Int. Dairy J.* **2016**, *53*, 29–36. [CrossRef]
35. *ISO 8589:2007/Amd 1:2014*; Sensory Analysis—General Guidance for the Design of Test Rooms. International Organization for Standardization: Geneve, Switzerland, 2014.
36. Khoshgozaran, S.; Azizi, M.H.; Bagheripoor-Fallah, N. Evaluating the effect of modified atmosphere packaging on cheese characteristics: A review. *Dairy Sci. Technol.* **2011**, *92*, 1–24. [CrossRef]
37. Costa, C.; Lucera, A.; Lacivita, V.; Saccotelli, M.A.; Conte, A.; Del Nobile, M.A. Packaging optimisation for portioned Canestrato di Moliterno cheese. *Int. J. Dairy Technol.* **2016**, *69*, 401–409. [CrossRef]
38. Atallah, A.A.; El-Deeb, A.M.; Mohamed, E.N. Shelf-life of Domiati cheese under modified atmosphere packaging. *J. Dairy Sci.* **2021**, *104*, 8568–8581. [CrossRef]
39. O' Callaghan, K.A.M.; Papkovsky, D.B.; Kerry, J.P. An Assessment of the Influence of the Industry Distribution Chain on the Oxygen Levels in Commercial Modified Atmosphere Packaged Cheddar Cheese Using Non-Destructive Oxygen Sensor Technology. *Sensors* **2016**, *16*, 916. [CrossRef] [PubMed]
40. Kristensen, D.; Orlien, V.; Mortensen, G.; Brockhoff, P.; Skibsted, L.H. Light-induced oxidation in sliced Havarti cheese packaged in modified atmosphere. *Int. Dairy J.* **2000**, *10*, 95–103. [CrossRef]
41. Kirkin, C.; Gunes, G.; Kilic-Akyilmaz, M. Preservation of precut white cheese by modified atmosphere packaging. *Int. J. Dairy Technol.* **2013**, *66*, 576–586. [CrossRef]
42. Solomakos, N.; Govari, M.; Botsoglou, E.; Pexara, A. Effect of modified atmosphere packaging on physicochemical and microbiological characteristics of Graviera Agraphon cheese during refrigerated storage. *J. Dairy Res.* **2019**, *86*, 483–489. [CrossRef] [PubMed]
43. Rodríguez-Alonso, P.; Centeno, J.A.; Garabal, J.I. Biochemical study of industrially produced Arzúa-Ulloa semi-soft cows' milk cheese: Effects of storage under vacuum and modified atmospheres with high-nitrogen contents. *Int. Dairy J.* **2011**, *21*, 261–271. [CrossRef]
44. Dermiki, M.; Ntzimani, A.; Badeka, A.; Savvaidis, I.N.; Kontominas, M.G. Shelf-life extension and quality attributes of the whey cheese "Myzithra Kalathaki" using modified atmosphere packaging. *LWT Food Sci. Technol.* **2008**, *41*, 284–294. [CrossRef]
45. Pintado, M.E.; Malcata, F.X. Optimization of modified atmosphere packaging with respect to physicochemical characteristics of Requeijao. *Food Res. Int.* **2000**, *33*, 821–832. [CrossRef]
46. Alves, R.M.V.; Sarantopoulos, C.; Van Dender, A.; Faria, J.d.A. Stability of sliced mozzarella cheese in modified-atmosphere packaging. *J. Food Prot.* **1996**, *59*, 838–844. [CrossRef]
47. Diezhandino, I.; Fernandez, D.; Sacristan, N.; Combarros-Fuertes, P.; Prieto, B.; Fresno, J.M. Rheological, textural, colour and sensory characteristics of a Spanish blue cheese (Valdeon cheese). *LWT Food Sci. Technol.* **2016**, *65*, 1118–1125. [CrossRef]
48. Di Marzo, S.; Di Monaco, R.; Iaccarino, T.; Cavella, S.; Masi, P. Determinazione della shelf life del Canestrato Pugliese in fette confezionato in film biodegradabili, Ricerche ed Innovazioni nell'Industria Alimentare. In *6° Congresso Italiano di Scienza degli Alimenti*; Chiriotti Editori: Pinerolo, Italy, 2004; pp. 414–419.

49. Martin, B.; Fedele, V.; Ferlay, A.; Grolier, P.; Rock, E.; Gruffat, D.; Chilliard, Y. Effects of grass-based diets on the content of micronutrients and fatty acids in bovine and caprine dairy products. Land use systems in grassland dominated regions. In Proceedings of the 20th General Meeting of the European Grassland Federation, Luzern, Switzerland, 21–24 June 2004; pp. 876–886.
50. Romani, S.; Sacchetti, G.; Vannini, L.; Pinnavaia, G.G.; Dalla Rosa, M.; Corradini, C. Storage stability of portioned packed Parmigiano Reggiano cheese [Emilia-Romagna]. *Sci. Tec. Latt. Casearia* **1999**, *50*, 273–290.
51. Avila, M.; Garde, S.; Nuñez, M. The influence of some manufacturing and ripening parameters on the colour of ewes' milk cheese. *Milchwissenschaft* **2008**, *63*, 160–164.
52. Singh, P.; Wani, A.A.; Karim, A.A.; Langowski, H.C. The use of carbon dioxide in the processing and packaging of milk and dairy products: A review. *Int. J. Dairy Technol.* **2012**, *65*, 161–177. [CrossRef]
53. Temiz, H. Effect of modified atmosphere packaging on characteristics of sliced kashar cheese. *J. Food Process. Preserv.* **2010**, *34*, 926–943. [CrossRef]
54. Agarwal, S.; Costello, M.; Clark, S. Gas-Flushed Packaging Contributes to Calcium Lactate Crystals in Cheddar Cheese. *J. Dairy Sci.* **2005**, *88*, 3773–3783. [CrossRef] [PubMed]
55. Hotchkiss, J.H.; Werner, B.G.; Lee, E.Y.C. Addition of Carbon Dioxide to Dairy Products to Improve Quality: A Comprehensive Review. *Compr. Rev. Food Sci. Food Saf.* **2006**, *5*, 158–168. [CrossRef]
56. Dybing, S.T.; Wiegand, J.A.; Brudvig, S.A.; Huang, E.A.; Chandan, R.C. Effect of processing variables on the formation of calcium lactate crystals on Cheddar cheese. *J. Dairy Sci.* **1988**, *71*, 1701–1710. [CrossRef]

foods

MDPI

Article

Vacuum-Assisted Block Freeze Concentration Studies in Cheese Whey and Its Potential in Lactose Recovery

Noelia Gil [1], Gisela Quinteros [1], Monica Blanco [2], Shafirah Samsuri [3], Nurul Aini Amran [3], Patrico Orellana-Palma [4], Elane Schwinden [5] and Eduardo Hernández [1,*]

1 Agri-Food Engineering and Biotechnology Department, Universitat Politècnica de Catalunya BarcelonaTech, Campus del Baix Llobregat, Edifici D-4 C/Esteve Terradas, 8, Castelldefels, 08860 Barcelona, Spain
2 Department of Mathematics, Universitat Politécnica de Catalunya BarcelonaTech, Parc Mediterrani de la Tecnologia Campus del Baix Llobregat, Edifici D-4 C/Esteve Terradas, 8, Castelldefels, 08860 Barcelona, Spain
3 Chemical Engineering Department, Universiti Teknologi PETRONAS, 32610 Seri Iskandar, Perak, Malaysia
4 Departamento de Ingeniería en Alimentos, Facultad de Ingeniería, Campus Andrés Bello, Universidad de La Serena, Avda. Raúl Bitrán 1305, La Serena 1720010, Chile
5 Departamento de Ciência e Tecnologia de Alimentos, Universidade Federal de Santa Catarina, Florianópolis 88036, Santa Catarina, Brazil
* Correspondence: eduard.hernandez@upc.edu; Tel.: +34-935521077

Abstract: Block freeze concentration (BFC) is considered an emerging technology which allows the acquiring of high quality organoleptic products, due to the low temperatures employed. In this study we have outlined how the vacuum-assisted BFC of whey was investigated. The effects of vacuum time, vacuum pressure, and the initial solids concentration in whey were studied. The results obtained show that the three variables significantly affect each of the following parameters analysed: solute yield (Y) and concentration index (CI). The best Y results were obtained at a pressure of 10 kPa, 7.5 °Bx, and 60 min. For CI parameter, the highest values were given at 10 kPa, 7.5 °Bx, and 20 min, respectively. In a second phase, by applying the conditions that provide higher solute yield to three different types of dairy whey, Y values of 70% or higher are reached in a single step, while that the CI of lactose are higher than those of soluble solids. Therefore, it is possible to recover, in a single step, at least 70% of the lactose contained in the initial whey samples. This suggests that vacuum-assisted BFC technology may be an interesting alternative for the recovery of lactose contained in whey.

Keywords: block freeze concentration; vacuum; whey; lactose

Citation: Gil, N.; Quinteros, G.; Blanco, M.; Samsuri, S.; Amran, N.A.; Orellana-Palma, P.; Schwinden, E.; Hernández, E. Vacuum-Assisted Block Freeze Concentration Studies in Cheese Whey and Its Potential in Lactose Recovery. *Foods* **2023**, *12*, 836. https://doi.org/10.3390/foods12040836

Academic Editors: Michele Faccia and Giuseppe Natrella

Received: 17 January 2023
Revised: 9 February 2023
Accepted: 11 February 2023
Published: 15 February 2023

1. Introduction

In Mediterranean countries, where the dairy sector (cheese making-oriented) has a marked traditional character, constraints in the utilization of fresh liquid whey are linked to high transportation costs of the bulky liquid and the low productivity of drying facilities [1]. Thereby, an excellent alternative to reduce operating time, maintenance costs, and storage spaces, among others, is the reduction of the volume (and the weight) of fresh liquid whey. For this purpose, it is important to concentrate these solutions, and thus, this process allows for the decreasing of the water activity, and in turn, to avoid any unwanted microbial growth [2]. Precisely, there are three main methods available to eliminate the part of water contained in objective solutions, allowing concentration in the liquid food products: membrane processes, vacuum evaporation, and freeze concentration. Although these technologies are effective, they also present certain disadvantages. For example, vacuum evaporation technologies are extremely expensive, and it has been found that the energy use is very high. In the same way, membrane technologies needs frequent maintenance because of membrane fouling [3]. Hence, when comparing the heat of evaporation (about 2260 kJ/kg under pressure of 0.1 MPa) with enthalpy of freezing (335 kJ/kg), the freeze concentration process seems to be cheaper than evaporation from the energy point of view.

Specifically, freeze concentration (FC) is considered an emerging technology in which liquid food is concentrated via partial or total water freezing, where the procedure involves a controlled decrease in the temperature of the liquid food. As a result of this process it is pushed below the freezing point, avoiding the eutectic temperature, where all the components of the product are frozen. Hence, FC process separates the ice fraction from the residual unfrozen solution [4]. Additionally, there are three techniques viable to applied freeze concentrate in liquid foods: suspension freeze concentration (SFC), progressive freeze concentration (PFC), and block freeze concentration (BFC). In BFC, also known as freeze concentration by freezing–thawing, the liquid is completely frozen and the temperature of the centre of the product is set below the freezing point. Subsequently, the block is thawed and the concentrated fraction is then separated from the ice fraction. The separation process is occasionally assisted by external forces such as centrifugation or vacuum, since these external forces increase the efficiency of the procedure [5]. Previous studies have suggested that the suction during the vacuum process by a pump takes advantage of the channels between the ice crystals [6]. Therefore, the result is an increase in the extraction of the concentrate fraction from the ice fraction, and an improvement of the efficiency and solute recovery. The use of vacuum as an assisted technique to enhance the freeze concentration performance has been studied in saline solutions, sucrose, coffee extracts, red wine, orange juice, and blueberry juice [6–12]. In this case, BFC could be an alternative technology for the concentration of by-products, such as whey. In dairy processing, BFC offers to minimize the thermal damage on sensitive milk components, such as proteins and flavours, among others. Thus, it provides an opportunity for producing dairy ingredients with enhanced functional and organoleptic qualities.

In particular, FC has been applied on dairy products [2,3,13–16], and thus, for the concentration of whey, the maximum concentration ranged between 25 and 35 wt.% [2,16]. Moreover, there has been a growing number of studies on the concentration of dairy products through SFC technology (it comprises of a primary phase formation of ice nuclei (nucleation), followed by a secondary phase growth of ice nuclei in the solution) [14,17], by layer crystallization [18,19] and BFC assisted by gravity and microwave processes [2,3,15,16]. Until now, no study on the application of vacuum-assisted BFC applied to whey is known.

On the other hand, whey contains at least half of the total solids present in the initial whole milk, and hence, it can be considered as a valuable by-product with several applications, especially, in the food industry [20]. Recent research stated that the components of whey are difficult to degrade, creating a major problem to any wastewater treatment plants [21]. Therefore, the use of a concentration technique could be applied in the development of new products, and at the same time, it may help to solve the industrial waste problem [22]. BFC assisted by vacuum suction, due to its cheaper capital and operating costs than other FC alternatives could be an attractive technology for dairy industries [23].

Taking into account the above, the objective of this study in a first phase was to research the concentration of vacuum-assisted BFC of fresh cheese whey, by studying the influence of the initial concentration (C_0), time (t), and vacuum pressure (V) on the response variables, concentration index (CI) and solute yield (Y). In a second phase, the best conditions of the previous stage (time and vacuum pressure) were applied to three types of whey, designated by: fresh cheese whey (CSW1), mató (a typical fresh cheese of Catalonia, Spain), cheese whey (CSW2), and blue cheese whey (CSW3). Additionally, the effect of vacuum assisted BFC on lactose content was also studied.

2. Materials and Methods

2.1. Material

Cheese whey was provided by a local supplier (Can Corder, Lliça d'Amunt, Barcelona, Spain) from cheese process. The three types of cheese was produced, namely, fresh cheese, mató, and blue cheese, use pasteurized cow's milk as raw material.

CSW1 from fresh cheese. Its manufacture is very simple and consists of two stages: coagulation is essentially lactic (with ferment) and normally lasts 24 h, while draining is never excessive and is carried out in the container itself by dividing the coagulum into portions. Whey without salt.

CSW2 from mató: It is an enzymatic coagulation cheese (with rennet), drained by draining, lightly pressed (self-pressed) by periodic turning. Therefore, it is soft paste. No ripening. Whey without salt.

CSW3 from blue cheese: It is a matured cheese produced by enzymatic coagulation. Salty milk whey.

2.2. Freezing and Vacuum Suction Procedure

The experimental procedure of vacuum-assisted BFC was carried out according to Petzold et al. [8]. Firstly, whey samples (45 mL) were placed in plastic tubes (internal diameter: 22 mm) and were frozen at −20 ± 2 °C for 48 h in a static freezer (Fricon Model THC 520, Portugal). Due to our availability of freezing equipment, other different conditions have not been considered. The tubes were covered with thermal insulation made of polystyrene foam (8 mm thickness, thermal conductivity: K = 0.035 W/mK) in order to facilitate axial heat transfer. The frozen whey tubes were removed from the freezer and transferred to a suction stage. The suction was generated by connecting a vacuum pump (Comecta, Spain; Pump rate: 3.6 m^3/h; Vacuum limit: 0.1 mbar) at the bottom of the frozen sample at room temperature (Figure 1). Subsequently, the concentrated solution was collected, and the concentration of the soluble solids within the solution as well as the concentration of the ice phase were measured by a refractometer (ATAGO DBX-55 Japan; Measurement range: 0.0–55 °Bx; Accuracy: 0.1 °Bx ± 0.1%). All measurements were taken in triplicate.

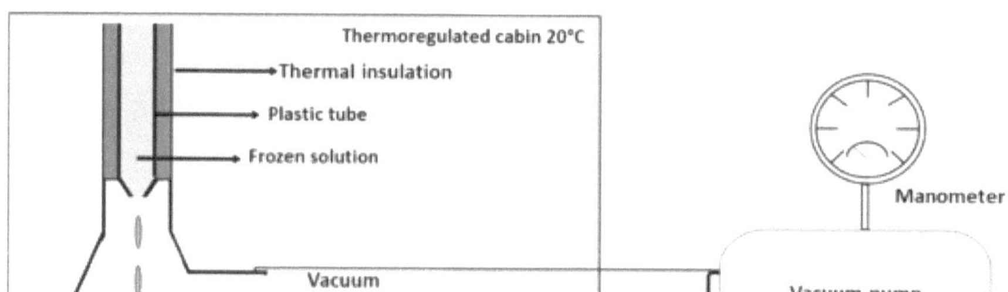

Figure 1. Vacuum unit for freeze concentration.

2.3. Experimental Design

2.3.1. Effect of Factors on Response Variables

A full factorial design (FFD) was used to study the effect of the following three independent factors: initial soluble solids concentration (C_0, °Bx), time (t, min), and vacuum pressure (V, kPa), on the response variables: solid yield (Y) and concentration index (CI). The independent factors and their levels have been selected based on previous studies [6,11,12], and according to the initial tests carried out in our laboratory. The experimental design comprised eight combinations of the independent factors, as shown in Table 1. All tests were carried out in triplicate.

Table 1. Experimental design.

	C_0 (°Bx)	t (min)	V (kPa)
Minimum	7.5	20	10
Maximum	19	60	70

According to previous tests, 20 and 60 min under suction vacuum were adopted to ensure a sufficient concentrated sample, and to avoid vacuum break in the ice column. In this study the pressures of 10 and 70 kPa, are absolute pressures (absolute atmospheric pressure 101 kPa) that correspond to 90 and 30 kPa of vacuum pressure, respectively.

2.3.2. Vacuum-Assisted BFC Tests on Three Types of Cheese Whey

In the second stage of experiments, the best conditions of time and vacuum pressure obtained in the previous phase with respect to the solute yield parameter (Y) were applied to the whey of three types of cheese: fresh cheese whey (CSW1), mató cheese whey (CSW2), and blue cheese whey (CSW3). Additionally, in these tests, the percentage of recovery of lactose with respect to the initial sample was determined.

2.4. Lyophilization Process

To carry out the freezing curves at different concentrations, as well as for the tests with whey at 19 °Bx, it was necessary to lyophilize the initial sample to obtain dry sample, which, once re-dissolved in water, allowed for the samples to be prepared at the desired concentration. Whey samples (approximately 25 mL) were frozen in Falcon tubes and placed horizontally in the freezer at −20 °C for 48 h. Then, the Falcon's cap was removed, and plastic wrap was placed to facilitate sublimation. The frozen samples were placed in the lyophilizer (CRYODOS FD-10 Series, Telstar Industrial S.L, Spain; Vacuum pump nominal flow: 6 m^3/h), for 48 h at −56.6 °C and a vacuum pressure of 4.7×10^{-2} mbar. Once the dried milk was obtained, the corresponding solutions were prepared with distilled water at 15, 25, and 35 °Bx. The same procedure was followed to prepare the samples at 19 °Bx.

2.5. Response Variables

Concentration index (CI) is defined as the relation between the concentration of soluble solids in the concentrated solution (C_f), and the concentration of soluble solids in the initial whey (C_0), according to Equation (1) [24]. The concentration index is also known as relative concentration [25].

$$\mathbf{CI} = \frac{\mathbf{C_f}}{\mathbf{C_0}} \quad (1)$$

Solute yield (Y) was calculated to analyze the soluble solids recovery. Y was defined as the relationship between the mass of soluble solids present in the separated liquid and the mass of soluble solids present initially in the initial solution. This can be seen in the following Equation (2) [26].

$$\mathbf{Y}\ (\%) = \frac{\mathbf{C_f \cdot m_f}}{\mathbf{C_0 \cdot m_0}} \cdot \mathbf{100} \quad (2)$$

where C_f and C_0 are the soluble solids content (°Bx) of the concentrate and initial whey, m_f and m_0 are the concentrated and the initial whey mass (g).

2.6. Validation of Results

For each experiment, an ice mass ratio W_{exp} (kg ice/kg initial) can be defined, as the amount of ice obtained, with respect to the initial amount of sample in each experiment. In FC systems, ice handling can be complex and can be a source of error. Alternatively, this ratio can be estimated through a mass balance, depending on the concentrations of the ice, concentrated and initial whey sample. The measurement of concentrations is much simpler and more reliable than the data obtained from ice handling. It is denoted as W_p, Equation (3) [27].

$$W_p = \frac{C_f - C_0}{C_f - C_i} \cdot 100 \quad (3)$$

The quality of the fit between experimental (W_{exp}) and predicted (W_p) values for N experimental points, i.e., the deviation between experimental and theoretical data, was tested by the Root Mean Square (RMS) as follows (Equation (4)).

$$RMS(\%) = 100\sqrt{\frac{\sum[(W_{exp} - W_p)/W_{exp}]^2}{N}} \quad (4)$$

2.7. Lactose Concentration

The lactose concentration of the initial, concentrated, and ice fractions were determined. The lactose content procedure was carried out according to Schuster-Wolff-Bühring et al. [28], with modifications. Hewlett Packard 1100 Series HPLC System (Agilent Technologies, Waldbronn, Germany) equipped with a Beckman 156 refractive index detector was used for determination. The separation was achieved with a tracer carbohydrate (250 × 4.6 mm, 5 μm) column (Teknokroma, Sant Cugat del Vallès, Barcelona, Spain). The volume injected was 20 μL and the mobile phase was a mixture of acetonitrile (Panreac Química SLU, Castellar del Vallès, Spain) and ultrapure water (75:25). The flow rate and column temperature were maintained as 1.3 mL/min and 28 °C, respectively. The detection was carried out with a refractive index detector, adjusting the zero against the mobile phase. Before the determinations, a portion of 1 mL samples was diluted with 8 mL of ultrapure water and mixed. Thus, 0.5 mL of Carrez Reagent 1 and 2 were added and mixed for 1 min. The mixture was allowed to settle for 15 min, and filtered by a nylon syringe filter (a pore diameter of 0.45 μm) (Agilent, Santa Clara, CA, United States). Each sample was prepared and injected in triplicate.

2.8. Statistical Analysis

The results obtained were statistically analysed using the application 'Minitab 18' for Windows (Minitab Inc., State Collage, PA, USA) and expressed as the mean ± standard deviation. To determine significant differences ($p < 0.05$) between results, one-way analysis of variance (ANOVA), and Tukey tests were used.

Full factorial design (FFD) with 3 factors and 2 levels (2^3) was applied, with $\alpha = 0.05$.

3. Results

3.1. Freezing Point Depression

The freezing point of a liquid depends on the concentration and type of solutes present in the solution [29], and thus, a high level of dissolved solids means a low freezing point. At the same mass concentration (%*w/w*), the solutes with low molecular weight (MW) have high molality, and therefore, a low freezing point. For cheese whey, the freezing point is influenced by the concentration of lactose, chlorides and other salts, and it is lower than that of pure water [30]. The freezing point was determined for initial and concentrates CSW1 whey with concentration of 15, 25 and 35 °Bx (Figure 2).

As expected, the freezing point determined through this experiment increased with a growing concentration of solids. The same trend was observed for whey [1,19,30] and milk [31]. For comparison, Figure 2 includes the results of salted cheese whey from the same manufacturer [19]. The trend line obtained is above the line of salted whey, and it may suggest a low concentration of salt in the tested whey. On the other hand, Figure 2 includes the freezing points of an ideal solution obtained from the Van't Hoff equation, taking 235 g/mol as the effective molecular weight for whey milk, suggested by Baschi et al. [30]. As can be seen, there is a very good correlation between the experimental (CSW1) and the calculated value for ideal solutions. This can be very useful for future freeze concentration works.

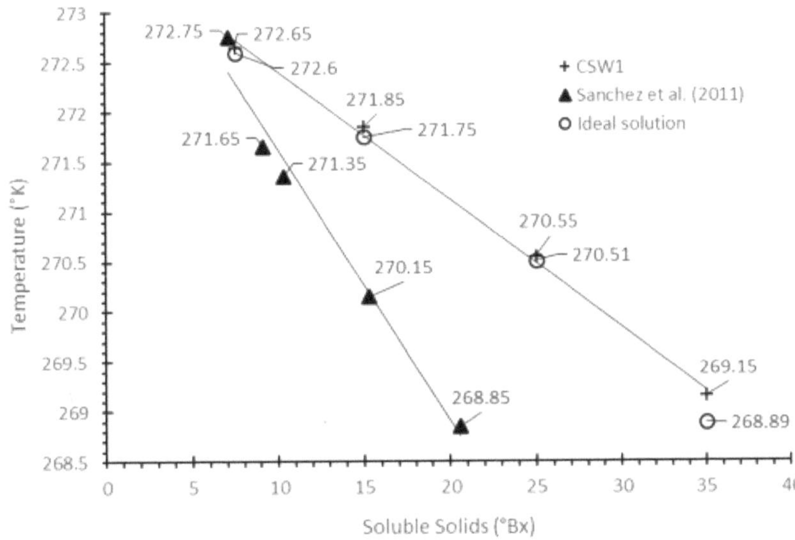

Figure 2. Freezing point (°K) of the CSW1 whey and an ideal solution, as a function of the soluble solids content (°Bx).

3.2. *Solute Yield and Concentration Index*

The responses obtained for Y and CI from the eight experiments are shown in Table 2. The best results for Y were obtained at a vacuum pressure of 10 kPa, an initial concentration of 7.5 °Bx and vacuum time of 60 min. For the CI parameter, the best condition was also 10 kPa, 7.5 °Bx, and 20 min of vacuum. The values suggest that it would be possible to improve the Y parameter by increasing the vacuum time.

Table 2. Full Factorial Design (FFD), responses of solute yield (Y) and concentration index (CI). Within a column, different lowercase letters denote significant differences ($p < 0.05$) between Y and CI.

V (kPa)	C_0 (°Bx)	t (min)	Y (%)	CI
10	7.5	20	28.4 ± 4.4 [b]	5.3 ± 0.4 [a]
10	7.5	60	65.5 ± 3.8 [a]	3.4 ± 0.2 [b]
70	7.5	20	9.4 ± 4.8 [cde]	4.5 ± 0.2 [a]
70	7.5	60	36.3 ± 10 [b]	3.2 ± 0.7 [b]
10	19	20	7.4 ± 3.9 [de]	2.6 ± 0.1 [bc]
10	19	60	24.3 ± 2.0 [bc]	1.7 ± 0.1 [cd]
70	19	20	0.26 ± 0.1 [e]	1.2 ± 0.3 [d]
70	19	60	21.9 ± 1.0 [bcd]	1.5 ± 0.2 [cd]

This is confirmed by the results of ANOVA presented in Table 3, since the p-values are shown in Table 3, indicating that all individual effects, as well as the interaction between C_0 and t, were significant ($p < 0.05$) on the solute yield (Y) and the concentration index (CI).

3.3. *Vacuum-Assisted BFC Tests on Three Types of Whey*

Based on the analysis of the results obtained in Section 3.2, vacuum-assisted BFC process was performed in three different whey samples, under conditions that allow for reaching the best values of solute yield (Y). These conditions correspond to an absolute pressure of 10 kPa and a time of 60 min under suction vacuum. The tests were performed in triplicate resulting in a total of 9 tests. Under these conditions, the Y values were close to 70%, as shown in Table 4.

Table 3. Results of ANOVA for solute yield (Y) and concentration index (CI). The asterisk (*) indicates a significant effect ($p < 0.05$).

	Y		CI	
	F-Value	***p*-Value**	**F-Value**	***p*-Value**
Vacuum (V)	39.08	<0.000 *	14.57	0.002 *
Concentration (C_0)	86.03	<0.000 *	197.47	<0.000 *
Time (t)	123.44	<0.000 *	32.34	<0.000 *
$V{\bullet}C_0$	17.68	0.001 *	0.53	0.479
$V{\bullet}t$	0.34	0.566	7.11	0.017 *
$C_0{\bullet}t$	7.57	0.014 *	16.79	0.001 *
$V{\bullet}C_0{\bullet}t$	2.60	0.127	0.57	0.461

Table 4. Results of concentration index (CI) and solute yield (Y) for the three types of serum subjected to 10 kPa for 60 min.

Whey Type	**C_0 (°Bx)**	**CI**	**Y (%)**
CSW1	6.6	3.5 ± 0.1	73 ± 1
CSW2	5.7	3.8 ± 0.2	69 ± 1
CSW3	6.4	3.8 ± 0.2	71 ± 4

Additionally, Figure 3 shows HPLC lactose chromatograms for the initial and concentrated of CSW1 whey.

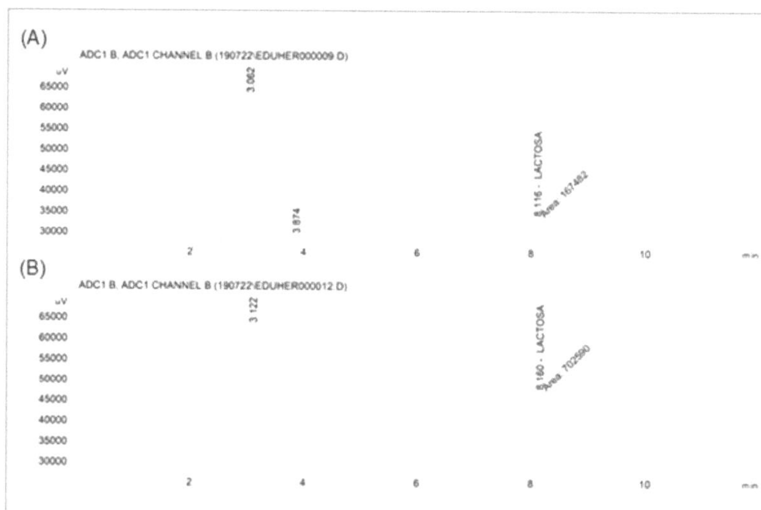

Figure 3. Lactose chromatograms of initial (**A**) and concentrated whey (**B**).

On the other hand, Figure 4 shows the lactose concentrations (g/L) in the initial whey, in the concentrate and ice fractions, of the three types of whey tested (CSW1, CSW2, and CSW3).

The results of the tests carried out with the three types of whey were validated, comparing the results calculated with the Equation (3) (W_p), with the measures of ice mass ratio obtained experimentally (W_{exp}). In Figure 5, the values W_p and W_{exp} of the 9 experiments are presented.

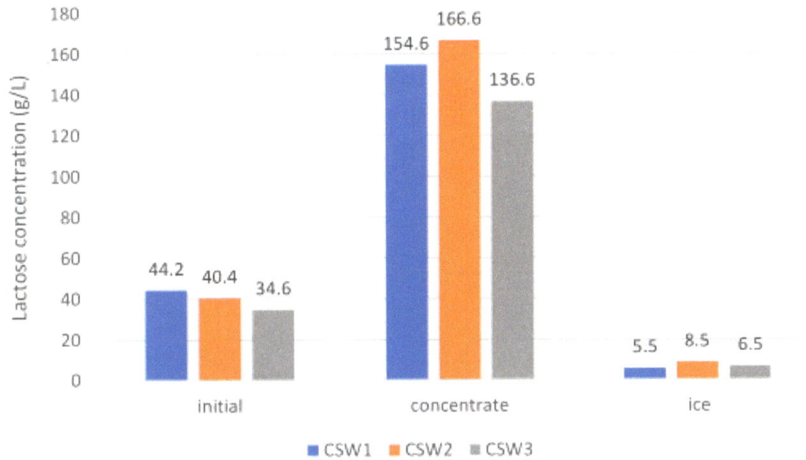

Figure 4. Lactose concentration of initial whey, concentrate and ice for three types of whey (CSW1, CSW2, and CSW3).

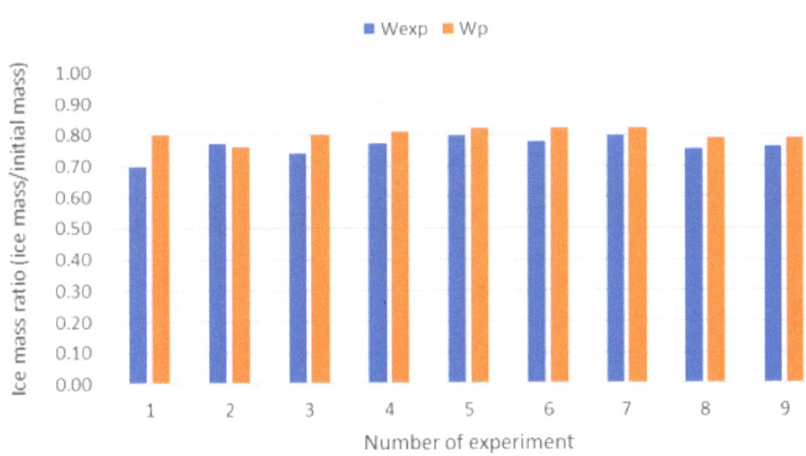

Figure 5. Experimental (W_{exp}) and predicted (W_p) values of ice mass ratio for CSW1 (experiment number 1, 2, and 3), CSW2 (experiment 4, 5, and 6), and CSW3 (experiment 7, 8, and 9).

The RMS value for the experiments performed with the three types of serum (CSW1, CSW2, and CSW3) is 6.5%.

4. Discussion

The concentration index (CI) showed values greater than 1 in all cases, with a downward trend as time increased. This suggests that the concentrated extract is mainly collected in the first thawed fractions. Similar behavior has been reported by other studies [32,33] in defrosting sugar solutions. These results may indicate that the solute in the frozen sample is recovered not only with fusion, but also with a diffusion of solute from the freeze concentrate phase (enhanced with vacuum). For Y, the behavior is the opposite, since the longer the vacuum time, the greater the amount of solutes recovered, even if they have a lower concentration.

Specifically, the factor that most influences Y is the vacuum time (t), while for CI, the main factor corresponds to the initial concentration (C_0). As can be seen in Table 2, the higher initial concentration acquired the higher concentrated solution, and in turn, the higher concentration solution obtained the higher increase in viscosity. According to Sánchez et al. [19], in the initial concentration range studied (7.5 to 19 °Bx), the viscosity of whey increased more than twice. In BFC, the capacity of ice separators is inversely proportional to the viscosity of the concentrate [32]. Therefore, as the viscosity increases, the efficiency and solute yield (Y) are reduced. Other authors [33] suggest that the increase of the initial concentration forms thinner and smaller ice structures. These small crystals of melted ice make it difficult to recover the concentrate fluid, and therefore, a low Y value is obtained with the highest concentration of whey. Table 3 shows that the C_0•t interaction is significant for both Y and CI. This suggests that centrifugation duration (t) has a greater effect at low initial concentrations (C_0) than at higher concentrations. This may also be related to the higher viscosity of the serum at higher concentrations, and therefore more difficult to extract under vacuum.

The freezing rate of the samples is also an important factor in the FC process. The freezing time of the 45 mL whey samples used in the BFC system has been estimated through applying the model proposed by Pham [34], obtaining an average value of 234 min, which is equivalent to an average freezing rate of 5.7 μm/s. This value is very close to previous works on BFC applied in orange juice [35]. In addition, this freezing rate is lower than the critical value (approximately 8 μm/s) provided by other authors [24,26]. These authors reported that for velocities higher than 8 μm/s, the ice occluded the solutes during the freezing stage, and it was not possible to expect a considerable separation of the concentrated solution from the ice fraction.

The vacuum-assisted BFC process has been applied on three types of whey (CSW1, CSW2, and CSW3) in a single stage, under the conditions that correspond to the best values of the previous Y tests (60 min and 10 kPa). The results obtained were 73%, 69%, and 71% for CSW1, CSW2, and CSW3, respectively. These results are better than those obtained in trials listed in Table 2 (65.5%). It may be due to the fact that in the previous trials the initial concentration (7.5 °Bx) was higher than that of the three dairy whey (between 5.7 and 6.6 °Bx), and as previously demonstrated, with a high initial concentration, there is a low yield of solutes. A work on PFC in whey [36] indicated a maximum of 76.4% (*w/w*) of solute yield in four freezing/thawing steps.

From Figure 4, CI values close to 3.5, 4.1, and 3.9 can be deduced for lactose of the three types of whey (CSW1, CSW2, and CSW3, respectively). A lactose concentration index of 3.5 has been reported by Korotkiy et al. [37] using a double FC system (two steps) in milk whey. The results of lactose CI in the present study are equal or higher than those of CI of soluble solids (Table 4), which seems to suggest that lactose tends to pass more easily to the concentrated phase than into the ice. Similar results have been informed by Aider and de Halleux [15], who studied the milk whey under BFC process in successive stages. In the first stage, the CI of the solids was 1.86, while the CI of lactose for that same stage was 2. This behavior may be due to the fact that lactose, a low molecular weight disaccharide compared to other whey molecules, mainly proteins, is easier to move and more difficult to trap in ice than other molecules with high molecular weight. This trend is consistent with Kawasaki et al. [38], since they found that lower molecular weight solutes were separated and concentrated more efficiently than higher molecular weight solutes. This corresponded well with the magnitude of the diffusion coefficient for each solute.

Results on BFC applied in whey have been reported, where thawing was performed by gravity [3,15], without the aid of vacuum. CI for solids and lactose around 2 were obtained. Therefore, the results of this study (CI between 3.5 to 3.8 for solids, and 3.5 to 4.1 for lactose) suggest that the application of vacuum in a single step can help to improve the recovery of the concentrated liquid fraction. A work of Lamkaddam et al. [36] related to the application of PFC in whey obtained a maximum content of 20.52% (*w/w*) of total solids in four steps.

The CI of total solids increases in each stage of FC and varies from 1.57 in the first stage, up to 3.42 in the fourth stage.

Based on the CI values of the soluble solids and of the lactose obtained in the present study, it seems possible that using the vacuum-assisted BFC technique, at least 70% of the lactose contained in the initial sample can be recovered, with a high CI (3.5–4.0). This performance can be increased if the formed ice is thawed and subjected to a new vacuum-assisted BFC stage.

Despite the difficulty in handling ice, a good agreement was observed between the experimental (W_{exp}) and predicted (W_p) ice mass ratio, as shown in Figure 5. An RMS of 6.5% was obtained, lower than the limit of 25% suggested by Lewicki [39] to consider an acceptable fit. These values were close to those reported in previous studies [6,11].

5. Conclusions

The vacuum-assisted BFC technique has been applied to whey samples. The results show that the variables studied (initial concentration, pressure, and vacuum time) have a significant effect on the process. The best solute yield (Y) results were obtained at a pressure of 10 kPa, 7.5 °Bx of initial concentration and 60 min of vacuum, while for the concentration index response variable (CI), the best conditions were 10 kPa, 7.5 °Bx, and 20 min. In a second phase of the work, by applying the conditions that provide higher solute yield (10 kPa and 60 min) to three different types of dairy whey, Y values around 70% or higher are obtained in a single stage. In all three types of whey, the CI for lactose are somewhat higher than those for soluble solids, suggesting that lactose tends to remain in the concentrated phase rather than on ice. In this way it is possible to recover, in a single step, at least 70% of the lactose contained in the initial whey. Vacuum-assisted BFC technology is postulated in this way, as an alternative for the recovery of valuable components contained in whey. Finally, the results are positive for the parameters tested, but it is necessary more studies to establish the effectiveness of BFC in the treatment of cheese whey.

Author Contributions: Conceptualization, E.H., M.B. and P.O.-P.; methodology, N.G., G.Q., E.H. and P.O.-P.; software, N.G., M.B. and E.H.; validation, N.G., G.Q., M.B. and E.H.; formal analysis, N.G., M.B., S.S., N.A.A., E.S. and E.H.; investigation N.G. and E.H.; resources, E.H.; data curation, G.Q., M.B., S.S., N.A.A., P.O.-P., E.S. and E.H.; writing—review and editing, E.H., G.Q., M.B., S.S., N.A.A., E.S. and P.O.-P.; visualization, N.G., G.Q., E.S. and E.H., supervision, M.B., E.S. and E.H. All authors have read and agreed to the published version of the manuscript.

Funding: The APC was financed by AGRUPS-2022 at Universitat Politècnica de Catalunya (UPC) and SGR-Cat 2021 of Departament de Recerca i Universitats (Generalitat de Catalunya).

Data Availability Statement: Data is contained within the article.

Acknowledgments: Author Eduardo Hernández thanks Pere Bassa Roca (Can Corder) for supplying milk whey for the tests.

Conflicts of Interest: The authors declare no conflict of interest.

References

1. Belén, F.; Raventós, M.; Hernández, E. Management of cheese whey by film freeze concentration. *Environ. Eng. Manag. J.* **2018**, *17*, 1373–1383. [CrossRef]
2. Aider, M.; de Halleux, D.; Akbache, A. Whey cryoconcentration and impact on its composition. *J. Food Eng.* **2007**, *82*, 92–102. [CrossRef]
3. Aider, M.; de Halleux, D.; Melnikova, I. Skim milk whey cryoconcentration and impact on the composition of the concentrated and ice fractions. *Food Bioproc. Tech.* **2009**, *2*, 80–88. [CrossRef]
4. Raventós, M.; Hernández, E.; Auleda, J.; Ibarz, A. Concentration of aqueous sugar solutions in multi-plate cryoconcentrator. *J. Food Eng.* **2007**, *79*, 577–585. [CrossRef]
5. Aider, M.; de Halleux, D. Production of concentrated cherry and apricot juices by cryoconcentration technology. *LWT-Food Sci. Technol.* **2008**, *41*, 1768–1775. [CrossRef]

6. Petzold, G.; Orellana, P.; Moreno, J.; Cerda, E.; Parra, P. Vacuum-assisted block freeze concentration applied to wine. *Innov. Food Sci. Emerg. Technol.* **2016**, *36*, 330–335. [CrossRef]
7. Hsieh, H.C. Desalinating process. *US Patent* **2008**, *7*, 467–526.
8. Petzold, G.; Niranjan, K.; Aguilera, J.M. Vacuum-assisted freeze concentration of sucrose solutions. *J. Food Eng.* **2013**, *115*, 357–361. [CrossRef]
9. Pardo, J.M.; Sánchez, R. Block freeze concentration intensification by means of vacuum and microwave pulses. *Eng. Compet.* **2015**, *17*, 143–151.
10. Moreno, F.L.; Raventós, M.; Hernández, E.; Ruiz, Y. Block freeze-concentration of coffee extract: Effect of freezing and thawing stages on solute recovery and bioactive compounds. *J. Food Eng.* **2014**, *120*, 158–166. [CrossRef]
11. Orellana-Palma, P.; Petzold, G.; Torres, N.; Aguilera, M. Elaboration of orange juice concentrate by vacuum-assisted block freeze concentration. *J. Food Process. Preserv.* **2017**, *2017*, e13438. [CrossRef]
12. Orellana-Palma, P.; Petzold, G.; Pierre, L.; Pensaben, J.M. Protection of polyphenols in blueberry juice by vacuum-assisted block freeze concentration. *Food Chem. Toxicol.* **2017**, *109*, 1093–1102. [CrossRef] [PubMed]
13. Qin, F.; Russel, A.; Chen, X.; Robertson, L. Ice fouling on a subcooled metal surface examined by thermo-response and electrical conductivity. *J. Food Eng.* **2003**, *59*, 421–429. [CrossRef]
14. Sung-Hee, P.; Jee-Yeon, K.; Geun-Pyo, H.; Hae-Soo, K.; Sang-Gi, M. Effect of ice recrystallization on freeze concentration of milk solutes in a lab-scale unit. *Food Sci. Biotechnol.* **2006**, *15*, 196–201.
15. Aider, M.; de Halleux, D.; Melnikova, I. Gravitational and microwave-assisted thawing during milk whey cryoconcentration. *J. Food Eng.* **2008**, *88*, 373–380. [CrossRef]
16. Aider, M.; de Halleux, M.; Melnikova, I. Skim acidic milk whey cryoconcentration and assessment of its functional properties: Impact of processing conditions. *Innov. Food Sci. Emerg. Technol.* **2009**, *10*, 334–341. [CrossRef]
17. Habib, B.; Farid, M. Freeze concentration of milk and saline solutions in a liquid-solid fluidized bed. part I. experimental. *Chem. Eng. Process. Process Intensif.* **2007**, *46*, 1400–1411. [CrossRef]
18. Chen, P.; Chen, X.D. A generalized correlation of solute inclusion in ice formed from aqueous solutions and food liquids on sub-cooled surface. *Can. J. Chem. Eng.* **2000**, *78*, 312–319. [CrossRef]
19. Sánchez, J.; Hernández, E.; Auleda, J.M.; Raventós, M. Freeze concentration of whey in a falling-film based pilot plant: Process and characterization. *J. Food Eng.* **2011**, *103*, 147–155. [CrossRef]
20. Mollea, C.; Marmo, L.; Bosco, F. Valorisation of cheese whey, a by-product from the dairy industry. In *Food Industry*, 1st ed.; Muzzalupo, I., Ed.; IntechOpen: London, UK, 2013.
21. Zandona, E.; Blažić, M.; Režek, A. Whey utilization: Sustainable uses and environmental approach. *Food Technol. Biotechnol.* **2021**, *59*, 147–161. [CrossRef]
22. Yadav, J.S.S.; Yana, S.; Pilli, S.; Kumar, L.; Tyagi, R.D.; Surampalli, R.Y. Cheese whey: A potential resource to transform into bioprotein, functional/nutritional proteins and bioactive peptides. *Biotechnol. Adv.* **2015**, *33*, 756–774. [CrossRef] [PubMed]
23. Prestes, A.A.; Helm, C.V.; Esmerino, E.A.; Silva, R.; da Cruz, A.G.; Prudencio, E.S. Freeze concentration techniques as alternative methods to thermal processing in dairy manufacturing: A review. *J. Food Sci.* **2022**, *87*, 488–502. [CrossRef] [PubMed]
24. Moreno, F.L.; Quintanilla-Carvajal, M.X.; Sotelo, L.I.; Osorio, C.; Raventós, M.; Hernández, E.; Ruiz, Y. Volatile compounds, sensory quality and ice morphology in falling-film and block freeze concentration of coffee extract. *J. Food Eng.* **2015**, *164*, 64–71. [CrossRef]
25. Khajehei, F.; Niakousari, M.; Eskandari, M.H.; Sarshar, M. Production of pomegranate juice concentrate by complete block cryoconcentration process. *J. Food Process Eng.* **2015**, *38*, 488–498. [CrossRef]
26. Nakagawa, K.; Maebashi, S.; Maeda, K. Freeze-thawing as a path to concentrate aqueous solution. *Sep. Purif. Technol.* **2010**, *73*, 403–408. [CrossRef]
27. Guerra-Valle, M.; Lillo-Perez, S.; Petzold, G.; Orellana-Palma, P. Effect of freeze crystallization on quality properties of two endemic Patagonian berries juices: Murta (*Ugni molinae*) and arrayan (*Luma apiculata*). *Foods* **2021**, *10*, 466. [CrossRef] [PubMed]
28. Schuster-Wolff-Bühring, R.; Michael, R.; Hinrichs, J. A new liquid chromatography method for the simultaneous and sensitive quantification of lactose and lactulose in milk. *Dairy Sci. Technol.* **2011**, *91*, 27–37. [CrossRef]
29. Raventós, M.; Hernández, E.; Auleda, J.M. Freeze concentration applications in fruit processing. In *Advances in Fruit Processing Technologies*, 1st ed.; Rodrigues, S., Narciso-Fernandes, F.A., Eds.; CRC Press: Boca Raton, FL, USA, 2012; pp. 263–286.
30. Bakshi, A.S.; Johnson, R.M. Calorimetric studies on whey freeze concentration. *J. Food Sci.* **1983**, *48*, 1279–1283. [CrossRef]
31. Muñoz, I.D.B.; Rubio, A.; Blanco, M.; Raventós, M.; Hernández, E.; Prudêncio, E.S. Progressive freeze concentration of skimmed milk in an agitated vessel: Effect of the coolant temperature and stirring rate on process performance. *Food Sci. Technol. Int.* **2019**, *25*, 150–159. [CrossRef]
32. Gulfo, R.; Auleda, J.M.; Moreno, F.L.; Ruiz, Y.; Hernández, E.; Raventós, M. Multi-plate freeze concentration: Recovery of solutes occluded in the ice and determination of thawing time. *Food Sci. Technol. Int.* **2013**, *20*, 405–419. [CrossRef]
33. Yee, P.L.; Wakisaka, M.; Shirai, Y.; Hassan, M.A. Effects of single food components on freeze concentration by freezing and thawing technique. *Japan J. Food Eng.* **2003**, *4*, 77–82. [CrossRef]
34. Pham, Q.T. Advances in food freezing/thawing/freeze concentration modelling and techniques. *Japan J. Food Eng.* **2008**, *9*, 21–32. [CrossRef]

35. Petzold, G.; Orellana, P.; Moreno, J.; Cuevas, C. Process parameters of vacuum-assisted freeze concentration. *Chem. Eng. Trans.* **2017**, *57*, 1789–1794. [CrossRef]
36. Lamkaddam, I.U.; Vega, E.; Colón, J.; Ponsá, S.; Llenas, L.; Mora, M. Progressive freeze concentration of cheese whey for protein and lactose recovery. *Int. Dairy J.* **2023**, *139*, 105572. [CrossRef]
37. Korotkiy, I.A.; Korotkaya, E.V.; Neverov, E.N.; Plotnikov, I.B.; Efremov, D.A. Separatory freezing and cryoconcentration of milk and whey. *IOP Conf. Ser. Earth Environ. Sci.* **2021**, *852*, 012052. [CrossRef]
38. Kawasaki, K.; Matsuda, A.; Kadota, H. Freeze concentration of equal molarity solutions with ultrasonic irradiation under constant freezing rate: Effect of solute. *Chem. Eng. Res. Des.* **2006**, *84*, 107–112. [CrossRef]
39. Lewicki, P.P. Raoult's law based food water sorption isotherm. *J. Food Eng.* **2000**, *43*, 31–40. [CrossRef]

MDPI AG
Grosspeteranlage 5
4052 Basel
Switzerland
Tel.: +41 61 683 77 34

Foods Editorial Office
E-mail: foods@mdpi.com
www.mdpi.com/journal/foods

www.ingramcontent.com/pod-product-compliance
Lightning Source LLC
Chambersburg PA
CBHW042010260926
51850CB00020B/2997
* 9 7 8 3 7 2 5 8 3 5 4 9 2 *